Novel Insights into Doppler Radar Observations

Novel Insights into Doppler Radar Observations

Edited by **Henry Collier**

New York

Published by Callisto Reference,
106 Park Avenue, Suite 200,
New York, NY 10016, USA
www.callistoreference.com

Novel Insights into Doppler Radar Observations
Edited by Henry Collier

International Standard Book Number: 978-1-63239-482-8 (Hardback)

The publisher's policy is to use permanent paper from mills that operate a sustainable forestry policy. Furthermore, the publisher ensures that the text paper and cover boards used have met acceptable environmental accreditation standards.

Trademark Notice: Registered trademark of products or corporate names are used only for explanation and identification without intent to infringe.

Printed in the United States of America.

Contents

Preface VII

Part 1 **Tropospheric Wind and
Turbulence Observations** 1

Chapter 1 **New Observations by Wind Profiling Radars** 3
Masayuki K. Yamamoto

Chapter 2 **Retrieving High Resolution 3-D
Wind Vector Fields from Operational Radar Networks** 27
Olivier Bousquet

Chapter 3 **Multiple Doppler Radar Analysis for Retrieving the
Three-Dimensional Wind Field Within Thunderstorms** 53
Shingo Shimizu

Chapter 4 **Synergy Between Doppler Radar
and Lidar for Atmospheric Boundary Layer Research** 69
Chris G. Collier

Part 2 **Weather Radar Quality
Control and Related Applications** 85

Chapter 5 **Quality Control Algorithms
Applied on Weather Radar Reflectivity Data** 87
Jan Szturc, Katarzyna Ośródka and Anna Jurczyk

Chapter 6 **Doppler Weather Radars and Wind Turbines** 105
Lars Norin and Günther Haase

Chapter 7 **Effects of Anomalous Propagation
Conditions on Weather Radar Observations** 127
Joan Bech, Adolfo Magaldi, Bernat Codina and Jeroni Lorente

Part 3   Advanced Techniques for
Probing the Ionosphere   153

Chapter 8   Incoherent Scatter Radar –
Spectral Signal Model and Ionospheric Applications   155
Erhan Kudeki and Marco Milla

Chapter 9   Aperture Synthesis Radar
Imaging for Upper Atmospheric Research   185
D.L. Hysell and J.L. Chau

Part 4   Other Advanced Doppler Radar Applications   205

Chapter 10   Doppler Radar Tracking Using Moments   207
Mohammad Hossein Gholizadeh and Hamidreza Amindavar

Chapter 11   Volcanological Applications of
Doppler Radars: A Review and Examples
from a Transportable Pulse Radar in L-Band   231
Franck Donnadieu

Permissions

List of Contributors

Preface

It is often said that books are a boon to humankind. They document every progress and pass on the knowledge from one generation to the other. They play a crucial role in our lives. Thus I was both excited and nervous while editing this book. I was pleased by the thought of being able to make a mark but I was also nervous to do it right because the future of students depends upon it. Hence, I took a few months to research further into the discipline, revise my knowledge and also explore some more aspects. Post this process, I begun with the editing of this book.

Doppler radar systems have been extremely helpful in enhancing our perception and monitoring capabilities of atmospheric phenomenon. Some of the topics discussed in this book are – extreme weather supervision, wind and turbulence retrievals, estimation of precipitation and nowcasting, and the volcanological applications of Doppler radar. This book is appropriate for graduate students who are seeking an initiation in the area or for experts wanting to refresh their knowledge and information on the same.

I thank my publisher with all my heart for considering me worthy of this unparalleled opportunity and for showing unwavering faith in my skills. I would also like to thank the editorial team who worked closely with me at every step and contributed immensely towards the successful completion of this book. Last but not the least, I wish to thank my friends and colleagues for their support.

Editor

Part 1

Tropospheric Wind and Turbulence Observations

New Observations by Wind Profiling Radars

Masayuki K. Yamamoto
Research Institute for Sustainable Humanosphere (RISH), Kyoto University,
Japan

1. Introduction

Wind profiling radar, also referred to as "radar wind profiler", "wind profiler", and "clear-air Doppler radar", is used to measure height profiles of vertical and horizontal winds in the troposphere. It receives signals scattered by radio refractive index irregularities (clear-air echo) and measures the Doppler shift of the scattered signals (Gage, 1990). Wind profiling radar measures wind velocities by steering its beam directions or using spaced receiving antennas (e.g., Larsen & Röttger 1989; May, 1990). The two methods are referred to as the Doppler beam swinging (DBS) technique and spaced antenna (SA) technique, respectively. Owing to its capability to measure wind velocities in the clear air with high height and time resolutions (typically a hundred to several hundreds of meters and less than several minutes, respectively), it is used for atmospheric research such as radio wave scattering, gravity waves, turbulence, temperature and humidity profiling, precipitation system, and stratosphere-troposphere exchange (STE) processes (Fukao, 2007; Hocking, 2011). Wind profiling radar is also utilized for monitoring wind variations routinely. In USA and Japan, a nationwide ultrahigh frequency (UHF) wind-profiling radar network is operated in order to provide upper-air wind data to numerical weather prediction (Ishihara et al., 2006; Stanley et al., 2004). In Europe, Cost Wind Initiative for a Network Demonstration in Europe (CWINDE), now renamed as the Co-Ordinated Wind Profiler Network in Europe, is also operated (Met Office, 2011).

For wind profiling radars, frequency range of 30-3000 MHz (i.e., very high frequency (VHF) and UHF bands) is generally used because the energy spectrum of atmospheric turbulence falls off rapidly with decreasing eddy size in the inertia subrange, and radar radio waves are scattered only from turbulent eddies at the Bragg scale (i.e., half the radar wavelength). For measurements from the ground to several thousand meters, UHF wind profiling radars are widely used because their small antenna size enables their easy installation and their quick switching time from transmission to reception is necessary for measurements near the ground. Such UHF wind profiling radars are referred to as the boundary layer radars. Because the minimum size of turbulent eddies increases exponentially with increasing altitude (e.g., Hocking, 1985), frequencies near 50 MHz are used for clear-air radars which measure the mesosphere, stratosphere and troposphere (MST radars) and those which measure the stratosphere and troposphere (ST radars). In the chapter, measurement results of VHF and UHF radars are presented.

Recent development in radar interferometry techniques provides means for enhancing radar resolution and improving data quality. In radar interferometry, spaced receiver antennas are used to improve angular resolution, and multiple carrier frequencies are used to improve range resolution. The former is referred to as coherent radar imaging (CRI) or spatial domain interferometric imaging (SDI; Palmer et al., 1998; Hassenpflug et al., 2008), and the latter is referred to as range imaging (RIM; Palmer et al., 1999) or frequency-domain interferometric imaging (FII; Luce et al., 2001). Hereafter the abbreviations CRI and RIM are used. Though development of the radar interferometry technique have decades of history (Hocking, 2011), CRI and RIM, which have been intensively developed for the last decade, are presented in section 2.

Wind profiling radars operated at approximately 50 MHz frequency (50-MHz wind profiling radars) are not sensitive for small-sized cloud particles. Therefore 50-MHz wind profiling radars are able to measure vertical and horizontal wind velocities in both the clear air and cloudy regions. Millimeter-wave radars, which use near 35-GHz or 95-GHz frequency (i.e., 8-mm or 3-mm wavelength) and hence are able to detect echoes scattered by small-sized cloud particles, are an indispensable means to measure microphysical properties of clouds (Kollias et al., 2007). Laser radars (lidars), which transmit laser light and receive echoes scattered by atmospheric molecules, aerosols, and hydrometeors, are useful to measure not only various physical quantities in the clear air but also particles and hydrometeors in the atmosphere (Wandinger, 2005). Recent measurements using collocated wind profiling radars and millimeter-wave radars/lidars have gained new insights of turbulence and cloud processes. The measurement results are presented in section 3.

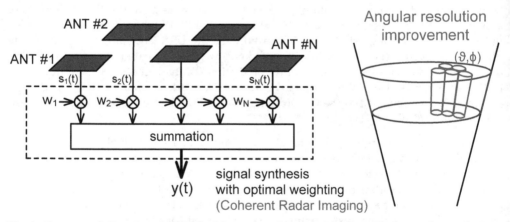

Fig. 1. Conceptual drawing of CRI. The blue-colored volume on the right shows the resolution without CRI (i.e., the angular resolution is determined by the antenna beam width), and the red-colored volumes show the angular resolution improvement attained by CRI.

2. Imaging techniques to enhance radar resolutions

Though development of the radar interferometry techniques have decades of history, CRI and RIM, which have been intensively developed for the last decade, are presented in section 2.1 and 2.2, respectively. The readers are recommended to refer Hocking (2011) for the thorough development history of the radar interferometry techniques. Further applications aiming at advanced probing of the atmosphere are presented in section 2.3.

2.1 Angular resolution enhancement using spaced receivers

2.1.1 Signal processing

Fig. 1. shows a conceptual drawing of CRI. In CRI, signals from the spaced receivers are synthesized with appropriate weights in order to steer the radar beam in certain directions with improved angular resolution. For CRI, the Capon method (Capon, 1969) is widely used because it satisfies both high angular resolution and simple calculation. Hereafter signal processing of CRI using the Capon method is described. The Capon method is described as the problem of finding optimal weights. The optimal weights used in order to calculate the weighted sum of signals which are received by the spaced receivers. $\mathbf{s}$ denotes a set of signals associated with the N spaced receivers at an arbitrary range gate and expressed by

$$\mathbf{s}(t) = (s_1(t), s_2(t), ..., s_N(t))^{\mathrm{T}}, \tag{1}$$

where t is the sampled time and T is the transpose operator. $\mathbf{w}$ denotes a set of weights for summation and is expressed by

$$\mathbf{w} = (w_1, w_2, ..., w_N)^{\mathrm{T}}. \tag{2}$$

The optimal weight vector is given by a solution that minimizes the resulting average power B. B is expressed by

$$B = \mathbf{w}^{\mathrm{H}} \mathbf{R} \mathbf{w}, \tag{3}$$

where H represents the Hermitian operator (conjugate transpose) and $\mathbf{R}$ is a covariance matrix given by

$$\mathbf{R} = \begin{pmatrix} R_{11} & R_{22} & \cdots & R_{1N} \\ R_{21} & R_{22} & \cdots & R_{2N} \\ \vdots & \vdots & \cdots & \vdots \\ R_{N2} & R_{N2} & \cdots & R_{NN} \end{pmatrix}, \tag{4}$$

R_{ij} is a covariance between s_i and s_j. The length of time used for calculating $\mathbf{R}$ should be determined by considering the accuracy of covariance value and the time resolution. $\mathbf{w}$ is constrained by the condition of constant gain to waves coming from the target volume, and the constraint is given by

$$\mathbf{e}^{\mathrm{H}} \mathbf{w} = 1, \tag{5}$$

where

$$\mathbf{e} = (e^{j\mathbf{k} \cdot \mathbf{D}_1}, e^{j\mathbf{k} \cdot \mathbf{D}_2}, ..., e^{j\mathbf{k} \cdot \mathbf{D}_N})^{\mathrm{T}}, \tag{6}$$

$\mathbf{k}$ represents the wavenumber vector of the focused direction with the zenith and azimuth angle of θ and ϕ, respectively ($\mathbf{k} = \dfrac{2\pi}{\lambda}[\sin\theta\sin\phi, \sin\theta\cos\phi, \cos\theta]$), and the vectors which represent the center of each receiving receiver are denoted by $\mathbf{D}_m$ for the m th receiver. $\mathbf{e}$ is referred to as the steering vector. The constrained minimization problem can be solved using the Lagrange method, and Palmer et al. (1998) describe details of solving the constrained minimization problem (see their appendix). As the solution of the constrained minimization problem, the optimal weight $\mathbf{w}_C(\mathbf{k})$ is given by

$$\mathbf{w}_C(\mathbf{k}) = \frac{\mathbf{R}^{-1}\mathbf{e}}{\mathbf{e}^H\mathbf{R}^{-1}\mathbf{e}} . \tag{7}$$

Using Equations (1) and (7), scalar output of the filter $y(t)$ is given by

$$y(t) = \mathbf{w}_C^H(\mathbf{k})\mathbf{s}(t) . \tag{8}$$

By calculating the Doppler spectrum of $y(t)$, brightness B_C (i.e., power density), radial Doppler velocity, and spectral width are able to be computed with improved angular resolution. B_C is able to be obtained without calculating $\mathbf{w}_C$ and given by

$$B_C(\mathbf{k}) = \frac{1}{\mathbf{e}^H\mathbf{R}^{-1}\mathbf{e}} . \tag{9}$$

When the brightness at arbitrary Doppler velocity needs to be calculated, $\mathbf{R}$ is replaced by the cross-spectral matrix of the N receiver signals. Palmer et al. (1998) showed a clear difference in angular distribution of brightness between positive and negative Doppler velocities (see their Plate 2). Brightness at the arbitrary Doppler velocity of received data is also able to be calculated by applying band-pass filtering to $\mathbf{s}$.

In the Fourier-based method, in which all the signals from receivers were synthesized with equal weight, a weight vector $\mathbf{w}_F(\mathbf{k})$ which steers the beam in the direction $\mathbf{k}$ is given by

$$\mathbf{w}_F(\mathbf{k}) = (e^{j\mathbf{k}\cdot\mathbf{D}_1}, e^{j\mathbf{k}\cdot\mathbf{D}_2}, \dots, e^{j\mathbf{k}\cdot\mathbf{D}_N})^T . \tag{10}$$

Scalar output of the filter is calculated by replacing $\mathbf{w}_C(\mathbf{k})$ in Equation (8) with $\mathbf{w}_F(\mathbf{k})$.

Other methods are able to be used for CRI. Details of multiple signal classification (MUSIC) method and maximum entropy method (MEM) are explained by Hélal et al. (2001) and Yu et al. (2000), respectively. For more general review of CRI, see Woodman (1997).

There are factors that affect the performance of CRI. Using numerical simulation, effects of receiver noise and turbulence distribution were evaluated by Yu et al. (2000) and Cheong et al. (2004). Further, Yu et al. (2000) evaluated relation between CRI performance and receiver arrangement. Effects of uncertainty of receiver gain and phase were evaluated for the case of turbulent Eddy Profiler (TEP), which was developed by the University of Massachusetts in order to carry out CRI measurement in the boundary layer (Mead et al. 1998).

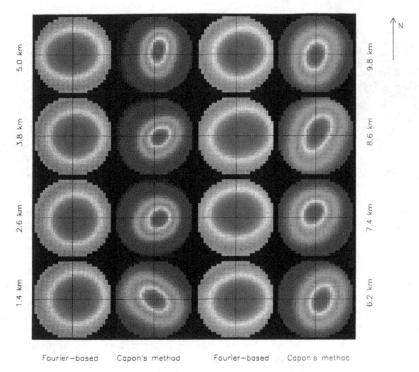

Fourier-based   Capon's method   Fourier-based   Capon's method

Fig. 2. Brightness distributions within ±4° from the center of beam direction at the altitudes from 1.4 km to 9.8 km (Palmer et al., 1998). Red color corresponds to highest brightness. Images on the left and right of each pair were obtained with the Fourier-based method and the Capon method, respectively.

2.1.2 Measurement results

An example of high-angle-resolution measurement using CRI is presented. Fig. 2 shows brightness distributions within ±4° from the center of beam direction measured by the Middle and Upper Atmosphere radar (MU radar). MU radar is a MST radar installed at Shigaraki MU Observatory, Japan (34.85°N, 136.10°E; Fukao et al., 1990). CRI successfully produces fine-scale angular distributions of backscattered clear-air echo power within the two-way half-power full beam width of 2.5°. Such high angular resolution cannot be attained without CRI. It is noted that the Capon method exhibits better resolution than the Fourier-based method. Using data collected by a VHF radar installed at Tourris, France (43.08°N, 6.01°E), Hélal et al. (2001) showed a fine-scale angular distribution of backscattered clear-air echo power using the Capon and MUSIC methods. Chau and Woodman (2001) also showed the angular distribution using the Fourier-based method, Capon method, MEM, and the fitting technique. Using the characteristic that raindrop fall velocity is much greater than vertical wind velocity, Palmer et al (2005) demonstrated that fine-scale angular distributions of backscattered power from clear air and that from raindrops are able to be obtained separately. From a CRI measurement by TEP, Pollard et al. (2000) demonstrated that horizontal distribution of refractive index structure function is able to be measured.

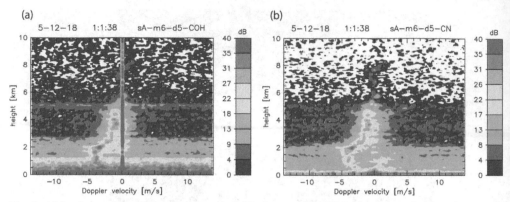

Fig. 3. Altitude profile of Doppler spectra (a) before and (b) after applying clutter mitigation algorithm (Nishimura et al., 2010). Colors show the backscattered power in an arbitrary unit.

CRI is also able to be used for clutter mitigation. Fig. 3 is an example of clutter mitigation. Measurement data were collected by the receiving antenna array of the Equatorial Atmosphere Radar (EAR), which were temporally installed for multistatic radar observations. The EAR is installed at West Sumatra, Indonesia and is operated with a center frequency of 47-MHz and peak output power of 100 kW (Fukao et al., 2003). Strong returns centered at 0 m s^{-1} in Fig. 3a are signals returned from the ground (i.e., ground clutter), and have to be removed in order to estimate spectral moments correctly. By applying the clutter mitigation algorithm developed by Nishimura et al (2010), the clutter signals are successfully removed (Fig. 3b). Nishimura et al. (2010) attained the clutter mitigation by combining the directional-constrained minimization of power with constrained norm (DCMP-CN; see Kamio et al. (2004) for details) and an algorithm that compensates electromagnetic coupling between antennas and the ground. The compensation was carried out because the electromagnetic coupling can cause a phase error of atmospheric echoes received by antenna arrays, and the phase error can lead to degradation of desired atmospheric echoes in the output of the adaptive clutter mitigation process.

In the case shown by Nishimura et al. (2010), each antenna element has an identical antenna gain. When a high-gain antenna is used for transmission and reception of scattering from atmospheric targets, using auxiliary antennas which are used only for receiving ground clutters is effective for clutter mitigation. In order to realize clutter mitigation using the main antenna of the MU radar and auxiliary antennas, Kamio et al. (2004) modified the DCMP-CN method. In the modified method, the weight of main antenna is kept to 1 in order to keep the main lobe pattern, and the weights of auxiliary antennas are optimized in order to minimize the received power from side lobes.

Moving biological targets like birds and insects can cause a large error of wind velocity measured by wind profiling radars (e.g., Vaughn, 1985; Wilczak et al., 1995). Using CRI, moving clutter is able to be suppressed. From the CRI measurement using TEP, Cheong et al. (2006) succeeded in separating clear-air echoes and the biological scattering which was moving in the grating-lobe region, and demonstrated that the separation of biological scattering greatly reduced the error of wind velocity estimates. Chen et al. (2007) also applied CRI to data measured by multiple antenna profiler radar (MAPR) of National

Center for Atmospheric Research (NCAR) in order to mitigate effects of bird contamination in wind velocity estimates.

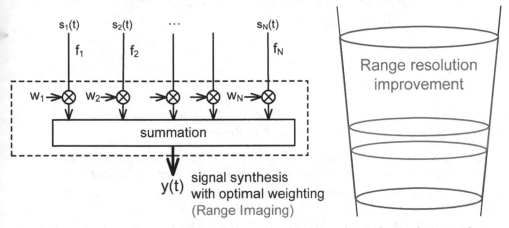

Fig. 4. Conceptual drawing of RIM. The blue-colored volume shows the resolution without RIM (i.e., range resolution is determined by the transmitted pulse width), and the red-colored volume shows the range resolution improvement attained by RIM.

2.2 Range resolution enhancement using multiple frequencies

2.2.1 Signal processing

Figure 4 shows a conceptual drawing of RIM. The width of range gate is determined by the sampling interval of analog-to-digital converter, and the range resolution without RIM is determined by the transmitted pulse width. In RIM, signals sampled from multiple frequencies are synthesized with appropriate weights in the subranges within a range gate. Using the synthesized signal, the first three spectral moments (brightness, Doppler velocity, and spectral width) at the subranges are estimated in order to attain range resolution improvement. Among various methods which can be used for RIM processing (Luce et al., 2001; Palmer, et al., 1998; Smaïni et al., 2002), the Capon method (Capon, 1969) is widely used because it satisfies both high range resolution and simple calculation. Hereafter signal processing of RIM using the Capon method is described. $\mathbf{s}$ denotes a set of signals collected by N carrier frequencies at an arbitrary range gate and is expressed by

$$\mathbf{s}(t) = (s_1(t), s_2(t), ..., s_N(t))^{\mathrm{T}} .\tag{11}$$

Frequencies are switched on a pulse-to-pulse basis in order to maximize the correlation of signals sampled by different frequencies. Using numerical simulation, Palmer et al. (1999) showed that using 3 or more frequencies are required for RIM measurements. For field measurements, four or five frequencies are used typically (e.g., Fukao et al., 2011; Palmer et al., 2001; Yu & Brown, 2004). The optimal weight vector is given as a solution that minimizes the resulting average power B_r. B_r is expressed by

$$B_r = \mathbf{w}^{\mathrm{H}} \mathbf{R} \mathbf{w} ,\tag{12}$$

where $\mathbf{w}$ denotes a set of weights for summation as expressed by Equation (2) and $\mathbf{R}$ is a covariance matrix as expressed by Equation (4). $\mathbf{w}$ is constrained by the condition of constant gain to waves coming from the target range, and the constraint is given by

$$\mathbf{e}^H\mathbf{w} = 1, \tag{13}$$

where $\mathbf{e}$ is a range steering vector and given by

$$\mathbf{e} = (e^{-j(2k_1 r_l - \phi_1)}, e^{-j(2k_2 r_l - \phi_2)}, \ldots, e^{-j(2k_N r_l - \phi_N)})^T, \tag{14}$$

where k_m denotes the wavenumber of m th frequency, r_l represents the range between the target and radar, and ϕ_m is the initial phase of m th frequency.

Because ϕ_m is determined not only by the total system delay throughout the transmitter and receiver chains but also by k_m, the values of $\phi_1, \ldots, \phi_N$ are different. Therefore the total system delay, from which the values of $\phi_1, \ldots, \phi_N$ are computed, needs to be known in order to determine $\mathbf{e}$ correctly. Chilson et al. (2004) and Palmer et al. (2001) measured the total system delay by leaking the transmitted signal back to the receiver through an ultrasonic delay line. In a practical manner, only the relative phase differences among the frequencies are necessary to correct the effects of system delay. Therefore, the correction is able to be attained by calculating the phase term of cross correlation between the two time series of received signals measured at different frequencies (Chen, 2004). Measurement results of phase correction using the clear-air echoes are shown by Chen et al. (2009, 2010) and Chen & Zecha (2009).

The optimal weight $\mathbf{w}_{rC}$ is given by

$$\mathbf{w}_{rC} = \frac{\mathbf{R}^{-1}\mathbf{e}}{\mathbf{e}^H \mathbf{R}^{-1} \mathbf{e}}. \tag{15}$$

Using Equations (11) and (15), scalar output of the filter $y(t)$ is given by

$$y(t) = \mathbf{w}_{rC}{}^H \mathbf{s}(t). \tag{16}$$

By calculating Doppler spectrum of $y(t)$, brightness B_{rC} (i.e., power density) and other spectral parameters are able to be computed with improved range resolution. B_{rC} is given by

$$B_{rC} = \frac{1}{\mathbf{e}^H \mathbf{R}^{-1} \mathbf{e}}. \tag{17}$$

Brightness at arbitrary Doppler velocity is also able to be calculated in the same manner as described in section 2.1.1.

Because of the limited width of transmitted pulse (i.e., wave form of transmitted pulse), received signal power within the range gate has range dependency (i.e., the received signal power decreases near the edge of range gate). Chen and Zecha (2009) proposed a practical method in order to correct the range weighting, and the correction results are also presented by Chen et al. (2009, 2010) and Chen & Zecha (2009).

Recently, Le et al. (2010) proposed a technique that improves received signal power by exploiting the temporal correlation difference between the desired signal and system noise. Le et al. (2010) showed that the technique has better performance of radar echo production than RIM in the low SNR regions.

2.2.2 Measurement results

In order to demonstrate that RIM is useful for improving range resolution, measurement results are presented. Fig. 5 shows an example of RIM measurement. The brightness profiles were produced with 6-m range intervals by applying RIM to received signals measured with the 2-µs transmitted pulse and four transmitted frequencies (914.0, 914.33, 915.33, and 916 MHz). The bandwidth of the RIM measurement was 2.5-MHz (2 MHz for the actual frequency spread and 0.5 MHz for the 2-µs transmitted pulse). During the experiment, the RIM measurement and single frequency measurement using the 0.5-µs transmitted pulse (i.e., corresponding to 75-m range resolution and 2-MHz bandwidth) were carried our alternatively. Although the frequency bandwidth difference between the two observation modes was as small as 500 kHz, it is clear that both the brightness produced by the Fourier-based method and by the Capon method show finer height variations than the backscattered power measured with the 75-m range resolution. Further, it is clear that the Capon method attains finer range resolution than the Fourier-based method. By applying RIM to the same dataset, Chilson et al. (2004) produced Doppler velocity with 15-m range intervals and showed that the Doppler velocity produced by RIM agreed well with that measured with the 0.5-µs transmitted pulse.

Fig. 6 shows an example of Kelvin-Helmholtz (KH) instability (i.e., shear instability) measured by the MU radar operated with a RIM observation mode (Fukao et al., 2011). The measurement was carried out using the new MU radar system upgraded in 2004 (Hassenpflug et al., 2008). Structure of KH billows is clearly seen in the brightness around 1.5 km from 00:35 to 00:53 (Fig. 6a). The resemblance to the evolution of KH vortices measured in the laboratory experiment (Patterson et al., 2006) is striking. Vertical wind velocity shows perturbations with magnitudes of 1 m s⁻¹ or more. Because accurate high-resolution vertical wind measurement is quite difficult for instruments other than clear-air Doppler radars (e.g., Fukao, 2007; Hocking 2011), measurements by wind profiling radars are indispensable to understand turbulence processes in the atmosphere. RIM measurements have revealed a fine structure of KH billows in the jet stream in the mid-latitudes (Luce et al., 2008) and upper-tropospheric easterly jet in the tropical region (Mega et al., 2010). Fukao et al (2011) carried out a statistical analysis of KH billows in order to quantify their occurrence frequency, spatial scales, energy dissipation rate, and vertical eddy diffusivity.

2.2.3 Advantage of range resolution enhancement using multiple frequencies

In the section, advantages of RIM over other methods are described. Radar range resolution is determined by the transmitted pulse width, and ranges typically a hundred to several hundreds of meters. However, for UHF wind profilers, a range resolution down to approximately 30 m is able be attained by transmitting shorter pulses (e.g., Wilson et al., 2005). Although the range resolution is able to be improved by transmitting shorter pulses, it requires not only wider bandwidth but also more transmitted power in order to keep the receiver sensitivity constant. RIM also contributes to efficient usage of frequencies. Chilson et al. (2003) pointed out that for the RIM measurement shown in Fig. 5, the spurious intensity of frequency power spectrum of transmitted pulses was smaller than the 0.4-µs transmitted pulse width which also uses 2.5-MHz bandwidth (see their Fig. 2). RIM is especially useful for VHF wind profiling radars because their frequency bandwidth allowed by license and that determined by antenna are limited.

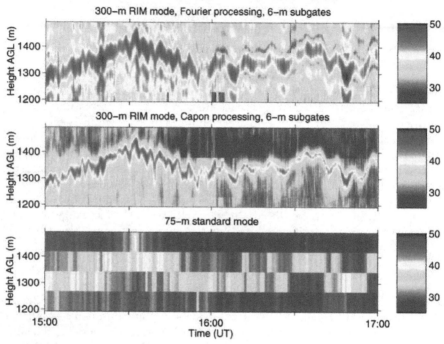

Fig. 5. Time-range plot of (top) brightness produced by the Fourier-based method, (center) brightness produced by the Capon method, and (bottom) backscattered power expressed in decibels (Chilson et al., 2003). The data was collected with the Platville 915-MHz tropospheric wind profiler. See text for details of the measurement.

Pulse compression using frequency-modulated continuous wave (FMCW) is also useful for UHF clear-air radars to improve their range resolution down to several meters or less (Eaton et al., 1995; Richter, 1969). However, a use of the FMCW pulse compression causes deterioration in data quality by range aliasing and by range ambiguity caused by Doppler shift of scatterers. Though range resolution in RIM depends on range distribution of scatterers, signal-to-noise ratio, and signal processing method used for RIM (e.g., Palmer et al., 1999; Luce et al., 2001; Smaïni et al., 2002), RIM does not suffer the drawbacks of using short transmitted pulse or FMCW pulse compression.

RIM contributes to reduce amount of on-line data size and computational complexity. In the case shown in Fig. 5, although the on-line sampling interval of the RIM measurement was 2 µs (i.e., 300-m range spacing), brightness data are able to be processed with finer (6-m) range spacing by off-line signal processing. On the other hand, on-line sampling interval in cases of short pulse transmission must be equivalent or shorter compared with the range resolution determined by transmitted pulse width. On-line sampling interval in cases of FMCW transmission also must be equivalent or shorter compared with the range resolution determined by sweep range of transmitted frequency.

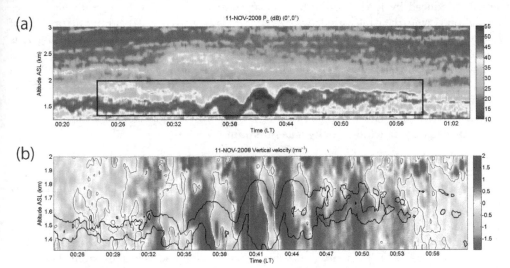

Fig. 6. Time-altitude plots of (a) brightness and (b) vertical wind velocity. Data were collected using vertically-pointing beam of the MU radar operated with the RIM observation mode (Fukao et al., 2011). The region within the black rectangular is plotted in panel (b). The thick black contours in panel (b) show 46 dB brightness level. The figure is reproduced with permission from the Royal Meteorological Society.

2.3 Further applications

2.3.1 High-resolution measurement using both spaced antennas and multiple frequencies

By using spaced receivers and multiple frequencies simultaneously, radar resolution is able to be improved both in angle and range, which leads to realize a three-dimensional (3-D) imaging. Signal processing procedures of the 3-D imaging are described by Yu & Palmer (2001). In 2004, the MU radar was upgraded for the 3-D imaging capability with 5 frequencies across a 1 MHz bandwidth and 25 intermediate frequency (IF) digital receivers (Hassenpflug et al., 2008). Using the 3-D imaging with the Capon method, Hassenpflug et al. (2008) showed a 3-D structure of radar echoes associated with billows of KH instability. Chen et al. (2008) applied the 3-D imaging to data collected by the MU radar in order to investigate relations between the angular distribution of clear-air echo power and tilted refractive-index layers caused by KH instability. The 3-D imaging technique is also able to be used for clutter mitigation. Using the MU radar, Yu et al. (2010) showed that the 3-D imaging provides comparable or better performance of both echo layer reconstruction and clutter mitigation compared to RIM.

The 3-D imaging technique is able to contribute to high-range-resolution wind measurement. Yu and Brown (2004) proposed a technique named RIM-SA that calculates wind velocity using both spaced antennas and multiple frequencies. In RIM-SA, first, RIM was separately applied to signals received by each of spaced antennas in order to produce high-range-resolution received signals. Next, SA technique was applied to the received signals produced by RIM in order to calculate horizontal wind velocity with high range resolution. Using data measured by MAPR, Yu and Brown (2004) produced profiles of horizontal wind velocity with 100-m intervals from a RIM-SA measurement using 2-μs transmitted pulse and four frequencies (914.667, 915.000, 916.000, and 916.667 MHz). Yu and Brown (2004) showed that the horizontal wind produced by RIM-SA agrees well with both wind velocity measured by a radiosonde and that measured by MAPR using the 0.67-μs transmitted pulse (i.e., 100-m range resolution).

2.3.2 Assessment of wind velocity measurement

Because DBS has been widely used for wind profiling radars, accuracy of wind velocity measured by DBS needs to be assessed. High resolution measurements using CRI and RIM provide the opportunity to assess the accuracy of wind velocity measured by DBS.

Wind field inhomogeneity within the scanning area of radar beams is a significant factor that produces errors of wind velocity measured by DBS. Cheong et al. (2008) carried out CRI measurement using TEP in order to obtain radial Doppler velocities from 490 beam directions, and used the radial Doppler velocity data in order to estimate how the wind field inhomogeneity affects the error of wind velocity measured by DBS. Cheong et al. (2008) concluded that optimal zenith angle of off-vertical radar beams is approximately 9-10° for minimizing the root-mean square (RMS) error in wind velocity measured by DBS, and that increasing number of off-vertical radar beams significantly reduces the RMS error in wind velocity measured by DBS.

Tilted refractive-index layers caused by KH instability deteriorate the measurement accuracy of vertical wind velocity because the tilted refractive-index layers cause contamination of horizontal wind velocity to the Doppler velocity of vertically-pointed radar beam, from which vertical wind is calculated in DBS (Muschinski, 1996; Yamamoto et al., 2003). Chen et al. (2008) applied the 3-D imaging to data collected by the MU radar in order to investigate relations between angular distribution of clear-air echo power and Doppler velocity measured by the vertically-pointed radar beam with improved range resolution. Chen et al. (2008) successfully showed the clear relation between the Doppler velocity bias measured by the vertically-pointed radar beam and the tilt of radar echo layers.

Though multistatic radar technique is not CRI, it is useful for measuring 3-D distribution of wind velocities. By installing two receiver arrays at approximately 1 km away from the westward and southward of the main antenna of the EAR, Nishimura et al. (2006) and (2010) realized the multistatic radar measurement of wind velocities. Their measurement results revealed 3-D wind perturbations down to the horizontal scale of 500 m.

2.3.3 High-resolution temperature measurement

Radio acoustic sounding system (RASS) is a radar remote sensing system for measuring profiles of temperature[1] with high time and height resolutions (e.g., May et al., 1990; Tsuda et al., 1989). RASS is also able to be used for monitoring humidity profiles (e.g., Furumoto et al., 2005; Tsuda et al., 2001). Using the MU radar, Furumoto et al. (2011) applied RIM to RASS measurement (RIM-RASS) in order to demonstrate that range resolution of temperature is improved down to approximately 60 m compared with the nominal range gate width of 150 m determined by the transmitted pulse width. Furumoto et al. (2011) pointed out that (1) sample time difference at an arbitrary range among data on different operational frequencies and (2) the Doppler shift bias due to the shape of range-gate weighting have to be corrected for RIM-RASS measurement, and developed an iteration algorithm that corrects (1) and (2). By applying CRI to temperature profiles measured by TEP with RASS, Dekker & Frasier (2004) retrieved virtual temperature structure function from horizontal distribution of temperature. The study by Dekker & Frasier (2004) indicates a usefulness of CRI to quantify turbulence intensity in the atmosphere.

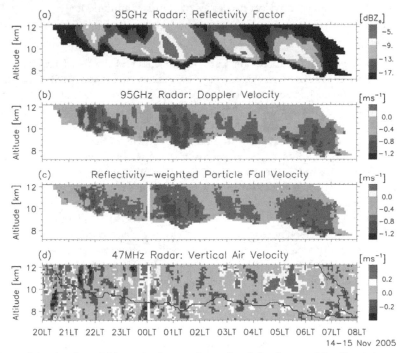

Fig. 7. Time-altitude plot of (a) equivalent radar reflectivity factor, (b) reflectivity-weighted particle fall velocity relative to the ground, (c) reflectivity-weighted particle fall velocity relative to the air, and (d) vertical wind velocity (Yamamoto et al., 2008). Equivalent radar reflectivity factor and reflectivity-weighted particle fall velocity relative to the ground were measured by a millimeter-wave radar. Vertical wind velocity was measured by the EAR. Horie et al. (2000) describe details of the millimeter-wave radar.

---

[1]To be exact, RASS measures a virtual temperature expressed by $Tv=T(1+0.61q)$, where Tv, T, and q are virtual temperature in K, tempeature in K, and water vapor mixing ratio in kg kg-1, respectively.

3. Multi-instrument measurements

Measurements using collocated wind profiling radar and millimeter-wave radar/lidar have been carried out in the last decade. In section 3, we focus on measurement results of vertical wind in and around clouds due to the following reasons. First, measurement of vertical wind in cloudy region with high resolution and accuracy is difficult by means other than wind profiling radars (e.g., Nishi et al., 2007). Though wind profiling radars measure only the area right above the radars, they are suitable for measuring wind motions with high time and vertical resolutions. Second, observations aiming at clarifying generation and maintenance mechanisms of clouds are indispensable. Clouds reflect a fraction of the solar radiation that would otherwise be absorbed at the Earth's surface. On the other hand, clouds also contribute to the blocking of outgoing radiation by the atmosphere (Wallace and Hobbs, 2006). However, the effects of clouds on the radiation budget of the earth–atmosphere system are not sufficiently quantified (Stephens, 2005). Measurements using collocated wind profiling radar and millimeter-wave radar/lidar provide an opportunity to observe dynamical and microphysical processes of clouds simultaneously.

3.1 Particle fall velocity

Fall velocity of cloud particles (hereafter particle fall velocity) is one of crucial factors that determine life time of clouds (e.g., Petch et al., 1997; Starr and Cox, 1985), and particle fall velocity relative to the air has been modeled in order to relate particle fall velocity to the size and shape of particles [e.g., Heymsfield & Iaquinta, 2000; Mitchell 1996; Mitchell and Heymsfield, 2005]. However, because fall velocity of cloud particles relative to the ground (i.e., a sum of particle fall velocity relative to the air and vertical wind velocity) is measured by millimeter-wave radars or lidars, measurement of vertical wind velocity is required to retrieve particle fall velocity relative to the air. Fig. 7 shows a measurement example. Reflectivity-weighted particle fall velocity relative to the ground, which was measured by a millimeter-wave radar, showed small-scale perturbations with a time scale less than several ten minutes (Fig. 7b). The small-scale perturbations were caused by vertical wind motions (Fig. 7d). By subtracting vertical wind velocity from the reflectivity-weighted particle fall velocity relative to the ground, reflectivity-weighted particle fall velocity relative to the air was retrieved (Fig. 7c). The small-scale perturbations caused by the vertical wind motions were not observed in the retrieved reflectivity-weighted particle fall velocity relative to the air. The results clearly demonstrate that vertical wind measurement by wind profiling radar is useful for measuring particle fall velocity accurately. Retrieval of particle fall velocity leads to data analysis to clarify cloud properties. Using data shown by Figs. 7a and c, Yamamoto et al. (2008) related particle fall velocity to cloud particle size. The collected data were also used to assess an algorithm that retrieves vertical wind velocity, effective diameter of cloud particles, and ice water content from millimeter-wave radar and lidar measurements (Sato et al., 2009). Using the MU radar and an X-band Doppler weather radar (Yamamoto et al., 2011a), Luce et al. (2010a) showed a similar retrieval result of particle fall velocity (see their Fig. 15).

3.2 Turbulence measurements in and around clouds

Fig. 8 presents a measurement result around the cloud bottom. Cloud bottom altitude estimated by the equivalent radar reflectivity factor showed protuberance structure. Such protuberances are referred to as mammatus cloud, whose generation mechanisms still remain to be discussed (Schultz et al., 2006). Brightness showed an increase around the cloud border due to the turbulence generation by KH instability. Vertical wind velocity showed oscillations exceeding ±3 m s^{-1} in magnitude due to KH instability, and played a role in producing protuberances of clouds through its downward motions. Luce et al. (2010a) suggested that the reduction of static stability at the interface between the clear air and cloud provided the favorable condition for triggering KH instabilities. Using the MU radar and a lidar (Behrendt et al., 2004), Luce et al. (2010b) also showed a correspondence between protuberances of clouds and vertical wind disturbances caused by atmospheric instability. Using the MU radar and a scanning millimeter-wave radar (Hamazu et al., 2003), Wada et al. (2005) showed a 3-D cell structure of cirrus clouds generated by KH instability, and the cell structures have greater vertical extension in the presence of upward vertical wind.

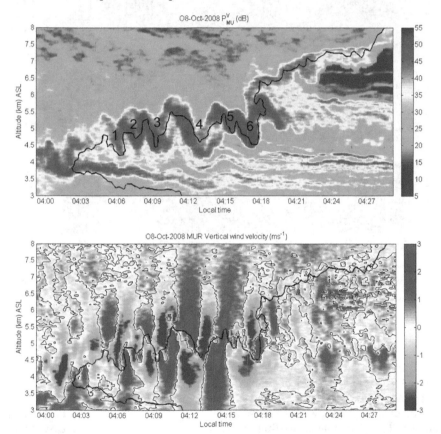

Fig. 8. Time-altitude plot of (upper) brightness and (lower) vertical wind velocity measured by the vertical beam of the MU radar operated with the RIM measurement mode (Luce et al., 2010a). Thick black curve shows a contour of -6 dBZ$_e$ equivalent radar reflectivity factor measured by a millimeter-wave radar. Numbers 1-6 shown in the upper panel indicate protuberances of clouds. Yamamoto et al. (2011b) describe details of the millimeter-wave radar.

Fig. 9 shows a measurement result around the cloud top. In the upper part of clouds, an increase of spectral width up to 0.5-0.7 m s^{-1}, which indicates the presence of small-scale turbulence triggered by convective instabilities, is observed (Fig. 9c). Further, in the top part of clouds (0-500 m below the cloud tops), downward wind up to 0.2-0.3 m s^{-1}, which was caused by radiative cooling, was observed. For further discussion on the generation mechanism of turbulence, see Yamamoto et al. (2009a). From a case study using data measured by the MU radar and Raman/Mie lidar, Yamamoto et al. (2009b) showed a clear relation between the cloud-top altitude of mid-latitude cirrus and the bottom altitude of subtropical jet with high time and altitude resolutions (12 min and 150 m, respectively).

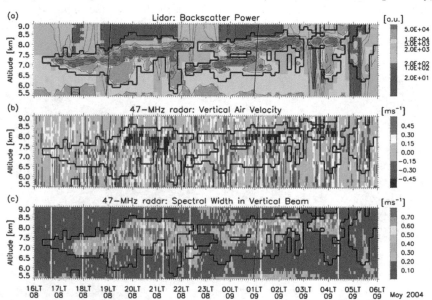

Fig. 9. Time-altitude plots of (a) backscattered power measured by a 532-nm Mie lidar, and (b) vertical wind velocity and (c) spectral width measured by the vertical beam of the EAR (Yamamoto et al., 2009a). Thick black curves in each panel indicate cloud boundaries estimated by the lidar backscattered power.

3.3 Other measurements

Measurements using wind profiling radars and other instruments are not limited to cloud researches. Measurement results of stratosphere-troposphere exchange (STE) processes have been reported in the last decade. Using observation network of wind profiling radar and lidar, Bertin et al. (2001) showed details of turbulent generation above and below the jet axis associated with a tropopause folding in the middle latitude. Using measurement data collected by the MU radar and ozonesonde, Gavrilov et al. (2006) compared distribution of turbulent diffusivity with vertical ozone flux. It is well known that clear-air echo power showed a vertical increase around the tropopause due to the increase of static stability or turbulence intensity (see section 14 of Hocking, 2011). Using data collected by the intensive observation of wind profiling radars and ozonesonde at Canada, Hocking et al. (2007) showed that stratospheric ozone has impacts on tropospheric ozone by its downward transportation to the ground, and demonstrated that tropopause altitude determined by wind profiling radar with high time resolution can be used to infer the possibility of ozone intrusions, as well as to represent tropopause motions in association with stratosphere–troposphere transport. Using the EAR and radiosondes, Fujiwara et al. (2003) showed the turbulence generation around the tropical tropopause caused by the Kelvin wave breaking.

Using results of RIM measurement by the MU radar and water vapor measurement by the Raman lidar (Behrendt, et al., 2004), Luce et al. (2010c) demonstrated that vertical gradient of humidity causes enhancement of backscattered intensity of radar echo by generating refractive index irregularities at the Bragg scale. Further coordinated observations would lead to clarify the radio scattering and propagation mechanisms in the clear air.

4. Conclusion

In the Chapter, new observations by wind profiling radars in the last decade were reviewed. In section 2, the signal processing and measurement results of radar interferometry techniques (CRI and RIM) were described. Resolution enhancements attained by CRI and RIM will be useful not only for scientific researches aiming at clarifying atmospheric processes but also practical radar utilization through improvement in data quality (i.e., clutter mitigation) and early detection of turbulence associated with storm or wind shear. Further assessments aiming at quantifying their usefulness under various weather conditions are desirable.

In section 3, it was demonstrated that multi-instrument measurement using wind profiling radar and millimeter-wave radar/lidar is useful to clarify phenomena related to cloud processes. In order to clarify interactions among dynamics, cloud physics, and radiation, simultaneous remote sensing and in-situ measurements are highly desirable. In-situ measurements of cloud particles, temperature, humidity, and radiation using balloons and/or aircrafts will contribute to quantify phenomena measured by wind profiling radar, millimeter-wave radar, and lidar. Numerical simulation is also important to assess the interactions. The author hopes that efforts to realize further coordinated studies are executed continuously.

5. Acknowledgment

This work was supported by the research grant for Exploratory Research on Sustainable Humanosphere Science from Research Institute for Sustainable Humanosphere (RISH), Kyoto University.

6. References

Behrendt, A., Nakamura, T. & Tsuda, T. (2004). Combined Temperature Lidar for Measurements in the Troposphere, Stratosphere, and Mesosphere, *Applied Optics*, Vol. 43, No.14, pp.2930-2939, doi:10.1364/AO.43.002930.

Bertin, F., Campistron, B., Caccia, J.L. & Wilson, R. (2001). Mixing Processes in a Tropopause Folding Observed by a Network of ST Radar and Lidar, Annales Geophysicae, Vol.19, No.8, pp.953-963, doi:10.5194/angeo-19-953-2001.

Capon, J. (1969). High-Resolution Frequency-Wavenumber Spectrum Analysis, *Proceedings of the IEEE*, Vol.57, No.8, pp.1408 -1418, doi:10.1109/PROC.1969.7278.

Chau, J.L. & Woodman, R.F. (2001). Three-Dimensional Coherent Radar Imaging at Jicamarca: Comparison of Different Inversion Techniques, *Journal of Atmospheric and Solar-Terrestrial Physics*, Vol.63, No.2-3, pp.253-261, doi:10.1016/S1364-6826(00)00142-5.

Chen, J.-S. (2004). On the Phase Biases of Multiple-frequency Radar Returns of Mesosphere-Stratosphere-Troposphere Radar, *Radio Science*, Vol.39, Art.no.RS5013, doi:10.1029/2003RS002885.

Chen, J.-S., Hassenpflug, G. & Yamamoto, M. (2008). Tilted Refractive-index Layers Possibly Caused by Kelvin–Helmholtz Instability and Their Effects on the Mean Vertical Wind Observed with Multiple-receiver and Multiple-frequency Imaging Techniques, *Radio Science*, Vol.43, Art.no.RS4020, doi:10.1029/2007RS003816.

Chen, J.-S., Su, C.-L., Chu, Y.-H., Hassenpflug, G. & Zecha, M. (2009). Extended Application of a Novel Phase Calibration Approach of Multiple-Frequency Range Imaging to the Chung-Li and MU VHF Radars. *Journal of Atmospheric and Oceanic Technology*, Vol.26, No.11, pp.2488-2500, doi:10.1175/2009JTECHA1295.1.

Chen, J.-S. & Zecha, M. (2009). Multiple-Frequency Range Imaging Using the OSWIN VHF Radar: Phase Calibration and First Results, *Radio Science*, Vol.44, Art.No.RS1010, doi:10.1029/2008RS003916.

Chen, J.-S., Furumoto, J.-I. & Nakamura, T. (2010). Effects of Radar Beam Width and Scatterer Anisotropy on Multiple-Frequency Range Imaging Using VHF Atmospheric Radar, *Radio Science*, Vol.45, Art.no.RS5001, doi:10.1029/2009RS004267.

Chen, M.-Y., Yu, T.-Y., Chu, Y.-H., Brown, W.O.J. & Cohn, S.A. (2007). Application of Capon Technique to Mitigate Bird Contamination on a Spaced Antenna Wind Profiler, *Radio Science*, Vol.42, Art.no.RS6005, doi:10.1029/2006RS003604.

Cheong, B.L., Hoffman, M.W., Palmer, R.D., Frasier, S. J. & López-Dekker, F.J. (2004). Pulse Pair Beamforming and the Effects of Reflectivity Field Variations on Imaging Radars, *Radio Science*, Vol.39, RS3014, doi:10.1029/2002RS002843.

Cheong, B.L., Hoffman, M.W., Palmer, R.D., Frasier, S.J. & López-Dekker, F.J. (2006). Phased-Array Design for Biological Clutter Rejection: Simulation and Experimental Validation, *Journal of Atmospheric and Oceanic Technology*, Vol.23, No4, .pp.585-598, doi:10.1175/JTECH1867.1.

Cheong, B.L., Palmer, R.D., Yu, T-Y., Yang, K-F., Hoffman, M.W., Frasier, S.J. & Lopez-Dekker F.J. (2008). Effects of Wind Field Inhomogeneities on Doppler Beam Swinging Revealed by an Imaging Radar, *Journal of Atmospheric and Oceanic Technology*, Vol.25, No.8, pp.1414–1422, doi:10.1175/2007JTECHA969.1.

Chilson, P.B., Yu, T.-Y., Strauch, R.G., Muschinski, A. & Palmer, R.D. (2003). Implementation and Validation of Range Imaging on a UHF Radar Wind Profiler, *Journal of Atmospheric and Oceanic Technology*, Vol.20, No.7, pp.987-996, doi:10.1175/1520-0426(2003)20<987:IAVORI>2.0.CO;2.

Chilson, P.B. (2004). The Retrieval and Validation of Doppler Velocity Estimates from Range Imaging. *Journal of Atmospheric and Oceanic Technology*, Vol.21, No.7, pp.1033–1043, doi:10.1175/1520-0426(2004)021<1033:TRAVOD>2.0.CO;2.

Dekker, P.L. & Frasier, S.J. (2004). Radio Acoustic Sounding with a UHF Volume Imaging Radar. *Journal of Atmospheric and Oceanic Technology*, Vol.21, No.5, pp.766–776, doi:10.1175/1520-0426(2004)021<0766:RASWAU>2.0.CO;2.

Eaton, F.D., McLaughlin, S.A. & J. R. Hines (1995). A New Frequency-Modulated Continuous Wave Radar for Studying Planetary Boundary Layer Morphology, *Radio Science*, Vol.30, No.1, pp.75–88, doi:10.1029/94RS01937.

Fujiwara, M., Yamamoto, M.K., Hashiguchi, H., Horinouchi, T. & Fukao, S. (2003). Turbulence at the Tropopause due to Breaking Kelvin Waves Observed by the Equatorial Atmosphere Radar, *Geophysical Research Letters*, Vol.30, No.4, 1171, doi:10.1029/2002GL016278.

Fukao, S., Sato, T., Tsuda, T., Yamamoto, M., Yamanaka, M.D. & Kato, S. (1990). MU radar: New Capabilities and System Calibrations, *Radio Science*, Vol.25, No.4, doi:10.1029/RS025i004p00477.

Fukao, S., Hashiguchi, H., Yamamoto, M., Tsuda, T., Nakamura, T., Yamamoto, M. K., Sato, T., Hagio, M. & Yabugaki, Y. (2003). Equatorial Atmosphere Radar (EAR): System Description and First Results, *Radio Science*, Vol.38, Art.No.1053, doi:10.1029/2002RS002767.

Fukao, S. (2007) Recent Advances in Atmospheric Radar Study, *Journal of the Meteorological Society of Japan*, Vol.85B, pp.215-239, doi:10.2151/jmsj.85B.215.

Fukao, S., Luce, H., Mega, T. & Yamamoto, M.K. (2011), Extensive Studies of Large-Amplitude Kelvin-Helmholtz Billows in the Lower Atmosphere with the VHF Middle and Upper Atmosphere radar (MUR), *Quarterly Journal of the Royal Meteorological Society*, Vol.137, No.657, 1019-1041, doi:10.1002/qj.807.

Furumoto, J.-I., Iwai, S., Fujii, H., Tsuda, T., Xin, W., Koike, T. & Bian L. (2005). Estimation of Humidity Profiles with the L-Band Boundary Layer Radar-RASS Measurements, *Journal of the Meteorological Society of Japan*, Vol.83, No.5, pp.895-908, doi:10.2151/jmsj.83.895.

Furumoto, J.-I., Shinoda, T., Matsugatani, A. & Tsuda T (2011). Measurements of Detailed Temperature Profiles within the Radar Range Gate Using the Range Imaging Technique, *Journal of Atmospheric and Oceanic Technology*, Vol.28, No.1, pp.22-36, doi:10.1175/2010JTECHA1506.1.

Gage, K.S. (1990). Radar Observations of the Free Atmosphere: Structure and Dynamics, In : *Radar in Meteorology*, D. Atlas, (Ed.), pp.534-565, American Meteorological Society, ISBN-10: 0933876866, Boston, Massachusetts, USA.

Gavrilov, N. M., Fukao, S., Hashiguchi, H., Kita, K., Sato, K., Tomikawa, Y. & Fujiwara, M. (2006), Combined MU Radar and Ozonesonde Measurements of Turbulence and Ozone Fluxes in the Tropo-stratosphere over Shigaraki, Japan, *Geophysical Research Letters*, Vol.33, Art.no.L09803, doi:10.1029/2005GL024002.

Hamazu, K., Hashiguchi H., Wakayama T., Matsuda T., Doviak R. J. & Fukao, S. (2003). A 35-GHz Scanning Doppler Radar for Fog Observations. *Journal of Atmospheric and Oceanic Technology*, Vol.20, No.7, pp.972-986, doi:10.1175/1520-0426(2003)20<972:AGSDRF>2.0.CO;2.

Hassenpflug, G., Yamamoto, M., Luce, H. & Fukao, S. (2008). Description and Demonstration of the New Middle and Upper Atmosphere Radar Imaging System: 1-D, 2-D, and 3-D Imaging of Troposphere and Stratosphere, *Radio Science*, Vol.43, Art.no.RS2013, doi:10.1029/2006RS003603.

Hélal,   D., Crochet, M., Luce, H. & Spano, E. (2001). Radar Imaging and High-Resolution Array Processing Applied to a Classical VHF-ST Profiler, *Journal of Atmospheric and Solar-Terrestrial Physics*, Vol.63, No.2-3, doi:10.1016/S1364-6826(00)00135-8.

Heymsfield, A.J. & Iaquinta, J. (2000). Cirrus Crystal Terminal Velocities, *Journal of the Atmospheric Sciences*, Vol.57, No.7, pp.916–938, doi:10.1175/1520-0469(2000)057<0916:CCTV>2.0.CO;2.

Hocking, W.K. (1985). Measurement of Turbulent Energy Dissipation Rates in the Middle Atmosphere by Radar Techniques: A Review, *Radio Science*, Vol.20, No.6, pp.1403-1422, doi:10.1029/RS020i006p01403.

Hocking, W.K., Carey-Smith T., Tarasick, D.W., Argall P.S., Strong, K., Rochon, Y., Zawadzki, I. & P. A. Taylor (2007). Detection of Stratospheric Ozone Intrusions by Windprofiler Radars, *Nature*, Vol.450, No.7167, pp.281-284, doi:10.1038/nature06312.

Hocking, W.K. (2011). A Review of Mesosphere–Stratosphere–Troposphere (MST) Radar Developments and Studies, Circa 1997–2008, *Journal of Atmospheric and Solar-Terrestrial Physics*, Vol.73, No.9, pp.848-882, doi:10.1016/j.jastp.2010.12.009.

Horie, H., Iguchi, T., Hanado, H., Kuroiwa, H., Okamoto, H. & Kumagai, H. (2000). Development of a 95-GHz Airborne Cloud Profiling Radar (SPIDER) - Technical aspects -, *IEICE Transactions on Communications*, Vol.E83-B, No.9, pp.2010-2020.

Ishihara, M., Kato, Y., Abo, T., Kobayashi, K. & Izumikawa, Y. (2006). Characteristics and Performance of the Operational Wind Profiler Network of the Japan Meteorological Agency, *Journal of the Meteorological Society of Japan*, Vol.84, No.6, pp.1085-1096, doi:10.2151/jmsj.84.1085.

Kamio, K., Nishimura, K. & Sato, T. (2004). Adaptive Sidelobe Control for Clutter Rejection of Atmospheric Radars, *Annales Geophysicae*, Vol.22, No.11, pp.4005-4012, doi:10.5194/angeo-22-4005-2004.

Kollias, P., Clothiaux, E.E., Miller, M.A., Albrecht, B.A., Stephens, G.L. & Ackerman T. P. (2007). Millimeter-Wavelength Radars: New Frontier in Atmospheric Cloud and Precipitation Research, *Bulletin of the American Meteorological Society*, Vol.88, No.10, pp.1608-1624, doi:10.1175/BAMS-88-10-1608.

Larsen, M.F. & Röttger, J. (1989). The Spaced Antenna Technique for Radar Wind Profiling, *Journal of Atmospheric and Oceanic Technology*, Vol.6, No.6, 920–938, doi:10.1175/1520-0426(1989)006<0920:TSATFR>2.0.CO;2.

Le, K.D., Palmer, R.D., Cheong, B.L., Yu, T.-Y., Zhang, G. & Torres, S.M. (2010). Reducing the Effects of Noise on Atmospheric Imaging Radars Using Multilag Correlation, *Radio Science*, Vol.45, RS1008, doi:10.1029/2008RS003989.

Luce, H., Yamamoto, M., Fukao, S., Helal, D. & Crochet M. (2001). A Frequency Domain Radar Interferometric Imaging (FII) Technique Based on High-Resolution Methods, *Journal of Atmospheric and Solar-Terrestrial Physics*, Vol.63, No.2-3, pp.221-234, doi:10.1016/S1364-6826(00)00147-4.

Luce, H., Hassenpflug, G., Yamamoto, M., Fukao, S. & Sato, K. (2008). High-Resolution Observations with MU Radar of a KH Instability Triggered by an Inertia–Gravity Wave in the Upper Part of a Jet Stream, *Journal of the Atmosphric Sciences*, Vol.65, No.5, pp.1711–1718, doi:10.1175/2007JAS2346.1.

Luce, H., Mega, T., Yamamoto, M.K., Yamamoto, M., Hashiguchi, H., Fukao, S., Nishi, N., Tajiri, T. & Nakazato, M. (2010a). Observations of Kelvin-Helmholtz Instability at a

Cloud Base with the Middle and Upper Atmosphere (MU) and Weather Radars, *Journal of Geophysical Research*, Vol.115, Art.no.D19116, doi:10.1029/2009JD013519.

Luce, H., Nakamura, T., Yamamoto, M. K., Yamamoto, M. & Fukao, S. (2010b). MU Radar and Lidar Observations of Clear-Air Turbulence Underneath Cirrus, *Monthly Weather Review*, Vol.138, No.2, pp.438-452, doi:10.1175/2009MWR2927.1.

Luce, H., Takai, T., Nakamura, T., Yamamoto, M., Fukao, S. (2010c). Simultaneous Observations of Thin Humidity Gradients in the Lower Troposphere with a Raman Lidar and the Very High-Frequency Middle- and Upper-Atmosphere Radar, *Journal of Atmospheric and Oceanic Technology*, Vol.27, No. 5, pp.950–956, doi:10.1175/2010JTECHA1372.1.

May, P.T. (1990). Spaced antenna versus Doppler Radars: A Comparison of Techniques Revisited, *Radio Science*, Vol.25, No.6, pp.1111–1119, doi:10.1029/RS025i006p01111.

May, P.T., Strauch, R.G., Moran, K.P. & Ecklund, W.L. (1990). Temperature Sounding by RASS with Wind Profiler Radars: A Preliminary Study, *IEEE Transactions on Geoscience and Remote Sensing*, Vol.28, No.1, pp.19-28, doi:10.1109/36.45742.

Mead, J.B., Hopcraft, G., Frasier, S.J., Pollard, B.D., Cherry, C.D., Schaubert, D.H. & McIntosh, R.E. (1998). A Volume-Imaging Radar Wind Profiler for Atmospheric Boundary Layer Turbulence Studies, *Journal of Atmospheric and Oceanic Technology*, Vol.15, No.4, pp.849–859, doi:10.1175/1520-0426(1998)015<0849:AVIRWP>2.0.CO;2.

Mega, T., Yamamoto, M.K., Luce, H., Tabata, Y., Hashiguchi, H., Yamamoto, M., Yamanaka, M. D. & Fukao, S. (2010). Turbulence Generation by Kelvin-Helmholtz Instability in the Tropical Tropopause Layer Observed with a 47 MHz Range Imaging Radar, *Journal of Geophysical Research*, Vol.115, Art.no.D18115, doi:10.1029/2010JD013864.

Met Office (2011), CWINDE Network, Available from http://www.metoffice.gov.uk/science/specialist/cwinde/.

Mitchell, D.L. (1996). Use of Mass- and Area-Dimensional Power Laws for Determining Precipitation Particle Terminal Velocities. *Journal of the Atmospheric Sciences*, Vol.53, No.12, pp.1710-1723, doi:10.1175/1520-0469(1996)053<1710:UOMAAD>2.0.CO;2.

Mitchell, D.L. & Heymsfield, A.J. (2005). Refinements in the Treatment of Ice Particle Terminal Velocities, Highlighting Aggregates. *Journal of the Atmospheric Sciences*, Vol.62, No.5, pp.1637-1644, doi:10.1175/JAS3413.1.

Muschinski, A. (1996). Possible Effect of Kelvin-Helmholtz Instability on VHF Radar Observations of the Mean Vertical Wind, *Journal of Applied Meteorology*, Vol.35, No.12, pp.2210-2217, doi:10.1175/1520-0450(1996)035<2210:PEOKHI>2.0.CO;2.

Nishi, N., Yamamoto, M. K., Shimomai, T., Hamada, A. & Fukao, S. (2007), Fine Structure of Vertical Motion in the Stratiform Precipitation Region Observed by a VHF Doppler Radar Installed in Sumatra, Indonesia, *Journal of Applied Meteorology and Climatology*, Vol.46, No.4, pp.522-537, doi:10.1175/JAM2480.1.

Nishimura, K., Gotoh, E. & Sato, T. (2006). Fine Scale 3D Wind Field Observation with a Multistatic Equatorial Atmosphere Radar, *Journal of the Meteorological Society of Japan*, Vol.84A, pp.227-238, doi:10.2151/jmsj.84A.227.

Nishimura, K., Harada, T. & Sato, T. (2010). Multistatic Radar Observation of a Fine-Scale Wind Field with a Coupling-Compensated Adaptive Array Technique, *Journal of the Meteorological Society of Japan*, Vol.88, No.3, pp.409-424, doi:10.2151/jmsj.2010-309.

Palmer, R.D., Gopalam, S., Yu, T.-Y. & Fukao S. (1998). Coherent Radar Imaging Using Capon's Method, *Radio Science*, Vol.33, No.6, pp.1585–1598, doi:10.1029/98RS02200.

Palmer, R.D., Yu, T.-Y. & Chilson, P. B. (1999). Range Imaging Using Frequency Diversity, *Radio Science*, Vol.34, No.6, pp.1485-1496, doi:10.1029/1999RS900089.

Palmer, R.D., Chilson, P.B., Muschinski, A., Schmidt, G., Yu, T.-Y. & Steinhagen H. (2001). SOMARE-99: Observations of Tropospheric Scattering Layers Using Multiple-Frequency Range Imaging, *Radio Science*, Vol.36, No.4, pp.681–693, doi:10.1029/1999RS002307.

Palmer, R.D., Cheong, B.L., Hoffman, M.W., Frasier, S.J., López-Dekker, F.J. (2005). Observations of the Small-Scale Variability of Precipitation Using an Imaging Radar. *Journal of Atmospheric and Oceanic Technology*, Vol.22, No.8, pp.1122-1137, doi:10.1175/JTECH1775.1.

Patterson, M.D., Caulfield, C.P., McElwaine, J.N. & Dalziel, S.B. (2006). Time-dependent Mixing in Stratified Kelvin-Helmholtz Billows: Experimental Observations, *Geophysical Research Letters*, Vol.33, Art.no.L15608, doi:10.1029/2006GL026949.

Petch, J.C., Craig, G.C. & Shine, K.P. (1997). A Comparison of Two Bulk Microphysical Schemes and Their Effects on Radiative Transfer Using a Single column Model, *Quarterly Journal of the Royal Meteorological Society*, Vol.123, No.542, pp.1561-1580, doi:10.1002/qj.49712354206.

Pollard, B.D., Khanna, S., Frasier, S.J., Wyngaard, J.C., Thomson, D.W., & McIntosh, R.E. (2000). Local Structure of the Convective Boundary Layer from a Volume-Imaging Radar, *Journal of the Atmospheric Sciences*, Vol.57, No.15, pp.2281–2296, doi:10.1175/1520-0469(2000)057<2281:LSOTCB>2.0.CO;2.

Richter, J.H. (1969). High Resolution Tropospheric Radar Sounding, *Radio Science*, Vol.4, No.12, pp.1261–1268, doi :10.1029/RS004i012p01261.

Sato, K. & Coauthors (2009). 95-GHz Doppler Radar and Lidar Synergy for Simultaneous Ice Microphysics and in-cloud Vertical Air Motion Retrieval, *Journal of Geophysiscal Research*, Vol.114, Art.no.D03203, doi:10.1029/2008JD010222.

Schultz, D. M. & Coauthors (2006). The Mysteries of Mammatus Clouds: Observations and Formation Mechanisms. *Journal of the Atmospheric Sciences*, Vol.63, No.10, pp.2409–2435, doi:10.1175/JAS3758.1.

Smaïni, L., Luce, H., Crochet, M. & Fukao, S. (2002). An Improved High-Resolution Processing Method for a Frequency Domain Interferometric Imaging (FII) Technique. *Journal of Atmospheric and Oceanic Technology*, Vol.19, No.6, pp.954–966, doi:10.1175/1520-0426(2002)019<0954:AIHRPM>2.0.CO;2.

Stanley G.B., Schwartz, B.E., Koch, S.E., & Szoke, E.J. (2004). The Value of Wind Profiler Data in U.S. Weather Forecasting, *Bulletin of the American Meteorological Society*, Vol.85, No.12, pp.1871–1886, doi:10.1175/BAMS-85-12-1871.

Starr, D. O'C. & Cox, S.K. (1985). Cirrus Clouds. Part II: Numerical Experiments on the Formation and Maintenance of Cirrus, *Journal of the Atmospheric Sciences*, Vol.42, No.23, pp.2682–2694, doi:10.1175/1520-0469(1985)042<2682:CCPINE>2.0.CO;2.

Stephens, G.L. (2005). Cloud Feedbacks in the Climate System: A Critical Review, *Journal of Climate*, Vol.18, No.2, pp.237–273, doi:10.1175/JCLI-3243.1.

Tsuda, T. and Coauthors (1989). High Time Resolution Monitoring of Tropospheric Temperature with a Radio Acoustic Sounding System (RASS), *Pure and Applied Geophysics*, Vol.130, No.2-3, pp.497–507, doi:10.1007/BF00874471.

Tsuda, T., Miyamoto, M. & Furumoto, J.-I. (2001). Estimation of a Humidity Profile Using Turbulence Echo Characteristics, *Journal of Atmospheric and Oceanic Technology*, Vol.18, No.7, pp.1214–1222, doi:10.1175/1520-0426(2001)018<1214:EOAHPU>2.0.CO;2.

Vaughn, C.R. (1985). Birds and Insects as Radar Targets: A Review, *Proceedings of the IEEE*, Vol.73, No.2, pp. 205 -227, doi:10.1109/PROC.1985.13134.

Wallace, J.M. & Hobbs P.V. (2006). *Atmospheric Science: An Introductory Survey*, 2nd Edition, pp.447-450, Academic Press, ISBN-10: 012732951X, San diego, California, USA.

Wada, E., Hashiguchi, H., Yamamoto, M.K., Teshiba, M. & Fukao, S. (2005). Simultaneous Observations of Cirrus Clouds with a Millimeter-Wave Radar and the MU Radar, *Journal of Applied Meteorology*, Vol.44, No.3, pp.313-323, doi:10.1175/JAM2191.1.

Wandinger, U. (2005). Introduction to Lidar, In : *Lidar : Range-Resolved Optical Remote Sensing of the Atmosphere*, C. Weitkamp, (Ed.), pp.1-18, Springer, ISBN-10: 0387400753, New York, USA.

Wilson, R., Dalaudier, F. & Bertin, F. (2005). Estimation of the Turbulent Fraction in the Free Atmosphere from MST Radar Measurements, *Journal of Atmospheric and Oceanic Technology*, Vol.22, No.9, pp.1326–1339, doi:10.1175/JTECH1783.1.

Wilczak, J. M. & Coauthors (1995). Contamination of Wind Profiler Data by Migrating Birds: Characteristics of Corrupted Data and Potential Solutions. *Journal of Atmospheric and Oceanic Technology*, Vol.12, No.3, pp.449-467, doi:10.1175/1520-0426(1995)012<0449:COWPDB>2.0.CO;2.

Woodman, R. (1997). Coherent Radar Imaging: Signal Processing and Statistical Properties, *Radio Science*, Vol.32, No.6, pp.2373-2391, doi:10.1029/97RS02017.

Yamamoto, M.K., Fujiwara M., Horinouchi, T., Hashiguchi, H. & Fukao S. (2003). Kelvin-Helmholtz Instability around the Tropical Tropopause Observed with the Equatorial Atmosphere Radar, *Geophysical Research Letters*, Vol.30, No.9, Art.no.1476, doi:10.1029/2002GL016685.

Yamamoto, M.K. & Coauthors (2008). Observation of Particle Fall Velocity in Cirriform Cloud by VHF and Millimeter-Wave Doppler radars, *Journal of Geophysical Research*, Vol.113, Art.no.D12210, doi:10.1029/2007JD009125.

Yamamoto, M.K. & Coauthors (2009a). Vertical Air Motion in Midlevel Shallow-Layer Clouds Observed by 47-MHz Wind Profiler and 532-nm Mie lidar: Initial Results, *Radio Science*, Vol.44, Art.no.RS4014, doi:10.1029/2008RS004017.

Yamamoto, M.K. & Coauthors (2009b). Wind Observation around the Tops of the Midlatitude Cirrus by the MU radar and Raman/Mie lidar, *Earth Planets Space*, Vol.61, e33-e36 (Available from
http://www.terrapub.co.jp/journals/EPS/abstract/6107/6107e033.html).

Yamamoto, M.K. & Coauthors (2011a). Doppler Velocity Measurement of Portable X-band Weather Radar Equipped with Magnetron Transmitter and IF Digital Receiver, *IEICE Transactions on Communications*, Vol.E94-B, No.6, pp.1716-1724, doi:10.1587/transcom.E94.B.1716.

Yamamoto, M.K. & Coauthors (2011b). Assessment of Radar Reflectivity and Doppler Velocity Measured by Ka-band FMCW Doppler Weather Radar, *Journal of Atmospheric Electricity*, Vol.31, No.2, pp.85-94.

Yu, T.-Y., Palmer, R.D. & Hysell, D.L. (2000). A Simulation Study of Coherent Radar Imaging, *Radio Science*, Vol.35, No.5, pp.1129-1141, doi:10.1029/1999RS002236.

Yu, T.-Y. & Palmer, R.D. (2001). Atmospheric Radar Imaging Using Multiple-Receiver and Multiple-frequency Techniques, *Radio Science*, Vol.36, No.6, pp.1493–1503, doi:10.1029/2000RS002622.

Yu, T.-Y. & Brown, W.O.J. (2004). High-Resolution Atmospheric Profiling Using Combined Spaced Antenna and Range Imaging Techniques, *Radio Science*, Vol.39, Art.no.RS1011, doi:10.1029/2003RS002907.

Yu, T.-Y., Furumoto, J.-I. & Yamamoto M. (2010). Clutter Suppression for High-Resolution Atmospheric Observations Using Multiple Receivers and Multiple Frequencies, *Radio Science*, Vol.45, Art.no.RS4011, doi:10.1029/2009RS004330.

2

Retrieving High Resolution 3-D Wind Vector Fields from Operational Radar Networks

Olivier Bousquet

Météo-France, Centre National de Recherches Météorologiques
France

1. Introduction

The ability to retrieve 3-D wind vector fields in a fully operational framework has a potentially wide-ranging impact on a variety of meteorological research and operational meteorological applications. This capability was recently evaluated by the French Weather Service in the course of an upgrade program aiming to introduce Doppler and dual-polarimetric technologies within its radar network. Starting in November 2006, real-time multiple-Doppler wind fields have been produced routinely every 15 minutes for 2 years within a 320x320 km² domain centered on Paris city (Bousquet et al. 2007, 2008a). The evaluation of wind fields synthesized in this framework was carried out from observations collected in a variety of weather situations including low-level cyclones, frontal systems and squall lines. Wind vectors retrieved in the greater Paris area were generally proved very realistic and have been found reliable enough to be used for research applications (mesoscale meteorology, statistical analysis) and numerical model verification (Bousquet et al. 2008b). In order to prepare for the field phase of the international Hydrological Mediterranean Experiment (HyMeX[1]), which will be conducted in 2012-2013, this analysis has later been successfully extended to regions of complex terrain located in the southern part of the country (Bousquet, 2009). The ability to perform operational wind retrieval in mountainous areas is an important step to improve our understanding of orographic precipitation developing in these usually poorly instrumented regions, and also demonstrates that operational real-time wind retrieval could potentially be carried out over the entire French territory – ground elevation exceeds 500 m over ~ 1 fifth of mainland France – which was the initial objective of the French Weather Service when this experiment was started. In 2009, the wind retrieval analysis was therefore extended to the full metropolitan radar network with the goal to implement an operational, nationwide, three-dimensional reflectivity and wind field mosaic to be ultimately delivered to forecasters and modelers, as well as automatic nowcasting systems for air traffic management purposes. In this composite analysis, which is expected to become operational in 2013, data collected by all 24 radars of the French radar network are concentrated, pre-processed and combined in real-time at a frequency of 15 minutes to retrieve the complete wind vector (u,v,w) and

---

[1] HyMeX is an international program that aims at a better understanding and quantification of the hydrological cycle and related processes in the Mediterranean, with emphasis on high-impact orographic weather events. Information about this program can be found on http://www.hymex.org

reflectivity fields at a horizontal resolution of 2.5 km within a domain of approximately 1000 x 1000 x 12 km³. This achievement, through the size of the retrieval domain, the number of Doppler radars (24) involved in the analysis, and the fact that retrieved three-dimensional winds rely exclusively on an operational infrastructure, represents an unprecedented breakthrough in operational applications of the Doppler information, which are so far generally limited to clutter filtering and Velocity Azimuth Display analysis (VAD, Browning and Wexler 1968).

The present study aims at examining the technical requirements needed to achieve real-time operational multiple-Doppler analysis in an operational framework, as well as to evaluate the performance and usefulness of three-dimensional wind composite retrieved from operational radar systems. After a recall of the principle of dual-Doppler wind retrieval and a description of the French radar network characteristics winds retrieved in this operational framework are evaluated using outputs produced during various high impact weather events that recently occurred over mainland France. This includes the extratropical storm Klaus, already referred to as the storm of the decade by many European forecasters, which stroke France with hurricane strength winds on 24 January 2009, as well as a heavy orographic precipitation event that occurred over the Massif Central Mountains in September 2010. The potential value of these unique datasets for both operational and a research application is also discussed with emphasis on the upcoming HyMeX program.

2. Wind retrieval

2.1 Principle of dual-Doppler wind retrieval

All current **dual-Doppler** analysis techniques originate from the seminal work of Armijo (1969) who demonstrates that it was possible to retrieve the three components of the wind field in precipitating area using i) the precipitation radial velocity data collected by 2 Doppler radars, ii) the anelastic air mass continuity equation and iii) empirical relationships between radar reflectivity and precipitation fallspeed (Z-R relationships). Among the numerous methods based on Armijo's methodology, three groups can be identified: i) analytical approaches (e.g. Scialom and Lemaitre 1990), which aim at retrieving the wind field under its analytical form, ii) coplane techniques (Lhermitte and Miller 1970, Chong and Testud 1996), which allow resolving the wind field in a cylindrical space and, iii) Cartesian methods, which aim to resolve the wind field in a Cartesian space by the mean of an iterative process between the radial velocity equations and the continuity equation (Heymsfield 1978). Among those 3 families, the latter is the most computationally efficient and the easiest to implement, making it by far the most popular method with researchers.

Despite its relative simplicity, the Cartesian method has nevertheless been an inexhaustible source of inspiration for radar scientists and led to more than 50 peer-reviewed publications over the last 30 years. Although highlighting a particular method among all available Cartesian techniques is difficult, one could mention the approaches proposed by Gamache (1995) – detailed information about this technique can be found in the appendix of Reasor et al. (2009) - Bousquet and Chong (1998) and Gao et al. (1999). All three techniques significantly improved the Cartesian approach by suppressing the iterative process traditionally associated with Cartesian retrieval algorithm (see below).

2.2 The MUSCAT analysis

The wind retrieval technique used by the French Weather Service is the Multiple-Doppler Synthesis and Continuity Adjustment Technique (MUSCAT), which is a variational algorithm allowing for a simultaneous and computationally efficient solution of the three Cartesian wind components (u, v, w). This method was originally proposed by Bousquet and Chong (1998) to overcome the drawbacks of iterative analysis techniques used to process data collected by airborne Doppler radars. It was later adapted to ground-based radars (Chong and Bousquet 2001) in order to analyze observations collected during the field phase of Mesoscale Alpine Programme (Bougeault et al. 2001) during which it was used in a semi-operational mode to guide research aircrafts in the field (Chong et al., 2000). The MUSCAT algorithm has been used successfully for more than 10 years in order to synthesize data collected by ground-based and mobile research radars (e.g. Bousquet and Chong 2000, Georgis al. 2003, Bousquet and Smull 2006), and has even been applied to wind lidars (Drechsel et al. 2009). In order to use this algorithm in an operational framework a modification of the initial MUSCAT formalism was proposed by Bousquet et al. (2008a) to take into account extensive radar separation distances prevailing in operational radar networks.

The current form of the MUSCAT algorithm is given hereafter. It consists in a global minimization, in a least-squares sense, of the function F:

$$F(u,v,w) = \int_S \left[A(u,v,w) + B(u,v,w) + C(u,v,w) + D(u,v) \right] dx\, dy \tag{1}$$

such that,

$$\frac{\partial F}{\partial u} \approx 0, \quad \frac{\partial F}{\partial v} \approx 0 \quad and \quad \frac{\partial F}{\partial w} \approx 0. \tag{2}$$

The expression of term A is given by:

$$A_{i,j}(u,v,w) = \frac{1}{N} \sum_{p=1}^{n_p} \sum_{q=1}^{n_q(p)} \omega_q \left[\alpha_q u + \beta_q v + \gamma_q (w + v_T) - V_q \right]^2 \tag{3}$$

where u, v, w, are the components of the wind field at grid point (i,j); V_T is the terminal particle fallspeed at grid point (i,j) evaluated from empirical relationships with pre-interpolated radar reflectivity, subscript q defines the qth measurement of a total number n_q that is observed from the pth radar and that falls inside an ellipsoid of influence centered on the grid point (i,j); N is the total number of n_q's over the considered domain; ω_q is the Cressman weighting function depending on the distance between measurement q within the ellipsoid and the considered grid point; and n_p is the total number of radars covering grid point (i,j) (≥ 2).

This term represents the optimal least-squares fit of the observed radial Doppler velocities to the derived wind component. The Cressman distance-dependent weighting function accounts for non-collocated data and grid point values, and allows the interpolation of the radar data onto the Cartesian grid of interest, in the data fit. In the current framework, the interpolation is performed using a fixed horizontal influence radius of the Cressman weighting function R_H of 3 km and a variable vertical radius of influence R_V that matches the beamwidth of ARAMIS radars.

Term B is given by:

$$B(u,v,w) = \mu_1 \left(\frac{\partial u}{\partial x} + \frac{\partial v}{\partial y} + \frac{1}{\rho} \frac{\partial \rho w}{\partial z} \right)^2 \tag{4}$$

where ρ is the air density and μ_1 is a normalized weighting parameter that controls the relative importance of this term with respect to term A.

This term represents the least-squares adjustment with respect to mass continuity. It is formulated for each individual grid box in terms of mass flux throughout the faces of the considered box, which allows solving the wind field over both flat and complex terrains (Chong and Cosma 2000). In this formalism the wind components estimated at the previously investigated plane are used as input values to solve the wind at the current level. In order to initialize the wind synthesis, horizontal components are assumed constant between the surface and the first plane for which the solution for the wind components is searched.

Term C is given by:

$$C(u,v,w) = \mu_2 \left[J(u) + J(v) + J(w) \right]^2 \tag{5}$$

where J is a differential operator based on 2nd and 3rd derivatives of the wind components [see Bousquet and Chong (1998) for the detailed expression of this term].

It is a constraint that acts as a low pass filter. Term C allows decreasing small-scale wind variations through the minimization of the second- and third order derivatives. It is controlled by a weighting factor μ_2, which is a function of the cutoff wavelength of the filter. In addition to provide more regular fields, this term is also essential to obtain an objective solution in regions of ill-conditioned analysis through realizing a regular extrapolation in these regions from surrounding properly conditioned areas.

Finally, D is given by:

$$D(u,v) = \mu_3 \left(\frac{\partial (u \sin^2 \alpha_i - v \sin \alpha_i \cos \alpha_i)^2}{\partial x} + \frac{\partial (-u \sin \alpha_i \cos \alpha_i + v \cos^2 \alpha_i)}{\partial y} \right)^2 \tag{6}$$

where μ_3 is given by:

$$\mu_3 = \cos^4 \beta_m \tag{7}$$

and β_m defines the angle between the horizontal projection of the two radar beam axes.

Term D is a constraint that is applied in regions covered by only 2 radars. It allows minimizing the variation of the cross-baseline component of the wind. Its effect is maximum close to the radar baseline and weak in properly conditioned areas (the reader is referred to Bousquet et al. 2008b for more details about this specific term).

According to Equations 3-6, the three-dimensional wind field reconstructed by MUSCAT represents a least squares fit to the available observations and does not perfectly satisfy the mass continuity equation. In order to obtain a wind field that truly verifies this equation, a posteriori upward integration is needed to limit the errors in the vertical component. Many integration methods can be found in the literature to accomplish this last step. For operational applications the basic adjustment technique proposed by O'Brien (1970) was chosen. This method consists in adjusting the vertical velocity by forcing w to be zero at the bottom and top of any column and to linearly distribute the error throughout the column.

3. Operational radar data

3.1 Overview of the French radar network

The French metropolitan operational radar network ARAMIS (Application Radar a la Météorologie Infra-Synoptique) is composed of 16 C-band and 8 S-band Doppler radars. 10 of these radars are equipped with dual-polarization capabilities and we expect the network to be fully polarimetric by 2016. The ARAMIS network covers about 95 % of mainland France with radar baselines fluctuating from ~200 km in the northern part of the country to ~ 60 km in Southeastern France (Fig. 1). It is composed of 5 different types of radars ranging from 25-year old facilities, with limited workload and scanning capabilities, to state of the art dual-polarimetric radar systems. All systems are however equipped with the same radar processor, which allows producing harmonized products despite different hardware characteristics. The French Weather Service, Météo-France, also operates 8 S-band Doppler radars overseas, as well as 2 "gap filling" polarimetric X-band radars in the French Alps. Those 10 additional radars are not considered in this paper.

3.2 The triple PRT Doppler scheme

Doppler capabilities have been introduced in 2002 in the frame of an 8-year upgrade program aiming to modernize the network and fill some gaps in the radar coverage. During this period, all radars have been progressively equipped with the triple pulse rise time (PRT) Doppler scheme proposed by Tabary et al. (2006). This scheme, which is based on the approach proposed by Zrnic and Mahapatra (1985), consists in operating the radars at different pulse repetition frequencies (PRF) in order to mitigate the effect of the "Doppler dilemma" ensuing from the inverse relationship between the unambiguous range and the unambiguous velocity (Doviak and Zrnic 1993). The French scheme yields an extended Nyquist velocity of 60 m s⁻¹ up to a range of ~ 250 km. Its particularity lies in the fact that interleaved frequencies are rather low (379, 321, and 305 Hz). This allows this scheme to be indifferently implemented on old and new radars, but also to get rid of potential second trip returns.

The capacity to retrieve multiple-Doppler winds in an operational framework directly arises from the mitigation of the Doppler dilemma. Indeed, one of the main consequences of this long lasting issue is to limit operational Doppler measurements to short range (~ 100 km) so as to mitigate velocity ambiguities resulting from the aliasing of radial velocities outside of the Nyquist interval. Because operational Doppler radars are generally separated by hundreds of kilometers, this limitation in range dramatically impedes overlapping areas where airflow can be successfully reconstructed, and does not allow for adequate dual- or multiple-Doppler wind synthesis.

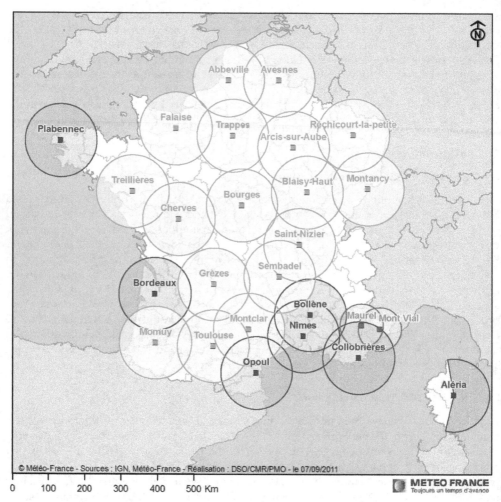

Fig. 1. Map of the French operational radar network ARAMIS. The 100 km (50 km for X-band) ranges of measurement associated with each radar are shown by circles. Green, purple and yellow colors correspond to C-, S-, and X-band radars, respectively.

In France, the ability to measure Doppler velocities at long range allows for a significant dual- and multiple-Doppler coverage over a large part of the country. The resulting coverage is quite heterogeneous due to the principle of ground-based radar measurements and is slightly superior in southern France due to the higher network density in this region. Between a height of 2 and 10 km (Fig. 2a) ~ 90% of the country is covered by at least 2 radar systems. Multiple-Doppler coverage (3 radars or more) is maximized between 2.5 and 9 km altitude, where ~80% of the French territory being covered by at least 3 radars. The radar overlapping near the surface is however quite limited due to both extensive radar baselines and beam blocking by terrain. The detailed maps of radar coverage at 2.5 and 5 km are shown in Fig. 2b-c.

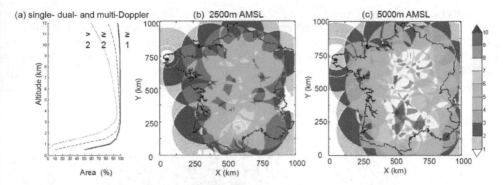

Fig. 2. (a) Radar overlapping within mainland France as a function of height and detailed overlapping map at (b) 2.5 and (c) 5 km AMSL.

3.3 Data processing

Reflectivity and radial velocity observations collected by all (24) ARAMIS radars are concentrated at the national center, in Toulouse, and automatically processed every 15′. Data consist in Cartesian, 512 km x 512 km, 1 km² in resolution, sweeps of radar observations, which are already exploited for current operational applications such as VAD analysis, quantitative precipitation estimates (Tabary et al. 2011) or data assimilation (Montmerle and Faccani 2009). Spurious reflectivity echoes are removed by the mean of a threshold on the pulse-to-pulse fluctuation of the reflectivity based on the work of Sugier et al. (2002). A 5x5 km² median filter is then applied to radial velocity measurements to discard potential spurious velocities resulting from dealiasing failures (Tabary et al. 2006). Finally, data are synchronized with respect to the ending time of the 15′ sampling period to account for the non-simultaneity of the measurements following the approach of Tuttle and Foote (1990). Once pre-processed, data are ingested in the MUSCAT analysis described in Section 2.

4. Examples of retrieved wind fields

A qualitative evaluation of the multiple-Doppler winds reconstructed in this framework is provided through the analysis of radar data collected during various rain events that occurred over mainland France between 2008 and 2010. This includes the extratropical cyclone Klaus, which stroke France on 24 January 2009 with hurricane force gusts, and a number of orographic convective precipitation events that produced large amount of rain over the Massif Central Mountains.

4.1 Extratropical cyclone "Klaus"

4.1.1 Overview

On 24 January 2009, an extratropical cyclone called "Klaus" made landfall over southwestern Europe with hurricane force gusts, causing widespread damage and many fatalities, especially across France and Spain. This event is considered the most intense storm affecting Western Europe since the infamous extratropical cyclones "Lothar"

(Wernli et al., 2002) and "Martin" in 1999. The heaviest damages occurred in southern France where millions of homes and commercial properties experienced power cuts and heavy damages due to falling trees. In Northern Spain and Southwestern France some of the most productive European forests have been profoundly impacted by the storm and will likely take decades to recover. In the French Landes department for instance, one estimates that 70 % of the pine forest – this forest, the largest of this kind in Europe, was accounting for about one third of France's lumber production - has been completely wiped in just a few hours.

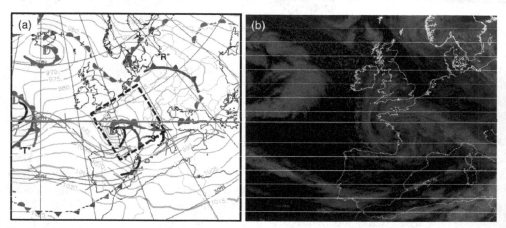

Fig. 3. (a) Surface analysis and satellite imagery over Western Europe valid 24 January 2009 at 6 UTC. The black square in (a) shows the area where wind retrieval is performed.

Klaus formed west of the Azores islands on 23 January 2009 near 00 UTC and made landfall about 30 hours later after crossing the Atlantic Ocean at a mean speed of 27 m.s[-1]. The life cycle of this system approximately follows the conceptual model proposed by Shapiro and Keyser (SK, 1990). It is characterized by an explosive intensification period, during which the sea level pressure (SLP) at the cyclone center deepened by ~ 36 hPa in 24 hours, ending up with a warm seclusion phase and a rapid decay. The SLP minimum (964 hPa) was reached at 00 UTC, 24 January, as the low-level vortex was located ~ 400 kilometers off the French coasts. Klaus made landfall slightly before 6 UTC on 24 January near Bordeaux, France (Fig. 3) with a minimum pressure of ~ 967 hPa (Fig. 4). The corresponding surface analysis (Fig. 3a) shows a warm frontal zone to the North of the cyclone center and a cold front extending far southward across the Pyrenees and Northern Spain. According to satellite images (Fig. 3b) the storm was elongated in the west-east direction along the warm front, which is in good agreement with the SK model theory (Schultz et al. 1998). After landfall, the system progressed eastward at a mean speed of about 15 m.s[-1] and reached Italy near 18UTC (Fig. 4). At this time the associated minimum pressure has already increased to 988 hPa. Maximum surface winds occurred in a region located approximately 300 to 350 km south of the cyclone center, along a 100 km swath oriented in a direction almost parallel to the cyclone trajectory. In France, wind gusts peaked at ~ 184 km.h[-1] near Opoul (Fig. 1), a value corresponding to category 3 hurricane winds on the Saffir-Simpson scale. The situation was even more impressive in Northern Spain where surface wind gusts over 200 km.h[-1] have been recorded at several locations.

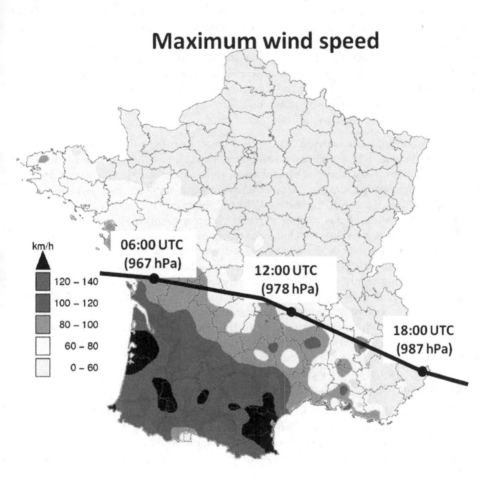

Fig. 4. Recorded maximum surface wind speed during the passage of Klaus over France. The trajectory and central pressure of the cyclone are also indicated.

4.1.2 Radar-derived wind and reflectivity fields

Observations collected between 03 and 9 UTC on 24 January 2009 are shown in Fig. 5 All radars performed nominally during this period with the exception of the Momuy radar, which was forced to cease operations near 05UTC due to heavy wind damages. Velocity data recorded just before the radar failure were in excess of 60 m.s^{-1} (216 km.h^{-1}) at 2000m MSL and ~ 45 m.s^{-1} (160 km.h^{-1}) close to the ground. Also note that the French radar network was not yet completely Doppler-ized at this time resulting in some gaps in the dual-Doppler radar coverage over northwestern and southern France.

Figure 5 presents horizontal cross-sections of wind and reflectivity fields over France at 2 km MSL and different stages of the storm evolution. At 3 UTC (Fig. 5a-b), a large part of the country was already affected by weak to moderate precipitation. The most intense rainfall occurred approximately 300-400 km in advance of the cyclone center along a SW-to-NE oriented rainband that marks the location of the cyclone associated cold front. Another area of intense precipitation associated with the cloud head (Browning 1999) could also be identified farther west. Both regions of intense precipitation were separated by an area of weaker precipitation associated with the dry intrusion identified in Fig.5a. Relatively intense winds in the range of 25- 30 m.s^{-1} could already be observed in the southeastern France (east of the cold front) and within the cloud head.

According to radar data, Klaus made landfall ~ 3 hours later near the city of Nantes (Fig. 5c-d). The closed circulation associated with the cyclone center can be clearly identified in Fig 5c. The diameter of the vortex deduced from radar observations was about 350 km. Severe winds reaching up to 50 m.s^{-1} have already penetrated deeply over land as seen by the patch of very intense winds located ~400 km of the cyclone center from each side of the dry slot. The location of this area of particularly strong winds is in good agreement with surface observations (Fig. 4). One can also notice the presence of several thin rainbands located to the west of the wind maximum, within the cloud head. This banded structure is consistent with that described by Browning (2004) and suggests the existence of multiple mesoscale slantwise circulations, which may have played an active role in strengthening the damaging winds (the investigation of processes at play during this event is outside the scope of this study). After landfall the patch of strong winds stretched out along a NW-SE axis more or less parallel to the orientation of the Pyrenees mountain chain and slowly progressed southeastward towards the Mediterranean coast. Strong winds remained active during the entire period. A wind maximum of up to 55 m.s^{-1} could be observed west of the Toulouse radar at 09UTC ($x \sim$ 400 km, $y \sim$ 200 km) in good agreement with surface observations (Fig. 4).

As there is currently no way to collect wind measurements at the space-time resolution achieved by ground-based Doppler radars, the validation of such operational wind data is extremely difficult. In order to evaluate these results, we propose to compare retrieved radar winds with those analyzed by the French operational numerical weather prediction system ALADIN (Aire Limitée Adaptation Dynamique Développement International; Radnóti et al. 1995). ALADIN is a limited area regional model that covers France and part of Western Europe at the horizontal resolution of 10 km. The comparison between radar-derived and analyzed horizontal winds at 1.5 km MSL at 6 UTC is shown in Fig. 7. Overall, the location of the cyclone center, the dimension of the vortex, and the intensity and direction of the winds at mid-level appear quite similar in both analyses. Some discrepancies can yet be seen along frontal boundaries (the wind shift associated with the cold front is for instance slightly more marked in the radar analysis) and to the North of the Massif Central Mountains, about 500 km east of the cyclone center. In the latter area the model produces a pronounced zonal wind component that is apparently not resolved in the radar analysis. This pronounced southerly component was nevertheless missing in all 15' analyses produced between 4 and 8 UTC. Although no strong conclusions can be inferred from this observation, this temporal consistency of the wind field may plead for an error in the analysis rather than in the retrieved winds.

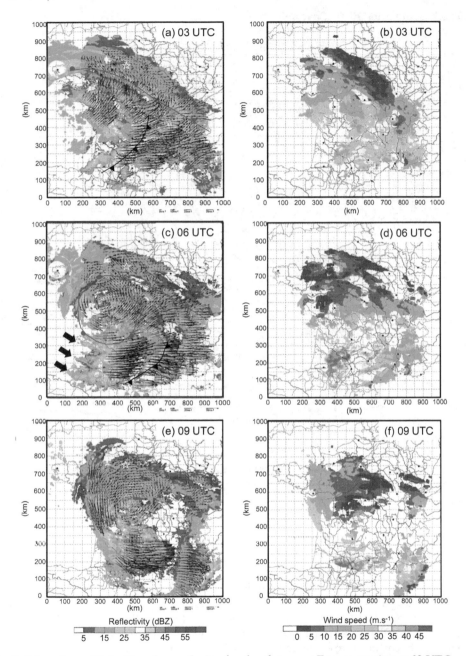

Fig. 5. 3-hourly multiple-Doppler analysis of radar data over France starting at 03 UTC on 24 January 2009. Left panel shows reflectivity (dBZ) superimposed on horizontal wind vector at 2.5 km AMSL. Right panel shows horizontal wind speed (m/s). One every sixth vector is plotted. Frontal boundaries deduced from surface analysis are shown in (a) and (c).

4.2 The 7 September 2010 "Cevenol" event

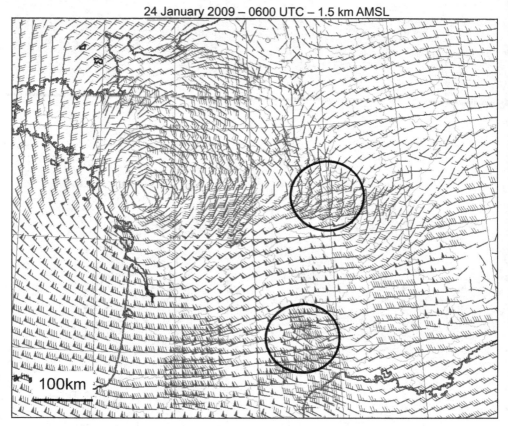

Fig. 6. Comparison between ALADIN operational (blue) and radar-derived winds (red) at 1500 AMSL on 24 January at 6 UTC. Black circles show area where the 2 analyses significantly differ.

4.2.1 Overview

During the fall season, the southeastern region of France is often affected by intense flash-flooding episodes ensuing from the formation of quasi-stationary mesoscale convectivesystems along the south-eastern flank of the Massif Central Mountains (Cevennes region). These systems can generate considerable amount of precipitation in relatively short periods of time. A well-known example of such systems is the so called "Gard case" (Delrieu et al. 2005) during which ~ 800 mm of rain fell down in less than 24 hours over the French Gard department, resulting in many fatalities and total damage amount of about 1.2 billion US dollars. Observing and understanding the dynamical and to some extent microphysical processes at play during these high impact weather events is critical to develop effective flood warnings systems and to improve their forecast. In the following we present examples of radar analyses produced during a heavy precipitation event that occurred in the Nimes area on 7 September 2010. Although this storm did not generate major damage, it locally produced rainfall accumulation in excess of 300 mm in just a few hours.

The 500 hPa analysis at 12 UTC, 7 September 2010 is shown in Fig. 7. Note that at this time convection over the Gard department has already started and was already well established. The upper level analysis indicates the presence of an upper-level cold low centered over the British Islands with an associated trough extending southward toward the Iberian Peninsula. This synoptic pattern generates a mid-to-upper level southwesterly flow over France and a low-level southerly flow over the Mediterranean Sea veering slightly south-easterly near the French coast.

The corresponding Deutscher Wetterdienst (DWD) low-level analysis (Fig. 8b) shows a surface low centered over Ireland and a main front extending meridionally through central France ahead of the trough which attained France on the 6 September (Fig. 8a). A secondary front can also be noticed a few hundred kilometers west of the main front, over the Atlantic Ocean. The latter is associated with a low-pressure anomaly located slightly north of the Spanish coast. In the following hours (Fig. 8c-d), this front and its associated surface low propagated rapidly eastwards to ultimately catch up with the leading front. The low surface pressure anomaly reached Western France by 18 UTC (Fig. 8c) and rapidly extended over a large part of the country (Fig. 8d).

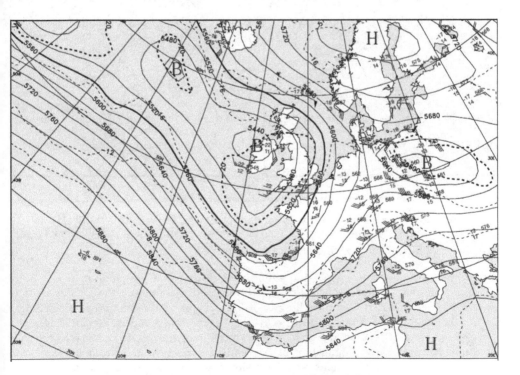

Fig. 7. Operational analysis of geopotential height at 500 hPa valid on 7 Sept 2010 at 12 UTC.

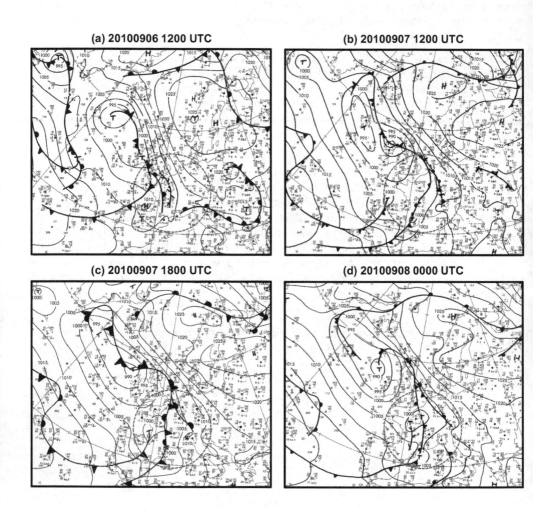

(a) 20100906 1200 UTC   **(b) 20100907 1200 UTC**

(c) 20100907 1800 UTC   **(d) 20100908 0000 UTC**

Fig. 8. DWD operational surface analyses valid at (a) 12 UTC 06 Sept 2010, (b) 12 UTC 07 Sept 2010, (c) 18 UTC 07 Sept 2010, and (d) 00 UTC 08 Sept 2010. The red circle indicates the area where multiple-Doppler wind retrieval is performed.

The 24-hour accumulated precipitation pattern (Fig. 9) shows a pronounced maximum of ~300 mm over the Gard department. The maximum rainfall amount was recorded close to the foothills of the Massif Central in the city of Conqueyrac where 308 mm of precipitation felt in 7 hours. Significant rainfall also occurred over the Massif Central as indicated by a well-defined band of accumulated precipitation in the range of 75-150 mm. This pattern is consistent with the slow propagation of the leading surface front noticed in Fig. 8 and suggests that frontal perturbations have been slowed-down and possibly enhanced by the Massif Central Mountains.

4.2.2 Radar-derived wind and reflectivity fields

Figure 10 presents multiple-Doppler analysis of radar data produced every 6 hours from 12 UTC, Sept 6 to 00 UTC, Sept 8 at a height of 2.5 km (left panel) and 6 km (right panel). The composite reflectivity patterns (right panel) show extensive frontal rainbands propagating eastwards in good agreement with the surface analyses shown in Fig. 8. Starting from 18 UTC, 6 September (Fig. 10c), one can notice the presence of widespread convective cells over southeastern France and northern Spain that seem to be triggered by the pronounced relief of the Pyrenees. These cells, which develop in a southwesterly midlevel level flow, are advected northeastwards towards the Massif Central Mountains before eventually aggregating into a stationary mesoscale convective system (MCS) along the flank of the mountains. The retrieved wind circulation at both 2.5 and 6 km altitude shows relatively uniform southwesterly wind, except near the Massif Central where the mid-level flow exhibits a more pronounced southerly component. This southerly flow, which tends to advect warm and moist air masses from the Mediterranean Sea toward the coast, impinges on the Massif Central Mountains and is responsible for the enhancement of convection over

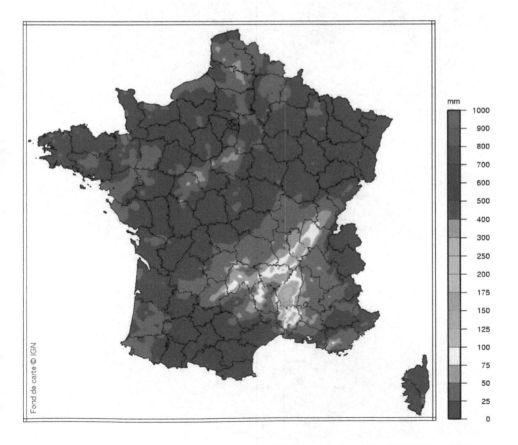

Fig. 9. 24-h accumulated precipitation (mm) over France starting at 00 UTC, 7 Sept 2010

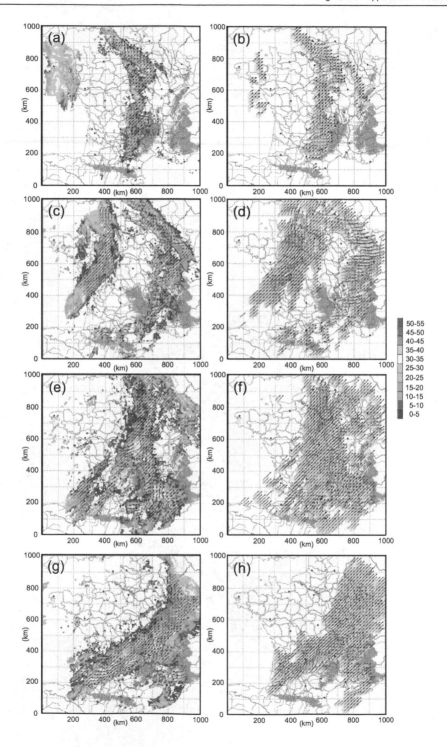

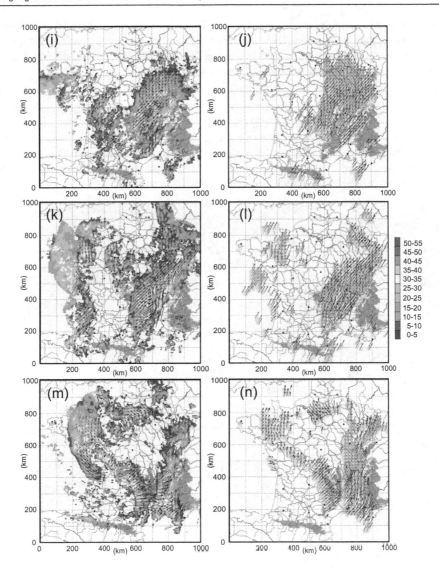

Fig. 10. 6-hourly multiple-Doppler analysis of radar data over France starting at 12 UTC on 6 Sept 2010. Left panel shows reflectivity (dBZ) superimposed on horizontal wind vector at 2.5 km AMSL. Righ panel shows vertical velocity (m/s) and horizontal wind vectors at 6 km AMSL. Vertical velocity values (m/s): dark green (w<-1) green (-1≤w<1), yellow (1≤w<2), orange (w≥2). One every sixth vector is plotted.

the Cevennes area. The convective activity in this region significantly intensified in the following 12 hours (Fig. 10e and g), as shown by radar reflectivity values increasing up to 55 dBZ. During this period, the MCS also became more organized and developed a well-defined stratiform region. At 12 UTC, 7 September, the MCS was absorbed by the frontal rainband associated with the cold front that had been moving across France for ~ 24 hours (Fig. 10i). The various interactions between air masses occurring during the merging process seemed to reinforce the convective activity along the southern flank of the Massif Central. The most intense convection was hence observed around 18 UTC (Fig. 10k).

In order to better understand the processes at play in the Cevennes region, Fig. 11 presents a zoomed view of low and midlevel flow at 00 and 21 UTC, 7 September within a 400 x 400 km² domain centered on Nimes. At 00 UTC (Fig. 11c), the low-level flow impinging on the Massif Central was from the SE and was oriented in a direction almost perpendicular to the mountains. The maximum convection was observed slightly off the slopes. This is likely due to the presence of a cold pool below the system, which acted to displace the triggering effect of the mountains farther south (cold dome effect; Reeves and Lin 2005).

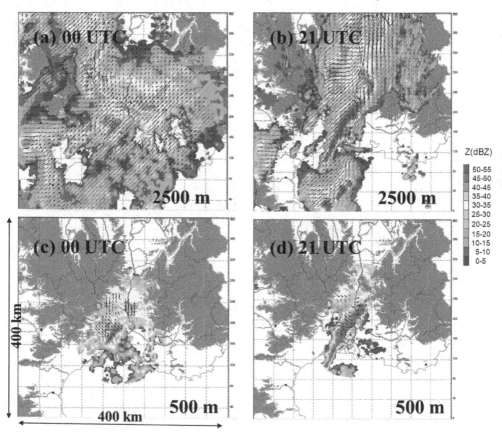

Fig. 11. Multiple-Doppler analysis of radar data within a domain of 400 km x 400 km centered near Nimes, France valid at 00 UTC and 21 UTC, 7 Sept 2010. Upper panels shows reflectivity (dBZ) superimposed on horizontal wind vector at 2.5 km AMSL. Lower panels shows panel shows reflectivity (dBZ) and horizontal wind vectors at 500m AMSL. Grey shading indicates Massif Central (left) and Alps (right) mountain chains. One every fourth vector is plotted.

At 21 UTC, one can note a profound reversal of the incident upslope-oriented flow, resulting in the formation of a northwesterly flow over the slopes of the Massif Central. Such pronounced return flow has already been observed over the Alps in the frame of the MAP experiment from airborne and mobile radar systems in response to negative buoyancy generated by both melting and evaporation of precipitation particles below the 0° C isotherm (Bousquet and Smull 2003, Steiner et al. 2003). The process generating the observed downslope flow by cooling from melting and evaporation of precipitation particles in the Massif Central is likely similar to that observed over the Alps in the late nineties. This is supported by the fact that stratiform precipitation lasted for several hours over the mountains before the formation of this downslope circulation (Fig. 10g,i,k). In this particular case, however, it seems that the downslope flow has had a strong impact on the convective activity by triggering new cells in the Rhone valley (Fig. 11d), whereas its effect was not found significant in previous studies.

After the merging, the frontal system remained blocked over the Massif Central and the eastern part of France for about 12 hours during which another 100 mm of rain felt over the Cevennes area. It finally passed the Cevennes near 00 UTC, 8 September as it was swept away by another frontal system approaching from the West. At this time, the low surface pressure anomaly identified in Fig. 8c had reached the French territory and had started to extend over a large part of the country. The associated cyclonic circulation (Fig. 10m) was well captured by the French radar network.

More information about the vertical structure of precipitation can be inferred from Fig. 12, which presents 6-hourly meridional cross-sections of the retrieved 3D composite reflectivity pattern along a 1000 km line ranging from the Golfe du Lion, in the Mediterranean Sea, to Belgium. These cross-sections provide a unique picture of the structure and evolution of precipitation over the entire country and can be very useful to quickly identify regions of intense rainfall, as well as to segregate between frontal and more convective precipitation. This time series, extending from 00 UTC to 18 UTC on 7 September, thus confirms that convection became significantly more intense after the frontal system reached the Massif Central Mountains. At 12 UTC (Fig. 12c) and 18 UTC (Fig. 12d) one can see that very deep convection was thus occurring over the Cevennes and the Rhone valley with convective cells reaching up to 35 dBZ at a height of 11 km.

In addition to horizontal wind fields, the MUSCAT analysis used by the French weather service also allows to retrieve accurate vertical velocities in the whole precipitating area within the 1000 km x 1000 km domain of analysis. Retrieved vertical motion fields at a height of 6 km, which is the altitude at which maximum vertical motion was observed, are displayed in Fig. 10. Overall, one can note a very good consistency between the location/intensity of updrafts and the position of the most active convective cells (left panel). Upward vertical motions are the most intense after the MCS has merged with the frontal rainband that is the moment when low level convergence was the most important. This observation is consistent with the vertical structure of precipitation deduced from Fig. 12 that indicates the presence of deep convection and shows a particularly impressive vertical extension of the convective cells.

During the Klaus storm, getting real-time or quasi real-time information about wind intensity would have been particularly useful to forecasters in order to trigger or cancel alerts, as well as to precisely monitor the propagation of the strong wind swath. On September 7 2010, a watch was ongoing for heavy precipitation and flash flood in the Cevennes area but forecasters were more interested in getting high resolution radar quantitative precipitation estimates in order to assess the hydrological risks.

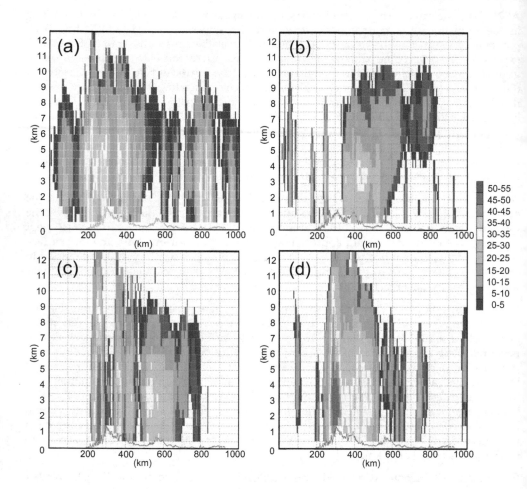

Fig. 12. Vertical cross-section of radar reflectivity (dBZ) at x=700 km (Fig. 10) at (a) 00 UTC, (b) 6 UTC, (c) 12 UTC and (d) 18 UTC on 7 Sept 2010.

Wind observations would nevertheless have been quite useful to anticipate the behavior of convective cells, especially in the Rhone valley where the intensity of the convection seemed directly related to the direction and strength of the wind. On the other hand, the benefit inferred from such products for research purposes is priceless. These wind fields could be used to develop new methods for model verification as well as more efficient nowcasting tools. Radar-derived wind information produced in a fully operational framework (i.e., in real time and automatically) could also be relied upon to evaluate numerical model output in real time (through identifying possible temporal or spatial phase shift in model output), as well as to build a weather database that would be used for statistical analysis purposes or more traditional case studies. Such capabilities, for instance, are at the heart of the upcoming HyMeX field phase, which will be held in Sept-Nov 2012.

4.3 20-22 October 2008 MCS Cevenol precipitation event

The last example consists in an isolated MCS (Fig. 13a) that developed along the south-eastern flank of the Massif Central Mountains on October 20th and remained stationary for about 15 hours. Again, this storm did not generate any significant damage, but it locally produced rainfall accumulation in excess of 250 mm in a just a few hours.

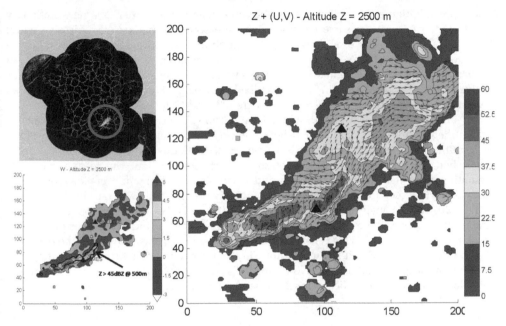

Fig. 13. Stationary mesoscale convective system observed on 20 October 2008 at 14 UTC. (Left, top): French operational radar reflectivity mosaic output. (Right): Horizontal wind vectors superimposed on radar reflectivity (shaded) at 2.5 km MSL within the 200 km x 200 km experimental domain, as derived from multiple-Doppler analysis of radar data. (Left, bottom): Associated retrieved vertical velocity. Triangles show the location of Nimes and Bollène radars. One every sixth vector is plotted.

The retrieved wind and reflectivity fields associated with this MCS at 14 UTC are shown in Fig. 13b within a domain of 200 x 200 km centered between the Nimes and Bollene radars (black triangles) at a horizontal resolution of 1 km. Note that only 4 radars (Bollene, Nimes, Montclar and Collobrieres, see Fig. 1) are used in this analysis. This setup will be used to produce high resolution wind fields in real-time during the first phase of the HyMeX program in fall 2012. Radar-derived wind and reflectivity fields will be used to guide both research aircraft and ground-based mobile radars systems towards the most interesting areas. Overall, the radar analysis shows a rather complex wind circulation resulting from the interactions between the incident flow originating from the Mediterranean Sea and the terrain. A region of strong convergence was observed near Nimes due to the interactions of westerly flow in the southern part of the massif with southerly-to-southwesterly flow to the east. As a consequence, strong convection could be observed in this region while more stratiform precipitation, associated with older cells, could be seen to the North. The vertical velocity field (Fig. 13, bottom right) in the convective region also shows intense updraft up to 8 m/s, which is consistent with observed reflectivity cores up to 60 dBZ. Note that retrieved vertical velocities were significantly higher than those retrieved within the 7 September 2010 system due to the much higher horizontal resolution of the wind field (1 km vs. 2.5 km).

A new precipitation event occurred over the same area the day after as a new stationary MCS formed at about the same location, along the southern flank of the Massif Central Mountains (Fig. 14). This time, however, the situation was quite different as a rainband associated with a southeastward propagating cold front merged with the isolated MCS at the end of the day and eventually swept it away. In essence, this new event is thus relatively similar to the 7 Sept 2010 case described previously. Figure 15 shows the evolution of the MCS between 0 UTC and 8 UTC on the 22nd of October. This time series begins slightly after the frontal system reached the Massif Central. The location of the cold front can be identified by the wind shift seen at both low and mid-levels. Note that the strong low-level southerly flow impinging on the barrier at 0 UTC quickly weakened as the front approached the eastern flank of the Massif, which acted to cut the feed of moisture originating from the Mediterranean Sea and prevented the formation of new convective cells. Convection over the area thus died very quickly and the MCS was rapidly swept away by the cold front. A major difference with the 2010 case is that the frontal system did not remain blocked over the relief and stratiform precipitation over the slopes thus only lasted for a short period of time. This is likely the reason why no reversal flow could be observed over the slopes of the Massif Central on this day as the cooling resulting from melting and evaporation of precipitation particles, which is responsible for the formation of the downslope flow, was not sufficient for this phenomenon to occur.

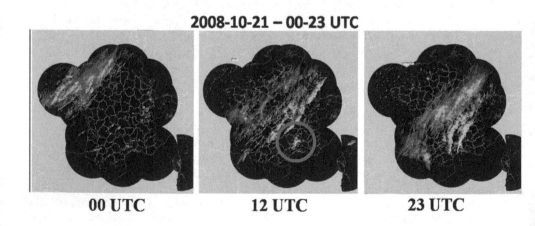

Fig. 14. French operational radar reflectivity mosaic outputs on 21 October 2008 at 00, 12, and 23 UTC. Red circle indicate the location of the isolated MCS.

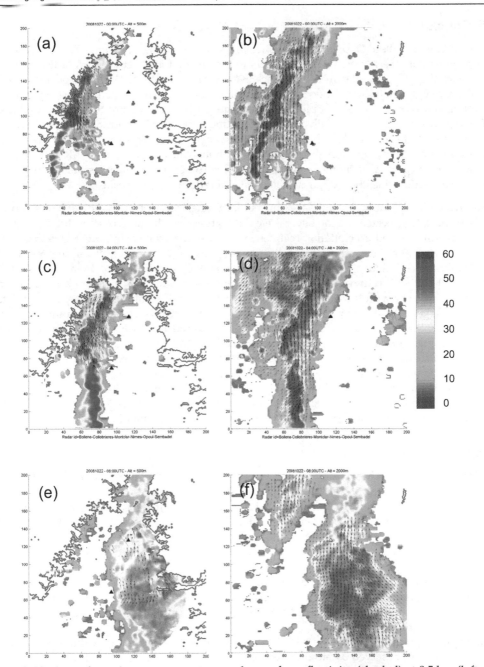

Fig. 15. Horizontal wind vectors superimposed on radar reflectivity (shaded) at 0.5 km (left panel) and 2.0 km (right panel) AMSL within a 200 km x 200 km experimental domain, as derived from multiple-Doppler analysis of radar data at (a,b) 00 UTC, (c,d) 04 UTC and (e,f) 08 UTC on 22 October 2008. One every sixth vector is plotted.

5. Conclusion

The ability to collect Doppler measurements up to long range resulting from the recent deployment of a new triple-PRT scheme within the French radar network ARAMIS allows to mitigate the Doppler dilemma and to achieve extensive multiple-Doppler coverage while collecting high quality radial velocities. This achievement brings new perspectives in terms of exploitation of operational Doppler measurements such as the ability to perform multiple-Doppler wind synthesis in a fully operational framework. In this context the French weather service has started to produce a new mosaic of wind and reflectivity covering the entire French territory. Reflectivity and radial velocity observations collected by the 24 ARAMIS radars, which are concentrated at the national center in Toulouse, are synthesized every 15' to produce a nationwide, three dimensional, wind and reflectivity composite.

An evaluation of multiple-Doppler wind fields synthesized in this framework was carried out from data collected during several weather situations characterized by fundamentally different airflow and precipitation regimes that is, the extratropical cyclone Klaus that stoke France with hurricane strength winds on 24 January 2009, as well as several heavy orographic precipitation events that occurred over the Massif Central Mountains in 2008 and 2010. Airflows retrieved at different horizontal resolutions ranging from 2.5 to 1 km are highly consistent with those documented earlier from high resolution research radar data, which suggests that multiple-Doppler winds retrieved in this operational framework are definitely reliable. Wind and reflectivity fields produced routinely in this framework are archived since about a year and are already available to researchers from many countries. This product will be used in a quasi-operational mode (real-time) to guide mobile research systems and evaluate numerical model outputs during the field phase of the Hydrometeorological Cycle in the Mediterranean Experiment (HyMeX) to be held in southern France in fall 2012. It should become fully operational near year 2013 after having been thoroughly evaluated by the forecasters of the French Weather Service.

6. References

Armijo, L., 1969: A Theory for the Determination of Wind and Precipitation Velocities with Doppler Radars. *J. Atmos. Sci.*, 26, pp. 570-573

Bougeault, P., P. Binder, A. Buzzi, R. Dirks, R. A. Houze Jr, R. Kuettner, R. B. Smith, R. Steinacker and H. Volkert, 2001: The MAP special observing period. *Bull. Am. Meteorol. Soc.*, 82, 433-462.

Bousquet, O., and M. Chong, 1998: A multiple Doppler and continuity adjustment technique (MUSCAT) to recover wind components from Doppler radar measurements. *J. Atmos. Oceanic Technol.*, 15, 343-359.

Bousquet, O., and M. Chong, 2000: The oceanic mesoscale convective system and associated mesovortex observed on 12 December 1992 during TOGA COARE. *Quart. J. Roy. Meteor. Soc.*, 126, 189-212.

Bousquet, O., and B. F. Smull, 2003: Airflow and precipitation fields within deep Alpine valleys observed by airborne radar. *J. of Appl. Meteorol.*, 42, 1497-1513

Bousquet, O., and B. F. Smull, 2006: Observed mass transports accompanying upstream orographic blocking during MAP IOP8, *Quart. J. Roy. Meteor. Soc.*, 132, 2393-2413.

Bousquet, O., P. Tabary, and J. Parent du Châtelet, 2007: On the value of operationally synthesized multiple-Doppler wind fields, *Geophys. Res. Lett., 34,* L22813, doi:10.1029/2007GL030464.

Bousquet, O., P. Tabary, and J. Parent du Châtelet, 2008a: Operational multiple-Doppler wind retrieval inferred from long range radar velocity measurements, *J. Appl. Meteorol. Clim., 47,* 2929-2945.

Bousquet, O., T. Montmerle, and P. Tabary, 2008b: Using operationally synthesized multiple-Doppler winds for high resolution NWP model horizontal wind verification. *Geophys. Res. Lett., 35,* L10803, doi:10.1029/2008GL033975.

Bousquet, O., 2009: Dynamical and microphysical properties of high impact orographic mesoscale convective systems from high resolution operational multiple-Doppler and polarimetric radar data. *Preprints,* 34th Conference on Radar Meteorology, Williamsburg, USA

Browning, K.A., and R. Wexler, 1968: The determination of kinematic properties of a wind field using Doppler radar, *J. Appl. Meteorol., 7,* 105-113.

Browning, K.A., 1999: Mesoscale aspects of extratropical cyclones: An observational perspective. in *Life Cycles of Extratropical Cyclones,* American Meteorological Society, Boston, 265-283.

Browning, K.A., 2004: The sting at the end of the tail: Damaging winds associated with extratropical cyclones. *Quart. J. Roy. Meteor. Soc., 130,* 375-399.

Chong, M., and J. Testud, 1996: Three-Dimensional Air Circulation in a Squall Line from Airborne Dual-Beam Doppler Radar Data: A Test of Coplane Methodology Software. *J. Atmos. Oceanic Technol., 13,* 36-53.

Chong, M., and O. Bousquet, 2001: On the application of Muscat to a ground-based dual-Doppler radar system., *Meteorol. Atmos. Phys., 78,* 133-139.

Chong, M., J.-F. Georgis, O. Bousquet, S. R. Brodzik, C. Burghart, S. Cosma, U. Germann, V. Gouget, R. A. Houze Jr., C. N. James, S. Prieur, R. Rotunno, F. Roux, J. Vivekanandan, Z.-X. Zeng, 2000: Real–Time Wind Synthesis from Doppler Radar Observations during the Mesoscale Alpine Programme. *Bull. Amer. Meteor. Soc., 81,* 2953-2962.

Delrieu, G., and co-authors, 2005: The Catastrophic Flash-Flood Event of 8–9 September 2002 in the Gard Region, France: A First Case Study for the Cévennes–Vivarais Mediterranean Hydrometeorological Observatory. *Journal of Hydrometeorology, 6,* 34-52.

Doviak, R. J., and D. S. Zrnic, 1993: Doppler radar and weather observations (second edition), Academic Press, Inc.

Drechsel, S., G. J. Mayr, M. Chong, M. Weissmann, A. Dörnbrack, and R. Calhoun, 2009: Three-dimensional wind retrieval: application of MUSCAT to dual-doppler lidar. *J. Atmos. Oceanic Technol., 26,* 635-646.

Gamache, J. F., 1995: A three-dimensional variational method for determining wind velocities from Doppler data as applied to the TOGA COARE test case. Summary Report, TOGA COARE Int. Data Workshop TOGA, COARE International Project Office, UCAR, Boulder, CO, 106 pp.

Gao, J., M. Xue, A. Shapiro and K. K. Droegemeier, 1999: A Variational Method for the Analysis of Three-Dimensional Wind Fields from Two Doppler Radars. *Mon. Wea. Rev., 127,* 2128-2142.

Heymsfield, G. M., 1978: Kinematic and Dynamic aspects of the Harrah tornadic storm analyzed from dual-Doppler radar data. Mon. Wea. Rev., 106, 253-264.

Lhermitte, R. M., and L. J. Miller, 1970: Doppler radar methodology for the observation of convective storms. *Preprints, 14th Radar Meteorology Conf*, Tucson, Arizona, Amer. Meteorol. Soc, pp 133-138.

Georgis, J-F., F. Roux, M. Chong and S. Pradier, 2003: Triple-Doppler analysis of the heavy rain event observed in the Lago Maggiorre region during MAP IOP 2b. *Quart. J. Roy. Meteor. Soc.*, 129, 495-522.

Montmerle, T., and C. Faccani, 2009: Mesoscale assimilation of radial velocities from Doppler radar in a pre-operational framework. *Mon. Wea. Rev.*,137, 1939–1953.

O'Brien, J. J., 1970: Alternative solutions to the classical vertical velocity problem., *J. Appl. Meteor.*, 9, 197-203.

Radnóti, G., R. Ajjaji, R. Bubnovà, M. Caian, E. Cordoneanu, K. von der Emde, J.-D. Gril, J. Hoffman, A. Horànyi, S. Issara, V. Ivanovici, M. Janousek, A. Joly, P. Le Moigne, S. Malardel, 1995 : The spectral limited area model Arpege/ Aladin. *PWPR Report Series* n° 7, WMO T.D. N. 799, 111-118.

Reasor, P. D., M. D. Eastin, and J. F. Gamache, 2009: Rapidly intensifying Hurricane Guillermo (1997). Part I: Low-wavenumber structure and evolution. *Mon. Wea. Rev.*, 137, 603–631.

Reeves, H. D., and Y-L Lin, 2005: Effect of stable layer formation over the Po Valley on the development of convection during MAP IOP-8. *J. Atmos. Sci.*, 63, 2567-2584.

Scialom, G., and Y. Lemaître, 1990 : A New Analysis for the Retrieval of Three-Dimensional Mesoscale Wind Fields from Multiple Doppler Radar. *J. Atmos. Oceanic Technol.*, 7, 640-665

Shapiro, M. A., and D. Keyser, 1990: Fronts, jet streams and the tropopause. *Extratropical Cyclones, The Erik Palmén Memorial Volume*, C. W. Newton and E. O. Holopainen, Eds., Amer. Meteor. Soc., 167-191.

Schultz, D. M., D. Keyser and L. Bosart, 1998: The Effect of Large-Scale Flow on Low-Level Frontal Structure and Evolution in Midlatitude Cyclones. *Mon. Wea. Rev.*, 126, pp. 1767-1791

Steiner, M., O. Bousquet, R. A. Houze Jr, B. F. Smull, and M. Mancini, 2003: Airflow within major Alpine river valleys under heavy rainfall. *Quart. J. Roy. Meteor. Soc.*, 129, 411-432.

Sugier, J., J. Parent-du-Châtelet, P. Roquain, and A. Smith, 2002: Detection and removal of clutter and anaprop in radar data using a statistical scheme based on echo fluctuation. *Proc. Second European Radar Conf.*, Delft, Netherlands, Copernicus GmbH, 17–24.

Tabary, P., F. Guibert, L. Perier, and J. Parent-du-chatelet, 2006: An operational triple-PRT scheme for the French radar network. *J. Atmos. Oceanic Technol.*, 23, 1645-1656.

Tabary, P., A.A. Boumahmoud, H. Andrieu, R. J. Thompson, A. J. Illingworth, E. Le Bouar and J. Testud, 2011: Evaluation of two "integrated" polarimetric Quantitative Precipitation Estimation (QPE) algorithms at C-band, *Journal of Hydrology*, 405., 248,260.

Tuttle, J. D., and G. B. Foote, 1990: Determination of boundary layer airflow from a single Doppler radar. *J. Atmos. Oceanic Technol.*, 7, 218–232.

Wernli, H., Dirren, S., Liniger, M. A. and Zillig, M., 2002, Dynamical aspects of the life cycle of the winter storm 'Lothar' (24–26 December 1999). *Quart. J. Roy. Meteor. Soc*, 128, 405-429

Zrnic, D. S., and P. Mahapatra, 1985: Two methods of ambiguity resolution in pulsed Doppler weather radars., *IEEE Trans. Aerosp. Electron. Syst.*, 21, 470–483.

3

Multiple Doppler Radar Analysis for Retrieving the Three-Dimensional Wind Field Within Thunderstorms

Shingo Shimizu

National Research Institute for Earth Science and Disaster Prevention/ Storm, Flood, and Land-Slide Research Department
Japan

1. Introduction

Multiple Doppler radar analysis has been widely used to retrieve three-dimensional wind fields within thunderstorms and meso-scale convective systems (MCS) since the late 1960s. A number of countries have constructed dense operational radar networks, such as the Operational Programme for the Exchange of Weather Radar Information (OPERA; Köck et al., 2000), to monitor and forecast severe weather in metropolitan regions. Multiple Doppler radar analysis using such operational radar networks improves 1) understanding of the physical mechanisms behind heavy rainfall and severe wind, 2) detection and forecasting of hazardous weather phenomena, and 3) planning for mitigation of human and socioeconomic losses in metropolitan regions.

Early single Doppler radar measurements provided a basic understanding of storm morphologies and their three-dimensional structures, including concepts for single-cell, multicell, and supercell storms (Browning, 1964, 1965). Single Doppler radar observations can only provide information on the radial component of wind (i.e., velocity which is directed toward or away from the radar), rather than the full three-dimensional structure. Armijo (1969) formulated a method that allowed the deduction of the three-dimensional wind structure by combining the data from several Doppler radars. Improvements in this multiple Doppler radar analysis method were reported during the 1970s and 1980s, including the design of optimal radar networks (Ray et al., 1979, 1983), the development of alternative analysis schemes for solving the mass continuity equation (Ray et al., 1980), and the introduction of floating boundary conditions (Chong & Testud, 1983). These improvements were primarily motivated by the need to overcome errors in the estimation of vertical velocity using upward integration of the mass continuity equation (Doviak et al., 1976). Errors in estimates of vertical velocity tend to amplify during such upward integration because of the stratification of density in the atmosphere (Doviak et al., 1976; Ray et al., 1980). Theoretical demonstrations indicate that downward integration of the mass continuity equation could yield more accurate estimates of vertical velocity than those that can be obtained from upward integration (Ray et al., 1980). Many subsequent studies have therefore applied downward integration schemes to determine the three-dimensional structure of winds within severe storms (Kessinger et al., 1987; Biggerstaff & Houze, 1991; Dowell & Bluestein, 1997).

Downward integration requires observations at the storm top; however, typical radar scan geometries are configured for operational monitoring of low-level precipitation and severe low-level wind phenomena (e.g., downburst and tornado). Such configurations do not often provide detailed observations of the storm top, where the vertical velocities may be significantly different from zero during storm development (Mewes & Shapiro, 2002). Several studies have used low pass filters, such as the Leise filter (Leise, 1981), to apply high wavenumber adjustments to the lower boundary conditions prior to upward integration of the mass continuity equation. Such adjustments reduce noise in estimates of upper-level winds (Parsons & Kropfli, 1990, Wakimoto et al., 2003).

One alternative is to apply the anelastic mass conservation equation as a weak constraint (Gao et al., 1999, hereafter G99; Gao et al., 2004, hereafter G04). This method is based on a three-dimensional variational approach, and removes the need to explicitly integrate the anelastic continuity equation. This prevents the accumulation of severe errors in the vertical velocity and ensures that uncertainties in the upper and lower boundary conditions do not propagate vertically. Furthermore, multiple Doppler radar analysis is usually performed in a Cartesian coordinate system; Doppler velocity data are often interpolated into this Cartesian coordinate system using a Cressman filter (Cressman, 1959). The scheme introduced by G99 bypasses this step by allowing reverse linear interpolation (from the regularly spaced Cartesian grid to the irregularly spaced radar observation points) during calculation of the cost function. This reverse interpolation procedure preserves the radial nature of radar observations; however, as noted above, operational radar networks are often incapable of providing dense observations, especially at upper levels. In such cases, the G99 scheme requires accurate background information, such as sounding data, to fill in the data-void regions between successive elevation angles. It is frequently difficult to obtain accurate background information in these cases, due in part to the coarse temporal resolution of sounding data. If spatially continuous Doppler velocity data could be obtained in Cartesian coordinates through the careful use of Cressman filters, accurate vertical velocity could be obtained from operational radar scans without the need for additional information. Otherwise, additional information regarding upper-level winds is necessary to reduce errors in estimates of vertical velocity near the storm top.

This chapter presents a simplified version of the G99 scheme that applies a three-dimensional variational approach on a regular Cartesian grid. The accuracy of calculated winds and the dependence of this accuracy on the density of upper-level radar observations are investigated using a set of idealized data sampled from a simulated supercell storm. A detailed description of the structure of this simulated supercell has been provided by Shimizu et al. (2008). The objective of this chapter is to propose an optimal method for analyzing severe thunderstorms using typical configurations of current operational radar data (less than 20 Plan Position Indicators, or PPIs, within 5–6 minutes).

2. Analysis method and variational scheme

This section briefly reviews the variational scheme for multiple Doppler radar analysis; a detailed description has been provided by Gao et al. (1999). The variational technique minimizes a cost function (J), which is defined as the sum of squared errors due to discrepancies between observations and analyses and additional constraint terms:

$$J = Jo + Jd + Js + Jb \tag{1}$$

J_o represents the difference between the analysed radial velocity and the observed radial velocity, J_d is the mass continuity equation constraint term, J_s is the smoothness constraint term, and J_b is the background constraint term. The definition of J_o used here differs from that used by G99. J_o is defined as

$$J_o = \frac{1}{N} \sum_{i,j,k,m} C_o \, (Vr^{i,j,k,m} - Vr\_obs)^2 \tag{2}$$

Vr is the analysed radial velocity on a specified Cartesian grid, where i, j, and k indicate spatial location in the x, y, and z directions, respectively, and m indicates the mth radar in the network. N is the total number of observations, which is equal to the product of the number of grid cells and the number of radars in the network. C_o is the reciprocal of the mean squared error in the observations. Vr_obs is the radial velocity interpolated to the regular Cartesian grid. The cost function is evaluated at each grid point in the Cartesian coordinates, rather than in spherical coordinates.

Each constraint is weighted by a factor that accounts for its respective proportion of the reciprocal of the mean squared error. As noted by G99, it is usually difficult to obtain appropriate values for the weighting coefficients. In particular, the value of the weighting coefficient for the anelastic mass conservation constraint plays an important role in determining the vertical wind component. This study uses the coefficient value used by G99, although G04 introduced a more objective method for estimating this coefficient.

The variational method uses the derivative of J with respect to the analysis variables to obtain an optimal solution. The gradient of the cost function is derived with respect to the control variables, namely the two horizontal wind components (u, v) and the vertical wind component (w). The form of the gradient used here differs slightly from that used by G99 because the form of the observational constraint differs. The gradient of the observational constraint with respect to u is given by

$$\frac{\partial J_o}{\partial u} = C_o \frac{x}{r} (Vr^{i,j,k,m} - Vr\_obs) \tag{3}$$

where r is the distance between the radar and the grid point and x is the component of r in the x direction.

After the cost function is evaluated and its gradients are obtained, a quasi-Newton-type optimization scheme is used to update the control variables. This analysis uses a limited-memory Broyden–Fletcher–Goldfarb–Shanno (L–BFGS) method (Liu & Nocedal, 1989). For most meteorological applications, the L–BFGS method is more efficient than the conjugate gradient method (Navon & Legler, 1987). L–BFGS uses an approximation of the second-order derivative, so that an iteration of the L–BFGS method typically requires less computation than an iteration of the CG method. L–BFGS is therefore a better choice for optimizing a computationally expensive cost function.

3. Observational system simulation experiment and model description

The performance of the variational technique is evaluated in the context of an observational system simulation experiment (OSSE). This OSSE is conducted using numerical simulations of a supercell thunderstorm observed near Tokyo on 24 May 2000 (Shimizu et al., 2008). The numerical simulations are generated using the Cloud-Resolving Storm Simulator (CReSS; Tsuboki & Sakakibara, 2002). CReSS is a three-dimensional nonhydrostatic model. The microphysical and other parameterization schemes used in CReSS have been described in detail by Tsuboki & Sakakibara (2002). The model grid comprises 300 x 300 x 70 grid points, with grid intervals of 1 km in the horizontal directions. The vertical grid interval increases with height from 0.2 km near the surface to 0.37 km at the model top. The OSSE focuses on the three-dimensional distribution of wind within a 50 × 50 km domain around the simulated supercell at 1206 local standard time (LST), assuming that four Doppler radars are observing the storm (Fig. 1).

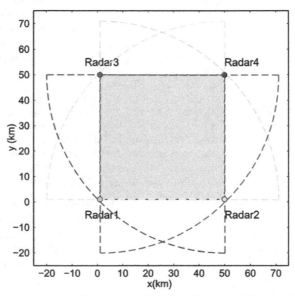

Fig. 1. Four radar locations around a simulated supercell storm. The simulated winds are compared with those derived from radar observations within the shaded region (50 × 50 km). The detection range of the radars is 70 km.

Updrafts associated with the simulated storm reached approximately 12 km above sea level (ASL). The locations of the four radars are chosen so that the distance between each radar and the storm is approximately 30–40 km (the supercell is located in the center of the shaded domain shown in Fig. 1). High elevation angle (~20°) radar scans are required to observe the storm top. Figure 2 shows the heights of the simulated radar beams. Three different volume scan strategies are considered in this chapter. The first strategy assumes that one volume scan consists of 17 PPIs. This strategy corresponds to X-band radar surveillance of a thunderstorm with echo top below 10 km ASL during an interval of 5–6 minutes. The elevation angles used in this strategy are 0.7°, 1.2°, 1.7°, 2.2°, 2.8°, 3.3°, 3.9°, 4.7°, 5.6°, 6.5°, 7.4°, 8.3°, 9.3°, 10.3°, 11.8°, 13.5°, and 15.6°. The second strategy adds three high elevation angles (16.7°, 17.8°, and 18.9°) to the previous volume scan (blue lines in Fig. 2). This strategy corresponds to a sector- or adaptive-scanning mode (Junyent et al., 2010) for a tall thunderstorm located near the radar. The third strategy adds nine additional high elevation angles (20.0°–32.0° spaced at 1.5° intervals) to the preceding volume scan (red lines in Fig. 2).

This corresponds to an ideal observing mode. The third scanning strategy appears to be impossible to implement at a rate of one volume scan every 5–6 minutes using current technology, but it may be possible using the high temporal resolution capabilities of next-generation phased array radars (Heinselman & Torres, 2011).

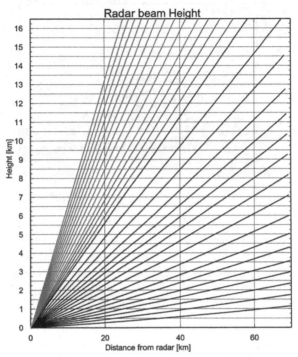

Fig. 2. Radar beam heights for the three analysed scanning strategies. Radar beam paths with elevation angles between 0.7° and 15.6° (17 PPIs) are shown as black lines, those with elevation angles between 16.7° and 18.9° are shown as blue lines, and those with elevation angles between 20.1° and 32° are shown as red lines.

4. Results

Figure 3 shows the mixing ratio of rain and the distribution of horizontal and vertical winds at an altitude of 4 km ASL. The Weak Echo Region (WER) within the strong updraft region was well simulated (the maximum updraft was 25 m s^{-1}). These strong updrafts were fed by southeasterly inflow below 1.5 km ASL (data not shown). Northwesterly wind was dominant at 4 km ASL, and advected the area of heavy precipitation toward the southeast (Fig. 3). Three downdraft cores were simulated at 4 km ASL. The first of these was located in the heavy rain region to the east of updraft, and was associated with precipitation loading. The second downdraft core was located in the light rain region to the southeast of the updraft, and was related to the melting and sublimation cooling of ice-phase precipitation (Shimizu et al., 2008). The third downdraft core was located in the non-precipitating region to the south of the updraft, and was associated with compensation for the nearby strong updraft. The strong updraft, first downdraft, and second downdraft cores were also simulated at 2 km ASL (Fig. 4). The maximum updraft speed exceeded 18 m s^{-1} at 2 km ASL. Anticlockwise wind rotation with a vertical vorticity of 0.08 s⁻¹ was simulated along with this strong vertical velocity (Fig. 4). At 2 km ASL, the downdraft in the heavy precipitation region (maximum velocity –3 m s⁻¹) covered a broad area to the northeast of the updraft at 2 km ASL.

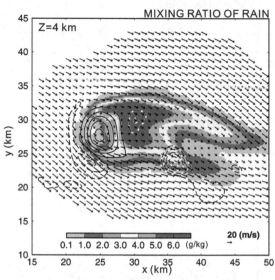

Fig. 3. CReSS model-simulated mixing ratio of rain at 4 km ASL. Updraft speeds are shown as solid contours with a contour interval of 5 m s⁻¹. Downdraft speeds are shown as dashed contours with a contour interval of 1 m s⁻¹. Winds are shown for all grid points where the vertically integrated mixing ratio of rain exceeded 0.0 kg kg⁻¹.

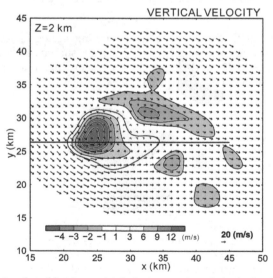

Fig. 4. CReSS model-simulated horizontal winds (vectors) and vertical velocity (shading) at 2 km ASL. Solid contours indicate updrafts at 3 m s⁻¹ contour intervals and downdrafts at 1 m s⁻¹ contour intervals. A vertical cross-section along the thick horizontal line is shown in Fig. 5.

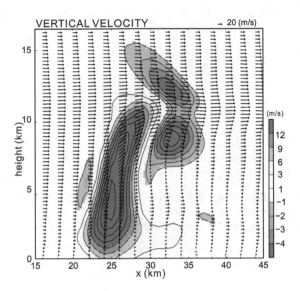

Fig. 5. Vertical cross-section of CReSS model-simulated vertical velocity at y = 26 km in Fig. 4. Contours are as in Fig. 4. Vectors indicate the speed and direction of wind flow along the vertical cross-section.

Figure 5 shows a vertical cross-section of the simulated storm along the y = 26 km transect in Fig. 4. The strong updraft extended upward to 12 km ASL, with a maximum updraft speed of 24 m s⁻¹ at 4 km ASL. Three downdraft cores flanked the strong updraft core. A weak compensating downdraft was located to the west of the strong updraft at approximately 6 km ASL. A second strong downdraft (6 m s⁻¹) formed at 8 km ASL, and was associated with a high graupel mixing ratio (data not shown). A third strong downdraft was simulated at heights above 11 km ASL in the non-precipitating region. This downdraft was likely related to gravity wave dynamics.

Four pseudo-radars are assumed to observe this simulated three-dimensional wind field (the locations of these pseudo-radars are shown in Fig. 1). The wind components are bilinearly interpolated from the model grid to sampling locations along the radar beams. Radial velocity is calculated from the three wind components interpolated to each radar grid point. The maximum range of detection is set to 70 km for all four pseudo-radars, as shown in Fig. 1. Each radar sweep observes a total of 90 azimuthal angles, with a gate spacing of 100 m and an azimuthal resolution of 1°. The radial velocities along each radar beam in the volume scan (see Fig. 2) are interpolated back onto the Cartesian coordinate system using a Cressman scheme with an influence radius R of 1.0 km. Some upper-level velocities will be lost because the highest elevation angle was less than 30° (Fig. 2). The robustness of the variational analysis method to noise is shown by adding random errors (mean 0 m s⁻¹, variance 1 m² s⁻²) to the radial velocities after interpolation back to the Cartesian coordinate system.

The accuracy of the multiple Doppler radar analysis is evaluated using root-mean-square errors in retrieved radial velocity (RMSE_VR) and vertical velocity (RMSE_W), defined as follows:

$$\text{RMSE\_VR}(k) = \frac{1}{N} \sum_{i,j,m} (Vr^{i,j,k,m} - Vr\_obs)^2 \tag{4}$$

$$\text{RMSE\_W}(k) = \frac{1}{N2} \sum_{i,j} (W^{i,j,k} - W^{true})^2 \tag{5}$$

Average values of RMSE_VR and RMSE_W are computed at each vertical level after L–BFGS optimization. W^{true} is the vertical velocity output by the CReSS model. N and $N2$ are the number of individual samples used to compute the averages at a given layer.

The L–BFGS optimization scheme is able to successfully minimize the cost function. Figure 6 shows the relationship between the value of the cost function and the number of iterations performed. The value of the cost function is effectively constant after 20 iterations.

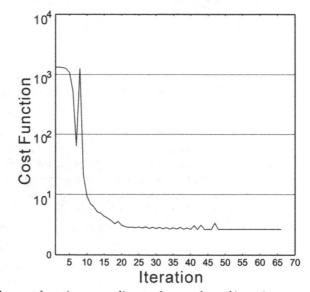

Fig. 6. Value of the cost function according to the number of iterations.

The results of the variational multiple Doppler radar analysis using volume scans with 30 PPIs (experiment name: EL30) is shown in Fig. 7 and Fig. 8. Figure 7 shows that the EL30 experimental setup successfully retrieves the strong updraft core shown in the original model output (Fig. 4). The EL30 results indicate a maximum updraft speed at 2 km ASL of 12 m s⁻¹, and successfully reproduce the anticlockwise rotation at this level. The downdraft region located in the area of heavy precipitation to the east of the strong updraft is retrieved by EL30, although the size of this downdraft region is too small (Fig. 4). The horizontal wind field retrieved by EL30 is similar to the original model output (Fig. 4), but there are two major discrepancies in vertical velocity. First, the strength of the updraft speed at 2 km ASL is underestimated by 3 m s⁻¹. Second, a spurious downdraft is identified to the southwest of the strong updraft. The estimated northwesterly winds are much stronger in this spurious downdraft region than in the original model output (Fig. 4). This stronger northwesterly wind causes stronger divergence, which in turn induces the spurious strong downdraft. The strength of the enhanced northwesterly winds at 2 km ASL is similar to the strong northwesterly winds at 3–4 km ASL (cf. Fig. 2). Vertical smoothing of the radial velocity field by the Cressman interpolation procedure likely plays a major role in the erroneous vertical velocity field. Errors in radial velocity from vertical interpolation tend to occur near the boundaries of the storm, as shown in Fig. 7. This concentration of errors near the storm boundaries occurs because the number of radial velocity samples measured at neighboring grid cells is limited, so that the relative influence of radial velocities measured at distant grid cells grows.

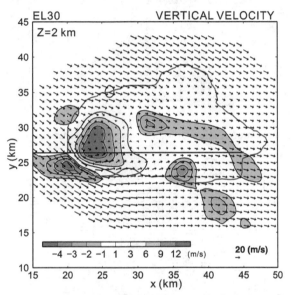

Fig. 7. Horizontal (vectors) and vertical wind velocity (shading) deduced by variational multiple Doppler radar analysis using 30 PPIs (EL30) at 2 km ASL. Solid contours indicate updrafts at 3 m s^{-1} contour intervals and downdrafts at 1 m s^{-1} contour intervals. The blue contour line outlines the region where the mixing ratio of rain exceeds 0.1 g kg^{-1}. A vertical cross-section along the thick horizontal line is shown in Fig. 8.

Figure 8 shows a vertical cross-section of the wind fields retrieved by EL30 along the y = 26 km transect in Fig. 7. Compared with the original model output (Fig. 5), the strong updraft and downdraft cores in the heavy precipitation region (at 8 km ASL) and the downdraft core associated with gravity wave dynamics (at 11 km ASL) are well retrieved. The maximum retrieved updraft speed is 21 m s^{-1}, and occurred at approximately 4 km ASL. The maximum retrieved downdraft speed is 7 m s^{-1}. The maximum updraft speed was underestimated by 3 m s^{-1}, while the maximum downdraft speed was overestimated by 1 m s^{-1}. The speed of the downdraft associated with gravity wave dynamics was underestimated by 1 m s^{-1} relative to the original model output (Fig. 5). Several spurious updrafts and downdrafts can be identified in the EL30 retrieval, especially at 7–12 km ASL. Figure 7 and Fig. 8 indicate that the EL30 pseudo-radar configuration provides a good estimation of vertical velocity. Errors in the retrieved vertical velocity are uniformly less than 3 m s^{-1}.

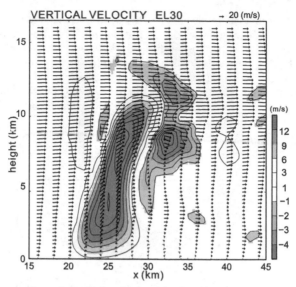

Fig. 8. Vertical cross-section of vertical velocity (shading) in EL30 at y = 26 km in Fig. 7. Contours are as in Fig. 5.

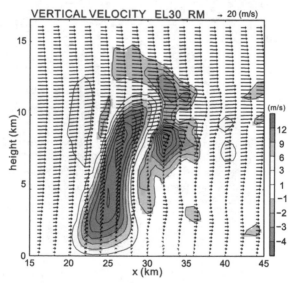

Fig. 9. As in Fig. 8 but with radial velocity observations contaminated by random noise.

As noted above, the use of the variational scheme removes the need to explicitly integrate the mass continuity equation, thus eradicating the vertical propagation of errors in the estimated wind field. Figure 9 shows the same vertical cross-section as in Fig. 8, but with random errors added to the input radial velocities. The distribution of retrieved vertical velocity is nearly identical in the cases with and without random errors. By contrast, the use of an upward integration scheme without random noise generates an erroneous vertical velocity field (too strong), especially at upper levels (Fig. 10). The addition of random noise to the upward integration method exacerbated these errors (data not shown).

These experiments reveal 1) that the variational approach provides more realistic estimates of vertical velocity than the upward integration method, and 2) that observational noise does not propagate upward when the variational method is used. These results have been obtained using a volume scan with 30 PPIs (experiment EL30). This choice of volume scan is currently unrealistic because it would take more than 10 minutes using typical current antenna rotation speeds (a few rotations per minute). For operational use, volume scans with 15–20 PPIs are realistic given the need for rapid updates (less than 400 seconds). Two further experiments are performed to mimic volume scan data with 20 PPIs (experiment name: EL20) and 17 PPIs (experiment name: EL17). The analysis method is the same for the EL20 and EL17 experiments as for the EL30 experiment. In EL20, the highest elevation angle is 18.9° (Fig. 2). A beam at this elevation angle reaches 14 km ASL 40 km away from radar. This is sufficient to ensure a valid upper boundary condition (w = 0) in this case, because the storm top is located at 12 km ASL. In EL17, however, the highest elevation angle is 15.7° (Fig. 2). This beam does not reach 12 km ASL within the detection range of the radar.

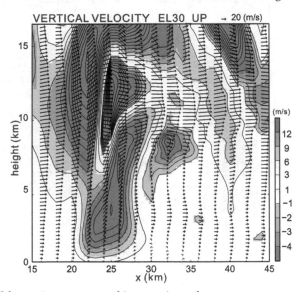

Fig. 10. As in Fig. 8 but using an upward integration scheme.

Figure 11 is the same as Fig. 8 but for EL20 rather than EL30. Vertical velocities retrieved by EL20 are similar to those retrieved by EL30 (Fig. 8) below 8 km ASL; however, the downdraft associated with gravity wave dynamics is not retrieved by EL20 because the pseudo-radar configuration does not observe that location. The maximum updraft was 20 m s^{-1}, a 1 m s^{-1} underestimate of the EL30 retrieval. As with EL30 (Fig. 8), a spurious updraft and downdraft are retrieved near 8–12 km ASL to the west of the strong updraft (Fig. 11). The area of this spurious downdraft is larger in EL20 than in EL30. This implies that EL20 is not capable of fully observing the storm top, so that erroneous upper boundary conditions induced the spurious downdraft.

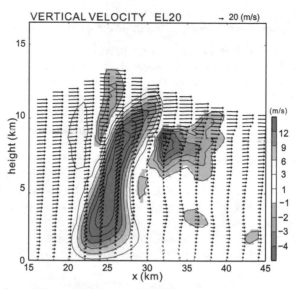

Fig. 11. As in Fig. 8 but with a volume scan of 20 PPIs.

Figure 12 is the same as Fig. 11 but for EL17 rather than EL20. The downdraft region associated with heavy precipitation to the east of the strong updraft is extended further downward in EL17. The maximum retrieved downdraft speed exceeds 9 m s^{-1}, a 3 m s^{-1} overestimate of the original model output (Fig. 5). These results indicate that even with the use of the variational approach, incomplete upper boundary conditions may lead to incorrect estimates of vertical velocity, especially near the storm top.

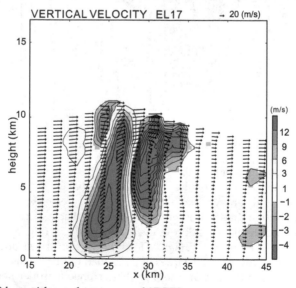

Fig. 12. As in Fig. 8 but with a volume scan of 17 PPIs.

Figure 13 shows root-mean-square errors in radial velocity (RMSE_VR) for the three experiments (EL30, EL20, and EL17). The EL30 RMSE_VR (black line in Fig. 13) was the smallest among the three experiments. The EL20 (red line) and EL17 (blue line) RMSE_VR both increased with height, while the RMSE_VR for all three experiments was small below 5 km ASL. The RMSE_VR and the root-mean-square error of the retrieved vertical velocity (RMSE_W) are closely related (Fig. 14). Differences in RMSE_W among the three experiments are relatively small below 5 km ASL and relatively large above 7 km ASL.

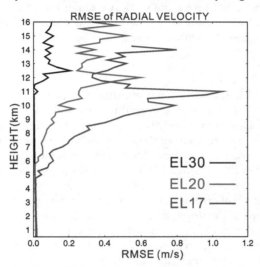

Fig. 13. Vertical profile of RMSE_VR for the EL30 (black), EL20 (red), and EL17 (blue) scanning strategies.

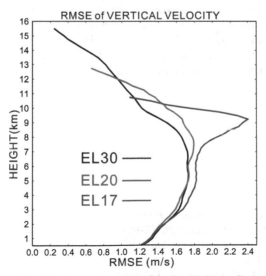

Fig. 14. Vertical profile of RMSE_W for the EL30 (black), EL20 (red), and EL17 (blue) scanning strategies.

For analyses of real thunderstorms the true value of vertical velocity is unknown, rendering the calculation of RMSE_W impossible; however, RMSE_VR can be calculated. Therefore, users could apply the variational multiple Doppler radar analysis approach to operational scans if the vertical profile of RMSE_VR is used to validate upper-level vertical velocities. The variational approach is a useful tool even for operational volume scans (less than 20 PPIs) because accurate three-dimensional winds can be retrieved at lower levels without contamination from sampling error in upper level or uncertainty in the upper boundary condition. These errors do not propagate vertically under the variational approach. Future work should examine the dependence of the retrieved wind field on scan strategy using observational datasets generated from real radar networks.

5. Conclusions

This chapter has introduced a variational multiple Doppler radar analysis for retrieving three-dimensional wind fields in a severe thunderstorm. A simplified version of the method presented by Gao et al. (1999) has been used to investigate the dependence of retrieved vertical velocity on scan strategy. Three volume scan strategies have been considered in this chapter: 1) a typical operational volume scan (17 PPIs), 2) a dense operational volume scan (20 PPIs), and 3) an extremely dense volume scan (30 PPIs). The variational approach has notable advantages over the upward integration method, particularly the avoidance of error accumulation during the upward integration; however, incomplete observations of the upper boundary condition can cause errors in estimates of vertical velocity near the storm top even when the variational approach is used. Users should limit their use of upper-level wind retrievals according to the root-mean-square error of radial wind, as described in this chapter. The variational method provides accurate estimates of the three-dimensional wind field at lower altitudes regardless of the upper boundary conditions. The density of operational radar networks in metropolitan regions has been increased in recent years to better monitor and forecast severe weather. Together with this increase in operational radar network density, the variational analysis method presented in this chapter will provide new information on the three-dimensional structure of wind within thunderstorms, and advance understanding of the physical mechanisms underlying heavy rainfall and severe winds.

6. Acknowledgment

The author thanks Dr. K. Iwanami of the National Research Institute for Earth Science and Disaster Prevention (NIED) for providing useful suggestions regarding multiple Doppler radar analysis. The model simulation was conducted at NIED using an SGI Altix 4700.

7. References

Armijo L. (1969). A theory for the determination of wind and precipitation velocities with Doppler radars, *J. Atmos. Sci.*, 26, 570-573
Byre, H. R. & Braham, R. R. (1949). *Thunderstorms,* , U.S. Government Printing Office, Washington D. C., 287 pp, USA.
Biggerstaff M. I. & R. A. Houze (1991). Kinematic and precipitation structure of the 10-11 June 1985 squall line, *Mon. Wea. Rev.*, 119, 3034-3065

Browning K. A. (1964). Airflow and precipitation trajectroies within severe local storms which travel to the right of the winds, *J. Atmos. Sci.*, 21, 634-639

Browning K. A. (1965). The evolution of tornadic storm, *J. Atmos. Sci.*, 22, 664-668

Chong M. J & J. Testud. (1983). Three-dimensinal wind field analysis from dual-Doppler radar data. Part III : the boundary condition : an optimal determination based on a variational concept, *J. Climate Appl. Meteor.*, 22, 1227-1241

Cressman G.(1959). An operatinal objective analysis system, *Mon. Wea. Rev.*, 87, 367-374

Doviak R. J. ; P. S. Ray ; R. G. Strauch & L. J. Miller (1976). Error estimation in wind fields derived from dual-Doppler radar measurement. *J. Appl. Meteor.*, 15, 868-878

Dowell D. C & H. B. Bluestein. (1997). The Arcadia, Oklahoma, storm of 17 May 1981 : Analysis of a supercell during tornadogenesis, *Mon. Wea. Rev.*, 125, 2562-2582

Gao J. ; M. Xue ; A. Shapiro & K. K. Droegemeier (1999). A variational method for the analysis of three-dimensional wind fields from two Doppler radras, *Mon. Wea. Rev.*, 127, 2128-2142

Gao J. ; M. Xue ; K. Brewster & K. K. Droegemeier (2004). A three-dimensional variational data analysis with recursive filter for Doppler radars, *J. Atmos. Oceanic Technol.*, 21, 457-469

Heinselman P. L. & S. M. Torres (2011). High-temporal-resolution capabilities of the national weather radar Testbed phased-array radar. *J. Appl. Meteor. Climatol.*, 50, 579-593

Junyent F. ; V. Chandrasekar ; D. McLaughlin ; E. Insanic & N. Bharadwaj (2010). The CASA Integrated Project I : Network radar system, *J. Atmos. Oceanic Technol*, 27, 61-78

Kessinger C. J. ; P. S. Ray & C. E. Hane (1987). The Oklahoma squall line of 19 May 1977. Part I : a multiple Doppler analysis of convective and stratiform structure, *J. Atmos. Sci.*, 44, 2840-2865

Köck K. ; T. Leitner ; W. L. Randeu ; M. Divjak & K. J. Schreiber (2000). OPERA: Operational Programme for the Exchange of Weather Radar information: First results and outlook for the future. *Phys. Chem. Earth*, 25B, 1147-1151

Leise J. E. ; E. F. Blick & R. R. Bensch (1981). A multidimensional scale-telescoped filter and data extrapolation package, NOAA Tech. Memo ERL. WPL-82, Wave Propagation Laboratory, 20 pp

Mewes J. J & A. Shapiro (2002). Use of the vorticity equation in dual-Doppler analysis of the vertical velocity field, *J. Atmos. Oceanic Technol.*, 19,543-567

Parsons D. B & R. A. Kropfli (1990). Dynamics and fine structure of a microburst, *J. Atmos. Sci.*, 47, 1674-1692

Ray P. S; K. W. Johnson & J. J. Stephens. (1979). Multiple Doppler network design, *J. Appl. Meteor.*, 18, 706-710

Ray P. S & K. L. Sangren. (1983). On multiple-Doppler radar network design, *J. Climate. Appl. Meteor.*, 22, 1444-1453

Ray P. S; C. L. Ziegler ; W. C. Bumgarner & R. J. Serafin. (1980). Single and multiple Doppler radar observations of tornadic storms, *Mon. Wea. Rev.*, 108,1607-1625

Shimizu S ; H. Uyeda ; Q. Moteki ; T. Maesaka ; Y. Takaya ; K. Akaeda & M. Yoshizaki (2008). Structure and formation mechanism on the 24 May 2000 supercell-like storm developing in a moist environment over the Kanto Plain, Japan, *Mon. Wea. Rev.*, 136, 2389-2407

Wakimoto R. ;   M. Hane ; V. Murphey ; D. C. Dowell & H. B. Bluestein (2003). The Kellerville Tornado during VORTEX : Damage survey and Doppler radar analyses, *Mon. Wea. Rev.*, 131, 2187-2221

4

Synergy Between Doppler Radar and Lidar for Atmospheric Boundary Layer Research

Chris G. Collier
National Centre for Atmospheric Science, University of Leeds
United Kingdom

1. Introduction

The principle of operation of radar and lidar is similar in that pulses of energy at wavelengths ranging from millimetres to metres for radar and 0.5 to 10 microns for lidar are transmitted into the atmosphere; the energy scattered back to the transceiver is collected and measured as a time-resolved signal. From the time delay between each outgoing transmitted pulse and the backscattered signal, the distance to the scatterer is inferred. The radial or line-of-sight velocity of the scatterers is determined from the Doppler frequency shift of the backscattered radiation. The systems use a heterodyne detection technique in which the return signal is mixed with a reference beam (i.e. local oscillator) of known frequency. A signal processing computer then determines the Doppler frequency shift from the spectra of the heterodyne signal. The energy content of the Doppler spectra can also be used to determine boundary layer eddy characteristics.

2. Characteristics of radar

The atmospheric boundary layer has been studied using weather radar extensively over the last forty years or so, such that networks of radars comprise systems sometimes operating unmanned in remote locations (see for example Atlas, 1990). Doppler radar operating at X, C or S-band (3cm, 5cm and 10cm wavelength respectively) has provided the opportunity to measure the reflectivity of target hydrometeors and the three dimensional wind structure of the lower parts of the atmosphere inferred from their motion (see for example Doviak and Zrnic, 1984). Typical parameters for a C-band radar are listed in Table 1. In addition, high power radar systems, such as the Oklahoma University Polarimetric radar for Innovations in Meteorology and Engineering (OU-PRIME) which operates at C-band and has a peak power of 1000 kW and a beamwidth of 0.45 degree (Palmer et al., 2011), provide information on the clear air structure of wind fields, and sometimes lower power systems may do the same at close range to the radar site. The detailed principle of operation has been described elsewhere in this book.

Millimetre-wave cloud radars exploit the fact that the echo intensity of Rayleigh scatterers increases with the inverse fourth power of the wavelength. These radars normally operate at 35 GHz and 94 GHz. UHF and VHF Doppler radar systems measure both wind speed and direction by detecting small irregularities in back scattered signals due to refractive index inhomogeneities caused by turbulence. In the lower troposphere the refractive index inhomogeneities are mainly produced by humidity fluctuations. The clear air Doppler shift provides a direct measurement of the mean radial velocity along the radar beam. Typically a UHF wind profiler operates at 1290 MHz or 915 MHz with a peak power of 3.5 kW and a beamwidth of 8.5 degrees.

Wavelength / frequency	5 cm (C-band) / 5430-5800 MHz
Pulse repetition frequency	250-1200 Hz selectable
Bandwidth single PRF	+ 15,9 m s^{-1}
dual PRF	± 63.8 m s^{-1}
Sampling frequency - IF	60 MHz
Peak power	250 kW
Minimum Detectable Signal	-111 dBm
Beamwidth	1 degree
Down range resolution/	75 m
Maximum range	200 km
Transmitter type	Coaxial magnetron

Table 1. Typical parameters of a C-band radar (from Selex Gematronik)

Generally radars used for weather forecasting other than wind profilers have a resolution of about 100 m with an antenna having a diameter of about 4 m. Some research radars, such as the Chilbolton radar in the UK, provide measurements with a resolution at 100 km range using a 0.25 degree beamwidth (25 m diameter antenna) of 0.4 km. These resolutions certainly improve our understanding of the structure and behaviour of boundary layer phenomena, and examples will be described in this chapter. Nevertheless, boundary layer turbulent eddies may exist with characteristic length scales from tens of metres to fractions of a metre close to the ground. Such length scales require different instrumentation. Doppler lidar is an instrument providing high resolution, clear air measurements with resolutions of around 30 m, albeit over much shorter ranges than available from radar systems.

3. Characteristics of lidar

Lidar has been developed which operate in various atmospheric windows, namely the 10, 2, 1.5 and 1 micron spectral regions. Hardesty et al. (1992) compared the transmission, backscatter, refractive turbulence and Doppler estimation characteristics of a 2 and a 10 micron Doppler lidar system. Whilst backscatter at 2 microns in the free atmosphere is 4-10 times higher than for a 10 micron Doppler lidar, the effects of turbulence on the 2 micron system beyond a few kilometres range are significant, with the signal to noise ratio being reduced by about 6 dB at 5 km range. However, early equipment operated with CO_2 lasers at 10.6 micron wavelength, but involved delicate optical systems (Post and Cupp, 1990; Mayor et al., 1997; Pearson and Collier, 1999). The advent of fibre optic technology has enabled compact, robust equipment to be developed and operated remotely at wavelengths of 1.5 microns (Pearson et al., 2002, 2009). Table 2 shows the parameters of this type of Doppler lidar. The range resolution of 30 m is considerably smaller than that used in CO_2 lidars of about 112 m.

Wavelength	1.5 µm
Pulse repetition rate	20 kHz
Bandwidth	± 14 m s^{-1}
Sampling frequency	30 MHz
Points per range gate	6
Number of pulses averaged	20 000
Δr	18 m
Δp	30 m
Averaging time	1 s

Table 2. Parameters of 1.5 micron Doppler lidar. Range gate parameters: Δr relates to pulse length and Δp is the down range extent of the range gate used in the signal processing (from Pearson et al., 2009).

4. Advantages and disadvantages of radar and lidar

Both radars and lidars have advantages and disadvantages, and in this chapter a review of the accuracy with which measurements of boundary layer winds from both instruments is given. However both instruments offer complimentary information. Lidar backscatter is from widely dispersed aerosol particles in the clear air, but their concentration decreases away from the surface. Although thin clouds also provide backscatter, hydrometeors strongly attenuate the lidar signal. Radar backscatter is generally from hydrometeors, and therefore the operating range is much greater than that of lidar although this depends upon the wavelength used as short wavelength signals are attenuated.

Topography strongly controls the flux of momentum and energy between the terrain surface and the boundary layer. In-situ instruments have minimum spatial coverage, and radar cannot make measurements very close to the ground due to beam side lobes which produce ground clutter. With beamwidths of 0.1-1 m. rad., lidar transverse resolution is 20-200 times finer than the one degree (17.5 m. rad.) of weather radars (Drechsel et al., 2009). Lidar provides wide area coverage, although not as extensive as radar, but does not suffer from ground clutter problems. Hence lidar offers the opportunity to improve our knowledge of flow over complex terrain close to the ground surface (Barkwith and Collier, 2011). However radar can provide detailed information in the boundary layer over very wide areas, particularly related to the development of convective systems.

5. Wind profiling

5.1 Single instrument measurements

Measurements of the vertical profile of wind may be made using Doppler lidar and both weather radars and UHF / VHF profilers. There are two modes that can be utilised for this. The beam can be scanned in a cone at fixed elevation and the resulting data fitted to a sine wave. This is known as the velocity-azimuth display (VAD) approach as described by Browning and Wexler (1968). This has been the preferred technique for radar systems although it has also been used with lidar systems. An alternative approach uses three fixed line of sights from which a vector analysis provides the three components of the wind (u, v and w) as described by Werner (2005).

Using the VAD technique Bozier et al. (2004) compared the average horizontal wind velocity difference and average standard deviation between a CO_2 Doppler lidar and tethered balloon borne turbulence probe measurements at a site in Eastern England as shown in Table 3. The in situ sensor captures a higher frequency turbulence component in the wind velocity data due to the higher measurement rate, 4 Hz, compared to the lidar data sampling rate of 0.1 Hz.

Measurement run	Height range (m)	Wind velocity difference (m s^{-1})	Lidar standard deviation (m s^{-1})	Balloon standard deviation (m s^{-1})
L1, B1	350–750	0.53	1.04	0.80
L2, B2	180–530	0.38	1.50	0.78
L3, B3	210–530	0.34	1.39	1.14
L4, B4	350–900	0.19	1.28	0.67

Table 3. Comparison of lidar and balloon borne instrument probe derived horizontal wind profiles (from Bozier et al., 2004)

Pearson et al. (2009) suggests that the three-beam technique may offer a better option in regions where the flow is not constant and laminar over the disc swept out by the VAD scan for example where topography influences the flow field. They showed examples of both approaches in a study of the performance of a 1.5 micron Doppler lidar. Based on 51 days continuous, unattended operation at a site in southern England, Figure 1 shows the results of an inter-comparison carried out on 13 September 2007 between a UHF radar and lidar data. There is no cloud cover. The wind speed data are less well correlated possibly due to ground clutter contamination of the radar velocity data, or a reduction in the Signal-to-Noise (SNR) leading to a larger degree of uncertainty in the sine-wave fit procedure. Similar results have been found using a 10.6 micron CO_2 Doppler lidar by Mayor et al. (1997), who compared vertical velocities derived from a 915 MHz radar profiler. A difference between the two instruments of -0.81 cm s^{-1} was found. It was noted that the difference may be due to the different sizes of the sampling volumes, the spatial separation of the two measurements and the different SNRs. Comparisons of VAD winds derived from Doppler lidar reported by Drechsel et al. (2009) deviated from wind profiler measurements by less than 1.0 m sec^{-1} and -1 degree, and only 0.1 m sec^{-1} and 2 degrees from radiosonde data.

5.2 Dual-doppler systems

A single Doppler radar or lidar measures the field of wind velocities that are directed towards or away from the instrument. Doviak and Zrnic (1984) describe how a second Doppler radar spaced far from the first produces a field of different radial velocities which can be vectorially synthesized to retrieve the two dimensional velocities in the plane containing the radials.

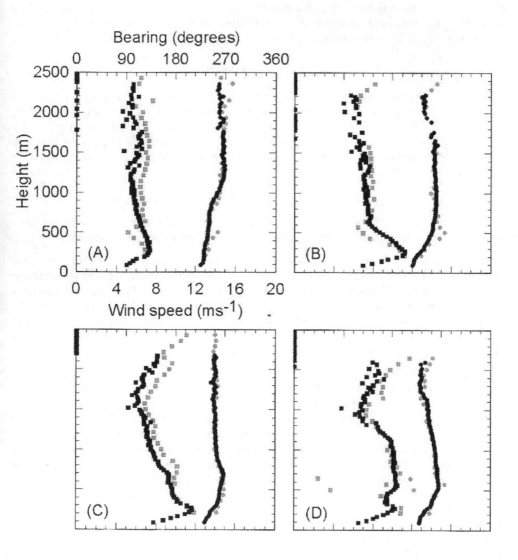

Fig. 1. Radar and lidar profiles from the evening of 13 Sept. 2007. The lidar scan took 9 min and the radar data are for a 10-min average. (A)-(D) Start times of 1900, 2100, 2200 and 2300 UTC respectively. The circles and squares show the bearing and speed data respectively (grey radar and black lidar) (from Pearson et al., 2009)

This is greatly simplified if this operation is performed in cylindrical coordinates with axis chosen to be in the line connecting the two radars 1 and 2, referred to as the baseline. The mean Doppler velocity is corrected for the reflectivity-weighted mean terminal velocity of the scatterers w_t. Hence the estimate of the radial component of air motion is:

$$v_{1,2} = v_{1,2}' + \overline{w_t} \sin \Theta_{e1,2}$$

where $v_{1,2}'$ are the mean Doppler target velocities measured by radars 1,2 at data points and w_t is positive. To estimate w_t an empirical expression such as that given by Atlas et al. (1973) can be used. The estimated radial velocities $v_{1,2}$ of the air can be interpolated to uniformly spaced grid points in planes at an angle α to the horizontal surface containing the baseline. The wind component w_α normal to the plane is obtained by solving the continuity equation in cylindrical coordinates:

$$1/r \, \partial/\partial r \, (r\rho w_r) + 1/r \, \partial/\partial x \, (\rho w_\alpha) + \partial/\partial s \, (\rho w_s) = 0$$

with the boundary condition $w = 0$ at the ground. Generally this approach covers a smaller area compared to the full dual-Doppler coverage and may contain residual errors.

An improved albeit similar approach has been developed by Bousquet and Chong (1998) for three dimensional wind retrieval from multiple airborne Doppler radars, called the Multiple Doppler Synthesis and Continuity Adjustment Technique (MUSCAT). It was extended for application over both flat or complex terrain by Chong and Cosma (2000), and for ground-based radar systems by Chong and Bousquet (2001). An alternative computationally inexpensive plane-to-plane solution known as the Multiple Analytical Doppler (MANDOP) system has been described by Tabary and Scialom (2001).

In 2003 two mobile Doppler lidars were sited at either end of a disused runway approximately 1.6 km apart at RAF Northolt in West London (Collier et al., 2005). The aim was to investigate the optimal lidar configuration to measure wind flow turbulence characteristics. Three dual-lidar configurations are shown in Figure 2. Figures 2a and 2c show data taken with the two lidars where the beams cross at a point, which is in the vertical plane defined by the line joining the two lidar positions. With each lidar system the radial velocities along the beams were measured every five seconds. The two radar computer clocks were first synchronised, and then for the different configurations, the time series of data were taken. Table 4 gives statistics derived from operating the lidar systems as in configuration 2c. The data were taken for a period of 700 seconds, and for different heights over a period of approximately 50 minutes. The errors in the vertical velocities from this dual-Doppler lidar deployment were analysed by Davies et al. (2005). It was found that the spread in vertical velocities due to the combined effects from instrumental errors of the two lidars can in some cases act to cancel each other out, although on other occasions this was not the case. We discuss the implications of this in the next section.

Drechsel et al. (2009) applied the MUSCAT processing system to dual-Doppler lidar data collected during the Terrain – induced Experiment (T-REX) in the spring of 2006. The flow pattern derived from 19 three dimensional wind fields revealed differences of wind speed and direction of less than 1.1 m sec[-1] and 3 degrees on average compared to radiosonde and wind profiler data. The average vertical motion from MUSCAT was -0.24 m sec[-1] compared to -0.52 m sec[-1] from wind profiler data and -0.32 m sec[-1] from radiosonde data.

Height (m)	Mean horizontal wind in direction of the lidar axis (m/s)	Mean vertical Wind (m/s)	Std. dev. of wind in direction of the lidar axis (m/s)	Std. Dev. of vertical wind (m/s)
100	-0.05	1.1	0.50	3.43
200	1.52	-0.90	0.59	1.45
400	0.77	-0.66	0.59	1.78
709	0.68	-0.83	1.49	1.13

Table 4. Means and standard deviations of the horizontal and vertical winds. The horizontal wind is the wind in the direction of the axis joining the two lidars. The heights of the data are determined by the crossing points of the two lidar beams as shown in the configuration of Figure 2c. The lidar data were taken from 1134-1224 UTC 23 July 2003 (from Collier et al. (2005)

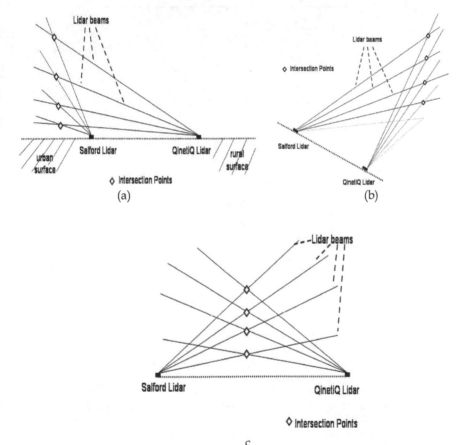

Fig. 2. Dual-lidar configurations. Diamonds denote beam intersection points (from Collier et al., 2005)

6. Measuring turbulent structure

A comprehensive review of the use of radar to measure the morphology of the boundary layer is provided by Gossard (1990), who also provides a review of the use of Doppler radar to measure turbulent velocity variance and covariance. VAD scans do not produce perfect sinusoids, and their derived time series are perturbed by random turbulent fluctuations at scales larger than the radar pulse volume, but much smaller than the diameter of the VAD. Wilson (1970) and Wilson and Miller (1972) developed a procedure for extracting quantitatively the variances and covariances of u, v and w. This was developed further by Kropfli (1984) who measured profiles of vertical momentum flux through the convective boundary layer as shown in Figure 3.

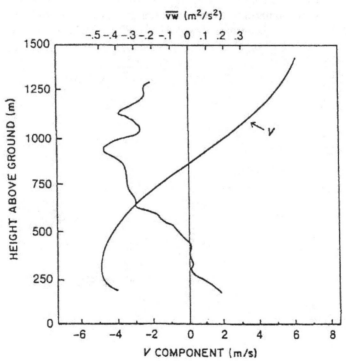

Fig. 3. Single radar measurement of vertical profile of momentum flux through the boundary layer by the method of Wilson (1970). Data are from a 100-min period beginning at 12-3 MDT during the PHOENIX experiment (from Kropfli, 1984)

The radar Doppler spectral width has also been used to extract small-scale turbulence information from spectral broadening due to wind shear, the radar antenna properties and the variance of the velocity component in the radial direction due to turbulence. In particular, the eddy dissipation rate can be measured, the first attempt to do this being by Gorelik and Mel'nichuk (1963). Table 5 is a summary of the literature on the measurement of turbulent energy dissipation (ε). Note that both radar and lidar have been used to measure this quantity.

Type	Typical Values of ε x 10^3 (m² sec⁻³)	Typical estimate	Reference
D_N	0.2	Model	Simmons and Hoskins, (1978)
D_{GW}	6 (Over major mountains)	Model	Shutts (2005)
D_C	1-10	Radar and aircraft observations	Meischner et al (2001)
D_F	1000 (within the surface layer) 2 (in shallow sloping frontal layers); 3 (close to the ground before surface front passes); 20 (jet stream level; clear air turbulence) 30 (jet stream level; clear air turbulence)	Hot-wire anemometer at 3m Radar observations Aircraft observations Radar observations	Piper and Lundquist (2004) Chapman and Browning (2001); Chapman and Browning (2001); Kennedy and Shapiro (1975, 1980) Gage et al (1980)
D_B (Urban areas)	Up to 4 (urban areas) 0.2 – 6.2 (urban areas)	Lidar observations at several hundred metres Lidar observations	Davies et al (2004); Davis et al (2008)

Table 5. Energy dissipation rates, ε (m² sec⁻³) (from Collier and Davies, 2009)

Generally the vertical velocity in the turbulent boundary layer is very small, and difficult to measure. It is clear that significant temporal averaging is necessary. Kropfli (1984) suggested averaging over 20 minutes for VAD radar data, whereas Davies et al. (2005) averaged lidar data over 10 to about 50 minutes. However, the errors in the wind velocity, particularly those in w were thought by Sathe et al. (2011) to prelude the use of Doppler lidar with the VAD technique from measuring boundary layer turbulence precisely. This conclusion is not necessarily appropriate when two lidars are used as demonstrated by Davies et al. (2005) and Pearson et al. (2009).

In order to investigate the detailed structure of the turbulent motion above the Urban Canopy Layer (UCL), Davies et al. (2004) compared Doppler lidar-measured turbulent structure functions with those derived using the Von Karman model of isotopic turbulence in the inertial sub-range. Making allowance for the spatial averaging of the lidar pulse volume, the correspondence is comforting (Fig. 4). Hence, estimates of the integral length-scale, the dominant spatial scale of the turbulence above the UCL, can be made from the fit of the model to the observations, giving a range from 250–400 m. In addition, measurements were made of the velocity covariance power spectra, and the corresponding eddy dissipation rates are shown in Fig. 5. The slope of the spectra within the inertial sublayer is usually −5/3, although this depends upon the presence of inversion layers and the strength of the turbulence. The fact that the spectrum falls off faster than −5/3 may indicate that the turbulence approaches isotropy locally in the inertial range (Lumley, 1965). However, in the case of the lidar measurements the fall off at less negative wave numbers is more likely to be due to the spatial averaging of the data over the beam. Climatologies of urban and rural eddy dissipation rates are needed to evaluate the urban effects.

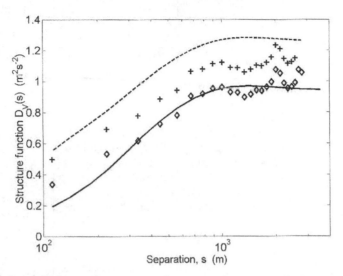

Fig. 4. Comparison of lidar measurements of structure function with the Von Karman model over Salford, Greater Manchester at 1317 UTC 2 May 2002. The upper dashed line represents the uncorrected model and the lower thick line the corrected model; crosses are uncorrected lidar measurements and diamonds are the corrected lidar measurements. (From Davies et al. 2004.)

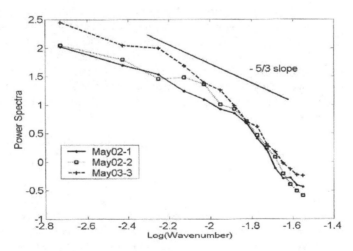

Fig. 5. Covariance power spectra and eddy dissipation rates, ε, over Salford, Greater Manchester, derived from line-of-sight measurements made during the SALFEX experiment for three datasets as indicated. The values of ε for the three curves range from 1.1 to 4.3 × 10^{-3} m²sec⁻³. (From Davies et al. 2004.)

Recently Krishnamurthy et al. (2011) reported Doppler lidar measurements of turbulent kinetic energy dissipation rate, integral length scale and velocity variance assuming a theoretical model of isotropic wind fields during the T-REX Project. Corrections to address the complications inherent in volumetric averaging of radial velocity over each range gate, noise in the data and the assumption made regarding the effects of smaller scales of motion were considered and tested. Comparisons between the lidar and tower measurements supported the soundness of the lidar measurements of boundary layer turbulence.

7. Doppler radar and lidar measurements of boundary layer phenomena

7.1 Storm outflows

One of the most difficult forecasting problems remains the identification of when and where convective cells develop. Two facets of this problem are (a) how first-generation convective cells are triggered in an environment which has been previously quiescent for a period, but which becomes more and more unstable; and (b) for convective cells that persist, how and where subsequent generations of cells are triggered by the propagation of cold outflows from existing cells. Over the years there have been many studies addressing these issues using radar data (see for example Bader et al., 1995). Indeed, it is well known that first generation convective cells may be triggered by orographic uplift, and by land surface heterogeneity caused by variations in the temperature and moisture fields.

Collier and Davies (2004) describe a study of the pre-storm environment for a case study using a Doppler lidar located at Northolt, North West London, a C-band weather radar sited at Chenies north of London and the S-band Chilbolton radar. It was noted that the Doppler lidar and the weather radar data complement each other. Figure 6 shows Chenies and Chilbolton radar images, a PPI from the Doppler lidar and a LDR PPI from the Chilbolton radar. An outflow boundary is evident in all the images. A reversal of the wind direction at low levels is shown near the lidar site. The Chenies radar shows a thin line of broken echoes about 12 km to the north and north west of the main area of convective rain. The Chilbolton radar Linear Depolarisation Ratio (LDR) suggests that the radar targets in the outflow region are probably not raindrops, but may be particulate matter (straw, dust). The Doppler radial velocities observed by Chilbolton are consistent with the lidar measurements in the figure.

Similar measurements of an outflow have been reported by Collier et al. (2008). This study illustrates the difficulty of measuring an outflow using a radar, in this case the DLR C-band radar. In Figure 7a an outflow from a thunderstorm over the Rhine Valley is partially observed by the radar, but the details are not clear as there is some confusion with ground clutter. Figure 7b shows Doppler lidar measurements of the vertical velocities made from Achern in the Rhine Valley. Here the outflow is clear. It is about 800 m deep, and a cap cloud is observed near the leading edge. The peak kinetic energy dissipation rate was calculated to be 0.18 $m^2 sec^{-3}$.

7.2 Observing smoke plumes

Combined observations of smoke plumes using lidar and radar have not been extensively reported. Such plumes may be generated from wild fires, or from prescribed (planned) burns. The plumes may contain lofted debri as the primary source of targets, although smoke and condensed water droplets may also be evident. Banta et al. (1992) used Doppler radar and lidar to observe both prescribed and wild fires. Figure 8 shows Doppler lidar measurements of plumes over Helsinki produced from forest fires in North West Russia. The correlation between the lidar backscatter data and the retrieved vertical velocities shows convective updrafts containing the particulate-laden air from the cleaner air contained in downdrafts.

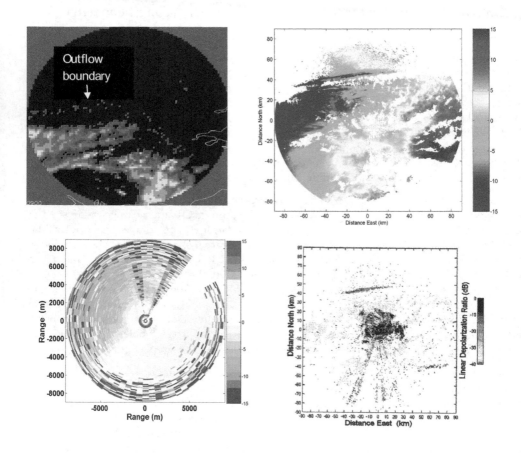

Fig. 6. (a) (top left) Chenies radar images on 16 July 2003 [Blue less than 1 mm h⁻¹; yellow 2-4 mm h⁻¹; grid 2 km x 2 km maximum range 75 km]; (b) (top right) Chilbolton unfolded radial velocities 16 July 0903 UTC in m sec⁻¹ [negative away from the radar]; (c) (bottom left) Doppler lidar PPI [10 deg.] radial velocities [m sec⁻¹] 16 July 2003.0850 UTC [negative towards the lidar]. Note the wind reversal at low levels; (d) (bottom right) Chilbolton radar Linear Depolarisation Ratio (LDR) 16 July 0903 UTC (from Collier and Davies, 2004).

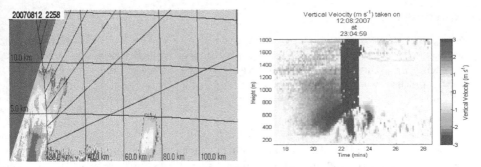

Fig. 7. Illustrating (a) RHI from the DLR Poldirad radar located at Waltenheim-sur-zorn towards the Supersites Achern (36 km), Hornisgrinde (46 km) and AMF (62 km) at 22.58 UTC 12 August 2007; and (b) vertical velocity measured at 23.04 UTC by Doppler lidar located at Achern on 12 August 2007 (from Collier et al., 2008).

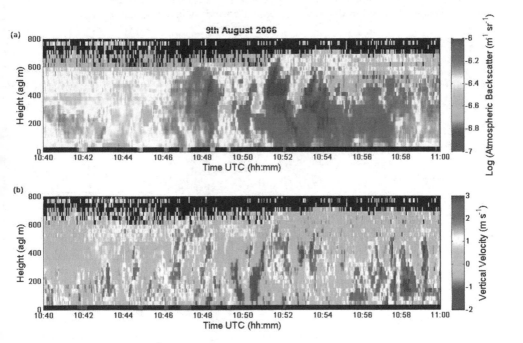

Fig. 8. Time series of lidar atmospheric backscatter and vertical velocity measurements taken on 9 August 2006 1040-1100 UTC in Helsinki, Finland. The log of the atmospheric backscatter is shown in (a) with the vertical velocity data from the lidar shown in (b). The vertical axis shows height above ground level (m) and the horizontal axis is time (UTC) (from Bozier et al., 2007).

7.3 Understanding convection

Skewness is a measure of asymmetry in the distribution of vertical velocity perturbations. Positive skewness at the surface suggests narrow intense updrafts from the surface and broad downdrafts (fair weather, clear). Negative skewness suggests sharp, narrow downdrafts and larger areas of weaker updraft, rather like "upside down" surface heating driven by turbulence on a cloudy day. Skewness may be calculated using,

$$S = w'^3 / (w'^2)^{3/2}$$

Bother Doppler radar and Doppler lidar are capable of measuring vertical velocity and they can measure skewness throughout the boundary layer. Knowing the skewness can help understand the structure of convection (see for example Hogan et al., 2009).

Although weather radars are available around the edges of tropical rain forests, there have as yet been only a few studies combining radar wind profilers and lidar (ceilometers) data (Grimsdell and Angevine, 1998). Vila-Guerau de Arellano et al. (2009) studied the isoprene fluxes in the tropical rain forest environment, but recommended the continued use of a radar wind profiler or Doppler lidar. Pearson et al. (2010) used a Doppler lidar to measure the diurnal cycle of the wind field in the tropical boundary layer, Sabah, Borneo.

8. Concluding remarks

Since radar and lidar provide measurements of backscatter and atmospheric motion based upon different targets, it is clear that much useful complementary information on atmospheric phenomena and processes can be obtained. Used together these instruments provide a powerful mechanism by which to enhance our knowledge of the atmosphere and develop improved forecasting procedures of a wide range of phenomena. Both technologies offer instrumentation capable of continuous unattended operation.

9. References

Atlas, D. (ED) (1990) *Radar in Meteorology*, Am. Met. Soc., 806pp

Atlas, D., Srivastava, R.C. and Sekhon, A.S. (1973). Doppler radar characteristics of precipitation at vertical incidence, *Rev. Geophys. Space Phys.*, 2, pp 1-35

Bader, M.J., Forbes, G.S., Grant, J.R., Lilley, R.B.E. and Waters, A.J. (1995). *Images in Weather Forecasting. A practical guide for interpreting satellite and radar imagery*, Cambridge University Press, 499pp

Banta, R.M., Olivier, L.D., Holloway, E.T., Kropfli, R.A., Bartram, B.W., Cupp, R.E. and Post, M.J. (1992). Smoke-column observations from two forest fires using Doppler lidar and Doppler radar, *J. Appl. Met.*, 31, 1328-1349

Barkwith, A. and Collier, C.G. (2011). Lidar observations of flow variability over complex terrain, *Meteor. Appl.*, 18, 372-382

Bousquet, O. and Chong, M. (1998). A Multiple-Doppler Synthesis and Continuity Adjustment Technique (MUSCAT) to recover wind components from Doppler radar measurements, *J. Atmos. Ocean. Tech.*, 15, pp 343-359

Bozier, K.E., Pearson, G.N. and Collier, C.G. (2007). Doppler lidar observations of Russian forest fire plumes over Helsinki, *Weather*, 62, no 8, pp 203-208

Browning, K.A. and Wexler, R. (1968). The determination of kinematic properties of a wind field using Doppler radar, *J. Appl. Met.*, 7, pp 105-113

Chong, M. and Cosma, S. (2000). A formulation of the continuity equation of MUSCAT for either flat or complex terrain, *J. Atmos. Ocean. Tech.*, 17, pp 1556-1565

Chapman, D. and Browning, K.A. (2001). Measurements of dissipation rate in frontal zones, *Quart. J. R. Met. Soc.*, 127, pp 1939-1959

Collier, C.G. and Davies, F. (2004). Observations of the development of the convective boundary layer using radar and Doppler lidar, *Proc. European Radar Conf. (ERAD)*, Copernicus GmbH, Visby, Gotland, Sweden, 6-10 Sept. 2004, ISBN 3-936586-29-2

Collier, C.G. and Davies, F. (2009). "Representing energy drain in Numerical Weather Prediction models induced by boundary layer sub-grid scale processes", *Atmos. Sci. Letters*, 10, Issue 3, DOI: 10.1002/asl.222, 146-151

Collier, C.G., Davies, F., Bozier, K.E., Holt, A., Middleton, D.R., Pearson, G.N., Siemen, S., Willetts, D.V., Upton, G.J.G. and Young, R.I. (2005). Dual Doppler lidar measurements for improving dispersion models, *Bull. Am. Met. Soc.*, 86, pp 825-838

Collier, C.G., Davies, F., Davis J.C., Pearson, G.N. and Hagen, M. (2008). Doppler radar and lidar observations of a thunderstorm outflow, *Proc. 5th European Conf. on Radar Met. and Hydrology (ERAD)*, Copernicus GmbH, June, Helsinki, ISSN 978-951-697-676-4

Davies, F., Collier, C.G., Pearson, G.N. and Bozier, K.E. (2004). Doppler lidar measurements of turbulent structure function over an urban area, *J. Atmos. Ocean. Tech.*, 21, pp 753-761

Davies, F., Collier, C.G. and Bozier, K.E. (2005). Errors associated with dual Doppler-lidar turbulence measurements, *J. Opt. A: Pure Appl. Opt.*, 7, pp S280-S289

Davis, J.C. Collier, C.G., Davies, F. and Bozier, K.E. (2008). Spatial variations of sensible heat flux over an urban area measured using Doppler lidar, *Meteor. Appl.*, 15, pp 367-380

Doviak, R.J. and Zrnic, D.S. (1984). *Doppler Radar and Weather Observations*, 2nd Ed., Academic Press, 562pp

Drechsel, S., Chong, M., Mayr, G.J., Weissmann, M., Calhoun, R. and Dornbrack, A. (2009). Three-dimensional wind retrieval: application of MUSCAT to dual-Doppler lidar, *J. Atmos. Ocean. Tech.*, 26, pp 635-646

Gage, K.S., Green, J.L. and Van Zandt, T.E. (1980). Use of a Doppler radar for the measurement of atmospheric turbulence parameters from the intensity of clear-air echoes, *Radio Sci.*, 15, pp 407-416

Gorelik, A.G. and Mel'nichuk, Yu.V. (1963). Radar study of dynamic process in the atmosphere, Tr. Vses. Nov. Meteor. Souesh. No5

Gossard, E.E. (1990). Radar research on the atmospheric boundary layer, in *Radar in Meteorology*, ed. D. Atlas, Am. Met. Soc., Boston, pp 477-527

Grimsdell, R.W. and Angevine, W.M. (1998). Covective boundary layer height measurement with wind profilers and comparison to cloud base, *J. Atmos. Ocean. Tech.*, 15, pp 1331-1338

Hardesty, R.M., Grund, C.J., Post, M.J., Rye, B.J. and Pearson, G.N. (1992). Measurements of winds and cloud characteristics: a comparison of Doppler lidar systems, Int. *Geoscience & Remote Sensing (Houston, IX, 1992)* Session TA-P, Paper 2

Hogan, R.J., Grant, A.L.M., Illingworth, A.J., Pearson, G.N. and O'Connor, E.J. (2009). Vertical velocity variance and skewness in clear and cloud-topped boundary layers as revealed by Doppler lidar, *Quart. J. R. Met. Soc.*, 135, pp 635-643

Kennedy, P.J. and Shapiro, M.A. (1975). The energy budget in a clear air turbulence zone as observed by aircraft, *Mon. Wea. Rev.*, 103, pp 650-654

Kennedy, P.J. and Shapiro, M.A. (1980). Further encounters with clear air turbulence in research aircraft, *J. Atmos. Sci.*, 37, pp 986-993

Krishnamurthy, A., Calhoun, R., Billings, B. and Doyle, J. (2011). Wind turbulence estimates in a valley by coherent Doppler lidar, *Meteor. Appl.*, 18, pp 361-371

Kropfli, R.A. (1984). Radar probing and measurement of the planetary boundary layer. Part II Scattering from particulates, in Probing the Atmospheric Boundary Layer, ed. D.H. Lenschow, Am. Met. Soc./NCAR, Chapters27a, 27b, pp 183-199

Mayor, S.D., Lenschow, D.H., Schwiesow, R.L., Mann, J., Frush, C.L. and Simon, M.K. (1997). Validation of NCAR 10.6µm CO_2 Doppler lidar radial velocity measurements and comparison with a 915 MHz profiler, *J. Atmos. Ocean. Tech.*, 14, pp 1110-1126

Meischner, P.F., Baumann, R., Holler, H. And Jank, T. (2001). Eddy dissipation rates in thunderstorms estimated by Doppler radar in relation to aircraft in situ measurements, *J. Atmos. Ocean. Tech.*, 18, pp 1609-1627

Palmer, R.D., Bodine, D., Kiemjian, M., Cheong, B., Zhang, G., Cao, Q., Bluestein, B., Ryzhkov, A., Yu, T-Y. and Wang, Y. (2011). Observations of the 10 May 2010 tornado outbreak using OU-PRIME, *Bull. Am. Met. Soc.*, 92, no7, pp 871-891

Pearson G.N. and Collier, C.G. (1999). A compact pulsed coherent CO_2 laser radar for boundary layer meteorology, *Quart J. R .Met. Soc.*, 125, 2703-2721

Pearson, G.N., Roberts, P.J., Eacock, J.R. and Harris, M. (2002). Analysis of the performance of a coherent pulsed fiber lidar for aerosol backscatter applications, *Appl. Opt.*, 41, pp 6442-6450

Pearson, G., Davies, F. and Collier, C. (2009). An analysis of the performance of the UFAM pulsed Doppler lidar for observing the boundary layer, *J. Atmos. Ocean. Tech.*, 26, pp 240-250

Pearson, G., Davies, F. and Collier, C. (2010). Remote sensing of the tropical rain forest boundary layer using pulsed Doppler lidar, *Atmos. Chem. Phys. Discuss.*, 10 , pp 5021-5049

Piper, M. and Lundquist, J.K. (2004). Surface layer turbulence measurements during a frontal passage, *J. Atmos. Sci.*, 61, pp 1768-1780

Post, M.J. and Cupp, R.E. (1990). Optimising a pulsed Doppler lidar, *Appl. Opt.*, 29, pp 4145-4158

Shutts, G. (2005). A kinetic energy backscatter algorithm for use in ensemble prediction systems, *Quart. J. R. Met. Soc.*, 131, pp 3079-3102

Simmons, A. and Hoskins, B.J. (1978). The life cycles of some nonlinear baroclinic waves, *J. Atmos. Sci.*, 35, pp 414-432

Vila-Guerau de Arellano, J., Van den Dries, K. And Pino, D. (2009). On inferring isoprene emission surface flux from atmospheric boundary layer concentration measurements, *Atmos. Chm. Phys.*, 9, pp 3629-3640

Werner, C. (2005). Doppler wind lidar, in *Lidar: Range-Resolved Optical Remote Sensing of the Atmosphere*, ed. C. Weitkamp, Series in Optical Sciences, Vol. 102, Springer, pp 339-342

Wilson, J.W. (1970). Integration of radar and gage data for improved rainfall measurement, *J. Appl. Met.*, 9, pp 489-497

Wilson, J.W. and Miller, L.J. (1972). Atmospheric motion by Doppler radar, in *Remote Sensing of the Atmosphere*, ed. V. E .Derr, Chapters 21a, 27a, 27pp

Part 2

Weather Radar Quality Control and Related Applications

5

Quality Control Algorithms Applied on Weather Radar Reflectivity Data

Jan Szturc, Katarzyna Ośródka and Anna Jurczyk
Institute of Meteorology and Water Management – National Research Institute
Poland

1. Introduction

Quality related issues are becoming more and more often one of the main research fields nowadays. This trend affects weather radar data as well. Radar-derived precipitation data are burdened with a number of errors from different sources (meteorological and technical). Due to the complexity of radar measurement and processing it is practically impossible to eliminate these errors completely or at least to evaluate each error separately (Villarini & Krajewski, 2010). On the other hand, precise information about the data reliability is important for the end user.

The estimation of radar data quality even as global quantity for single radar provides very useful and important information (e.g. Peura et al., 2006). However for some applications, such as flash flood prediction, more detailed quality information is expected by hydrologists (Sharif et al., 2004; Vivoni et al., 2007, Collier, 2009). A quality index approach for each radar pixel seems to be an appropriate way of quality characterization (Michelson et al., 2005; Friedrich et al., 2006; Szturc et al., 2006, 2008a, 2011). As a consequence a map of the quality index can be attached to the radar-based product.

2. Sources of radar data uncertainty

There are numerous sources of errors that affect radar measurements of reflectivity volumes or surface precipitation, which have been comprehensively discussed by many authors (e.g. Collier, 1996; Meischner 2004; Šálek et al., 2004; Michelson et al., 2005).

Hardware sources of errors are related to electronics stability, antenna accuracy, and signal processing accuracy (Gekat et al., 2004). Other non-meteorological errors are results of electromagnetic interference with the sun and other microwave emitters, attenuation due to a wet or snow (ice) covered radome, ground clutter (Germann & Joss, 2004), anomalous propagation of radar beam due to specific atmosphere temperature or moisture gradient (Bebbington et al., 2007), and biological echoes from birds, insects, etc. Next group of errors is associated with scan strategy, radar beam geometry and interpolation between sampling points, as well as the broadening of the beam width with increasing distance from the radar site. Moreover the beam may be blocked due to topography (Bech et al., 2007) and by nearby objects like trees and buildings, or not fully filled when the size of precipitation echo is relatively small or the precipitation is at low altitude in relation to the antenna elevation (so called overshooting).

Apart from the above-mentioned non-precipitation errors, meteorologically related factors influence precipitation estimation from weather radar measurements. Attenuation by hydrometeors, which depends on precipitation phase (rain, snow, melting snow, graupel or hail), intensity, and radar wavelength, particularly C and X-band, may cause the strong underestimation in precipitation, especially in case of hail. Another source of error is Z–R relation which expresses the dependence of precipitation intensity R on radar reflectivity Z. This empirical formula is influenced by drop size distribution, which varies for different precipitation phases, intensities, and types of precipitation: convective or non-convective (Šálek et al., 2004). The melting layer located at the altitude where ice melts to rain additionally introduces uncertainty into precipitation estimation. Since water is much more conductive than ice, a thin layer of water covering melting snowflakes causes strong overestimation in radar reflectivity. This effect is known as the bright band (Battan, 1973; Goltz et al., 2006). Moreover the non-uniform vertical profile of precipitation leads to problems with the estimation of surface precipitation from radar measurement (e.g. Franco et al., 2002; Germann & Joss, 2004; Einfalt & Michaelides, 2008), and these vertical profiles may strongly vary in space and time (Zawadzki, 2006).

Dual-polarization radars have the potential to provide additional information to overcome many of the uncertainties in contrast to situation when only the conventional reflectivity Z and Doppler information is available (Illingworth, 2004).

3. Methods for data quality characterization

3.1 Introduction

Characterization of the radar data quality is necessary to describe uncertainty in the data taking into account potential errors that can be quantified as well as the ones that can be estimated only qualitatively. Generally, values of many detailed "physical" quality descriptors are not readable for end users, so the following quality metrics are used as more suitable:

- total error level, i.e. measured value ± standard deviation expressed as measured physical quantity (radar reflectivity in dBZ, precipitation in mm h^{-1}, etc.),
- quality flag taking discrete value, in the simplest form 0 or 1 that means "bad" or "excellent" data,
- quality index as unitless quantity related to the data errors, which is expressed by numbers e.g. from 0 to 1.

Many national meteorological services provide quality information in form of flags to indicate where radar data is burdened with specific errors and if it is corrected by dedicated algorithms (Michelson et al., 2005; Norman et al., 2010). The flags are expressed as discrete numbers.

The quality index (QI) is a measure of data quality that gives a more detailed characteristic than a flag, providing quantitative assessment, for instance using numbers in a range from 0 (for bad data) to some value (e.g. 1, 100, or 255 for excellent data). The quality index concept is operationally applied to surface precipitation data in some national meteorological services (see review in Einfalt et al., 2010).

3.2 General description of *QI* scheme

An idea of quality index (*QI*) scheme is often employed to evaluate radar data quality. In this scheme the following quantities must be determined (Szturc et. al., 2011):

1. Quality factors, X_i (where $i = 1, ...n$) – quantities that have impact on weather radar-based data quality. Their set should include the most important factors that can be measured or assessed.
2. Quality functions, f_i – formulas for transformation of each individual quality factor X_i into relevant quality index QI_i. The formulas can be linear, sigmoidal, etc.
3. Quality indices, QI_i – quantities that express the quality of data in terms of a specific quality factors X_i:

$$QI_i = \begin{cases} 0 & \text{bad data} \\ 1 & \text{good data} \\ f_i(X_i) \in (0,1) & \text{other cases} \end{cases} \qquad (1)$$

4. Weights, W_i – weights of the QI_is. The optimal way of the weight determination seems to be an analysis of experimental relationships between proper quality factors X_i and radar data errors calculated from comparison with benchmark data (on historical data set).
5. Final quality index, QI – quantity that expresses quality of data in total, calculated using one of the formulae:

- minimum value:

$$QI = \min(QI_i), \qquad (2a)$$

- additive scheme (weighted average):

$$QI = \sum_{i=1}^{n}(QI_i \cdot W_i), \qquad (2b)$$

- multiplicative scheme (multiplication):

$$QI = \prod_{i=1}^{n}(QI_i \cdot W_i) \text{ or } QI = \prod_{i=1}^{n}QI_i. \qquad (2c)$$

The latter seems to be the most appropriate and its form is open (e.g. changes in set of quality indicators do not require the scheme parameterization).

4. Quality control algorithms for radar reflectivity volumes

Starting point in dealing with weather radar reflectivity data should be quality control of 3-D raw radar data. There are not many papers focused on quality characterization of such data. Fornasiero et al. (2005) presented a scheme employed in ARPA Bologna (Italy) for quality evaluation of radar data both raw and processed. The scheme developed in Institute of Meteorology and Water Management in Poland (IMGW) in the frame of BALTRAD project (Michelson et al., 2010) was described by Ośródka et al. (2010, 2012). Commonly employed groups of quality control algorithms are listed in Table 1.

Task	Correction algorithm	Quality factor	QC	QI
Evaluation of technical radar parameters	–	Set of technical radar parameters		x
Assessment of effects related to distance to radar site	–	Horizontal and vertical beam broadening		x
Ground clutter removal	Using Doppler filter or 3-D clutter map	Presence of ground clutter	x*	x
Removal of non-meteorological echoes	Analysis of 3-D reflectivity structure. Using dual-polarization parameters	Presence of the non-meteorological echoes	x	x
Beam blockage correction	Using topography map	Presence of beam blockage	x	x
Correction for attenuation in rain	Based on attenuation coefficient. Using dual-polarization parameters	Attenuation in rain along the beam path	x	x
Spatial variability evaluation	Analysis of 3-D reflectivity structure	Spatial variability of reflectivity field		x

* commonly the correction is made by built-in radar software.

Table 1. Groups of quality control algorithms (correction QC and characterization QI) for 3-D reflectivity (Z) data.

4.1 Technical radar parameters

This algorithm aims to deliver data quality metric only. A set of technical radar parameters that impact on data quality can be selected as quality factors. The parameters are for instance: operating frequency, beam width, pointing accuracy in elevation and azimuth, minimal detectable signal at 1000 m, antenna speed, date of last electronic calibration, etc. (Holleman et al., 2006). All the factors are static within the whole radar range and characterize quality of each particular radar so different radars can be compared in terms of their quality. The threshold values for which the quality index becomes lower than one should be set for all parameters according to the common standards.

4.2 Horizontal and vertical broadening of a radar beam

Radar measurements are performed along each beam at successive gates (measurement points in 3-D data space), which represent certain surrounding areas determined by the beam width and pulse length. Since the radar beam broadens with the distance to the radar site, the measurement comes from a larger volume and related errors increase as well. There is no possibility to correct this effect, however it can be quantitatively determined and taken into account in the total quality index.

The horizontal and vertical broadening of radar beam for each gate can be geometrically computed knowing its polar coordinates: elevation, azimuth, and radial distance to radar site, and two parameters of radar beam: beam width and radar pulse length. Related quality index may be determined from broadenings of the both beam cross section (Ośródka et al., 2012).

4.3 Ground clutter removal

The correction of radar data due to contamination by ground clutter is commonly made at a level of radar system software which uses statistical or Doppler filtering (e.g. Selex, 2010). In such situation the information about the correction is not available so generation of a ground clutter map for the lowest (and higher if necessary) scan elevation must be employed, e.g. using a digital terrain map (DTM). In order to determine areas contaminated by ground clutter a diagram of partial beam blockage values (*PBB*) is analysed. The *PBB* is defined as a ratio of blocked beam cross section area to the whole one.

Gates where ground clutter was detected should be characterized by lowered quality index. A simple formula for quality index QI_{GC} related to ground clutter presence can be written as:

$$QI_{GC} = \begin{cases} a & \text{ground clutter is detected} \\ 1 & \text{no clutter} \end{cases} \tag{3}$$

where a is the constant, e.g. between 0 and 1 in the case of $QI_i \in (0, 1)$. The quality index decreases in each gate with detected clutter even if it was removed.

4.4 Removal of non-meteorological echoes

Apart from ground clutter other phenomena like: specks, external interference signals (e.g. from sun and Wi-Fi emitters), biometeors (flock of birds, swarm of insects), anomalous propagation echoes (so called anaprop), sea clutter, clear-air echoes, chaff, etc., are considered as non-meteorological clutter. Since various types of non-precipitation echoes can be found in radar observations, in practice individual subalgorithms must be developed to address each of them. More effective removal of such echoes is possible using dual-polarization radars and relevant algorithms for echo classification.

Removal of external interference signals. Signals coming from external sources that interfere with radar signal have become source of non-meteorological echoes in radar data more and more often. Their effect is similar to a spike generated by sun, but they are observed in any azimuth at any time, mainly at lower elevations, and may reach very high reflectivity. The spurious spike-type echoes are characterized by their very specific spatial structure that clearly differs from precipitation field pattern (Peura, 2002; Ośródka et al., 2012): they are observed along the whole or large part of a single or a few neighbouring radar beams. Commonly reflectivity field structure is investigated to detect such echo on radar image (Zejdlik & Novak, 2010). Recognition of such echo is not very difficult task unless it interferes with a precipitation field: its variability is low along the beam and high across it. The algorithm removes it from the precipitation field and replaces by proper (e.g. interpolated) reflectivity values. In the algorithm of Ośródka et al. (2012) two stages of spike removal are introduced: for "wide" and "narrow" types of spikes.

Removal of "high" spurious echoes. "High" spurious echoes, not only spikes, are echoes detected at altitudes higher than 20 km where any meteorological echo is not possible to exist. All the "high" echoes are removed.

Removal of "low" spurious echoes. "Low" spurious echoes are all low-reflectivity echoes detected at low altitudes only. No meteorological echo can exist here. All the "low" echoes are removed. The algorithm can be treated as a simple method to deal with biometeor echoes (Peura, 2002).

Meteosat filtering. As a preliminary method for non-meteorological echo removal the filtering by Meteosat data on cloudiness can be used. A Cloud Type product, which is provided by EUMETSAT, distinguishes twenty classes of cloud type with the classes from 1 to 4 assigned to areas not covered by any cloud. All echoes within not clouded areas are treated as spurious ones and removed. Such simple technique can turn out to be quite efficient in the cases of anomalous propagation echoes (anaprop) over bigger areas without clouds (Michelson, 2006).

Speck removal. Generally, the specks are isolated radar gates with echo surrounded by non-precipitation gates. Number of echo gates in a grid around the given gate (e.g. of 3 x 3 gates) is calculated (Michelson et al., 2000). If a certain threshold is not achieved then the gate is classified as a speck, i.e. measurement noise, and the echo is removed. Algorithm of the reverse specks (i.e. isolated radar gates with no echo surrounded by precipitation gates) removal is analogous to the one used for specks.

Using artificial intelligence techniques. Artificial intelligence algorithms, such as neural network (NN), are based on analysis of reflectivity structure (Lakshmanan et al., 2007). The difference is that similarity of the given object pattern to non-meteorological one, on which the model was learned, is a criterion of spurious echo detection. For this reason NN-based algorithms are difficult to parameterize and control their running.

Using dual-polarization observations. The basis is the fact that different types of targets are characterized by different size, shape, fall mode and dielectric constant distribution. In general, different combinations of polarimetric parameters can be used to categorize the given echo into one of different types (classes). The fuzzy logic scheme is mostly employed for the combination. Such methods consider the overlap of the boundaries between meteorological and non-meteorological objects. For each polarimetric radar observable and for each class a membership function is identified basing on careful analysis of data. Finally, an object is assigned to the class with the highest value of membership function.

The most often horizontal reflectivity (Z_H), differential reflectivity (Z_{DR}), differential phase shift (Φ_{DP}), correlation coefficient (ρ_{HV}), and analyses of spatial pattern (by means of standard deviation) of the parameters are employed in fuzzy logic schemes. Radars operating in different frequencies (S-, C-, and X-band) may provide different values of polarimetric parameters as they are frequency-dependent. For that reason, different algorithms are developed for identification of non-meteorological echoes using different radar frequencies, see e.g. algorithms proposed by Schuur et al. (2003) for S-band radars and by Gourley et al. (2007b) for C-band. A significant disadvantage of such techniques is that they are parameterized on local data and conditions so they are not transportable to other locations.

Quality index. Quality index for the gates in which non-meteorological echoes are detected is decreased to a constant value using formula similar to Equation (3).

An example of algorithms running for spike- and speck-type echoes removal is depicted in Figure 2b (for Legionowo radar).

4.5 Beam blockage

Radar beam can be blocked by ground targets, i.e. places where the beam hits terrain. A geometrical approach is applied to calculate the degree of the beam blockage. This approach is based on calculation what part of radar beam cross section is blocked by any topographical object. For this purpose a degree of partial beam blocking (PBB) is computed from a digital terrain map (DTM). According to Bech et al. (2003, 2007), the PBB is calculated from the formula:

$$PBB = \frac{y\sqrt{a^2 - y^2} + a^2 \arcsin\frac{y}{a} + \frac{\pi a^2}{2}}{\pi a^2}$$ (4)

where a is the radius of radar beam cross section at the given distance from radar, y is the difference between the height of the terrain and the height of the radar beam centre. The partial blockage takes place when $-a < y < a$, and varies from 0 to 1 (see Figure 1).

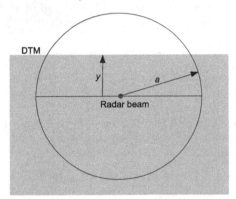

Fig. 1. Scheme of partial beam blockage PBB calculation using Bech et al. (2007) algorithm.

Quantity y in Equation 4 and Figure 1 is calculated as an altitude obtained from DTM for pixel located in radar beam centre taking into account altitude of radar antenna, the Earth curvature, and antenna elevation. Then the correction of partial beam blocking is made according to the formula (Bech et al., 2007):

$$Z_{cor} = Z + 10 \cdot \lg_{10}(1 - PBB)^{-1}$$ (5)

The correction is introduced if the PBB value is lower than 0.7. For higher PBB values "no data" (Bech et al., 2007) or reflectivity from neighbouring higher elevation (Ośródka et al., 2012) may be taken. A quality of blocked measurement dramatically decreases and can be expressed by:

$$QI_{PBB} = \begin{cases} 1 - PBB & PBB \leq a \\ 0 & PBB > a \end{cases}$$ (6)

where coefficient a can be set as 0.5 (Fornasiero et al., 2005) or 0.7 (Bech et al., 2007; Ośródka et al., 2012). If reflectivity in a specific gate has been replaced by reflectivity from higher elevation then QI_{PBB} is taken from the higher one multiplied by factor b set as e.g. 0.3 (Ośródka et al., 2012). An example of the algorithm running is presented in Fig. 2 for Pastewnik radar which is located near mountains.

4.6 Attenuation in rain

Attenuation is defined as decrease in radar signal power after passing a meteorological object, that results in underestimation of the measured rain:

$$A = 10 \cdot \log_{10} \frac{Z_{corr}}{Z} \tag{7}$$

where A is the specific attenuation (dB km^{-1}), Z_{corr} is the non-attenuated rain and Z is the measured one (mm^6 m^{-3}). Especially at C- and X-band wavelength the attenuation can considerably degrade radar measurements. The aim of the algorithm is to calculate the non-attenuated rain. Empirical formulae for determination of specific attenuation can be found in literature. Using 5.7-cm radar wavelength (C-band radar) for rain rate the two-way attenuation A in 18°C can be estimated from the formula (Battan, 1973):

$$A = 0.0044 \cdot R^{1.17} \tag{8}$$

Reflectivity-based correction made iteratively ("gate by gate") is a common technique of correction for attenuation in rain (Friedrich et al., 2006; Ośródka et al., 2012). For a given gate i the attenuation at distance between gate i-1 and gate i can be calculated taking into account underestimations calculated for all gates along the beam from the radar site up to the i-1 gate (based on Equation 8). Finally, corrected rain rate in the gate i is computed from the attenuation and underestimations in all previous gates.

In case of dual-polarization radars specific attenuation for horizontal polarization A_H and specific differential attenuation A_{DP} (in dB km^{-1}) can be calculated using different methods. For C-band radar typically specific differential phase K_{DP} is applied using a nearly linear relation between the attenuation and K_{DP}, e.g. (Paulitsch et al., 2009):

$$A_H = 0.073 K_{DP}^{0.99}, \; A_{DP} = 0.013 K_{DP}^{1.23} \tag{9}$$

or a linear one.

The iterative approach can lead to unstable results because it is very sensitive to small errors in both measurement and specific attenuation. Therefore, in order to avoid the instability in the algorithm, certain threshold values must be set to limit the corrections. For dual-polarization radar a ZPHI algorithm is recommended, in which specific attenuation is stabilized by differential phase shift Φ_{DP} (Testud et al., 2000; Gourley et al., 2007a).

Magnitude of the correction in precipitation rate can be considered as a measure of quality due to radar beam attenuation (Ośródka et al., 2012).

4.7 Spatial variability of reflectivity field

Small-scale variability of precipitation field is directly connected with uncertainty because heavy precipitation is more variable in space and time, as it can be especially observed in the case of small-scale convective phenomena. Moreover non-precipitation echoes, such as ground clutter, are often characterized by high variability that differs from that for stratiform precipitation echoes. Spatial variability can be quantified as 3-D reflectivity gradient (Friedrich & Hagen, 2004) or standard deviation in a certain spatial grid (Szturc et al., 2011) and should be taken into account in quality index determination.

4.8 Total quality index

Computation of the total quality index QI is the final step in estimation of radar volume data quality. If the individual quality indices QI_i characterizing data quality are quantitatively determined, then the total quality index QI is a result of all the individual values QI_i employing one of the formulas 2a – 2c.

Each elevation of raw reflectivity volume can be compared with final corrected field. A set of such data for the lowest elevation is presented in Fig. 2. In this Figure a strong impact of spike echoes is observed for Legionowo radar whereas ground clutter and related blockage on data from Pastewnik radar is evident. Both radars are included in Polish radar network POLRAD (Szturc & Dziewit, 2005).

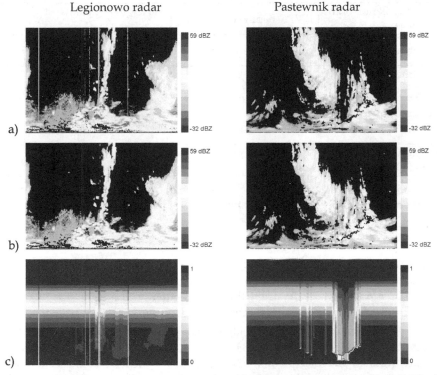

Fig. 2. Example of influence of all correction algorithms for the lowest elevation (0.5°): a) raw data Z (in dBZ); b) corrected data Z; c) total quality index QI (the left image for Legionowo radar, 10.05.2010, 15:30 UTC, the right for Pastewnik radar, 5.05.2010, 18:00 UTC; distance to radar up to 250 km). The panels represent range (y-axis) vs. azimuth (x-axis) displays.

5. Quality control algorithms for surface precipitation products

Corrections of 2-D radar data should constitute consecutive stages in radar data processing in order to get the best final radar products. These corrections include algorithms related to specific needs of the given product. The particular quality factors employed for calculating quality indices for 3-D data which also influence quality of 2-D data are not described here.

Many algorithms for surface precipitation field estimation from weather radar measurements applied in operational practice (e.g. Michelson et al., 2005) are described in this Section. For precipitation accumulation a different group of quality factors is applied. More common quality control algorithms employed in the practice are listed in Table 2.

Task	Correction algorithm	Quality factor	QC	QI
Z–R relationship estimation	Changeable Z–R relationship used	–	x	x
Bright band (melting layer) effect correction	VPR-based correction	Presence of melting layer	x	x
Data extrapolation onto the Earth surface	VPR-based correction	Height of the lowest radar beam	x	x
Orographic enhancement	Physical model	Magnitude of the enhancement	x	x
Adjustment with rain gauge data	Correction using rain gauge data	Radar precipitation – rain gauge differences	x	x
For accumulation: number of rate data	–	Temporal continuity of data (number of the products)		x
For accumulation: averaged QI for rate data	–	Quality of included data (averaged QI)		x

Table 2. Quality control algorithms (correction and characterization) for 2-D surface precipitation data (in order of implementation into the chain).

5.1 Estimation of Z–R relationship

The Z–R relationship ($Z = aR^b$) variability is one of the most significant error sources in precipitation estimation. Each hydrometeor contributes to the precipitation intensity roughly to 3.7th power of its diameter, thus assumption on the drop size distribution is needed as the integral intensity is measured. Nowadays, for a single polarization radar it is a common practice to apply a single (usually Marshall and Palmer formula $Z = 200 \cdot R^{1.6}$) or seasonally-dependent Z–R relationship. However, use of a fixed Z–R relation can lead to significant errors in the precipitation estimation, as it depends on precipitation type (stratiform or convective), its kind (rain, snow, hail), etc. There are approaches that use tuned Z–R relationships for different meteorological situations. It requires the different types of precipitation to be identified on the basis of dedicated algorithms, which is easier if disdrometer measurements are available (Tenório et al., 2010).

Improvement in precipitation rate R estimation is noticeable using dual-polarization parameters, especially for heavy rainfall. In addition to the horizontal reflectivity Z_H available for single polarization radar, the specific differential phase K_{DP} and the differential reflectivity Z_{DR} can be applied (Bringi & Chandrasekhar, 2001). Typical forms of relationships for precipitation estimation are as follows: $R = f(K_{DP})$, $R = f(Z_H, Z_{DR})$, $R = f(K_{DP}, Z_{DR})$, and $R = f(Z_H, K_{DP}, Z_{DR})$. These approaches for precipitation rate estimation are potentially unaffected by radar calibration errors and attenuation, unbiased by presence of hail, etc.

5.2 Bright band phenomenon

Vertical profile of reflectivity (VPR) provides very useful information for radar data quality control. An averaged VPR is suggested to be taken from radar pixels lying at distance between about from 30 to 80 km from radar site to obtain the profile valid for the whole range of heights (Franco et al., 2002; Germann & Joss, 2004; Einfalt & Michealides, 2008). The bright band is a phenomenon connected with the presence of the melting layer. It is assumed that the melting layer is placed in range from the 0°C isotherm down to 400 m below (Friedrich et al., 2006). The melting of ice precipitation into water drops and related overestimation of precipitation rate results in errors of ground precipitation estimation. The phenomenon is clearly visible in vertically pointing radar observations. For dual-polarization radar a vertical profile of correlation coefficient (ρ_{HV}) is investigated instead of reflectivity profile analysis (Tabary et al., 2006).

It is proposed that the relevant quality index equals 0 inside the melting layer due to bright band, and equals 0.5 for measurement gates above the layer (Friedrich et al., 2006). In the case when the melting layer does not exist (in winter season or within convective phenomena) the quality index equals 1.

5.3 Data extrapolation onto the Earth surface

Information available from VPR can be used for another quality correction algorithm, which is extrapolation of precipitation data from the lowest beam to the Earth surface, especially at longer distances over 80 km. The averaged VPR is estimated for distance to radar site in range from 30 to 80 km and then employed to extrapolate radar data from the lowest beam to the Earth surface (Šálek et al., 2004). A quality factor which describes the relevant quality index is the height of the lowest radar calculated from radar scan strategy, digital terrain map (DTM), and the radar coordinates. It strongly depends on terrain complexity and related radar beam blocking and is defined as a minimum height for which radar measurement over a given pixel is feasible.

5.4 Orographic enhancement (seeder-feeder effect)

Orographic enhancement is a result of so called seeder–feeder mechanism which is observed when ascent of air is forced by hills or mountains. The low-level clouds formed in this way (feeder clouds) provide a moisture source that is collected by drops falling from higher clouds (seeder clouds). Radar is not able to capture the enhancement, which occurs close to the ground, as the measurement is performed at certain height over the hill. This effect can be estimated by 3-D physical model taking account of information from numerical weather prediction model: wind speed, wind direction, relative humidity, temperature, as well as the topography of the region (Alpert & Shafir, 1989). Magnitude of such correction can be taken to determine related quality index.

5.5 Adjustment with rain gauge data

Weather radar-based precipitation may differ from "ground truth", which can be locally estimated from rain gauge measurements, especially in close vicinity of the gauge. It is assumed that rain gauge measures precipitation exactly as its correction can be calculated (Førland et al., 1996), whereas radar provides information about space distribution. The idea is to use rain gauge information to improve radar data, as so called adjustment. The following solutions are proposed (Gjertsen et al., 2004):

- Mean field correction is a simple method to make the radar measurements unbiased. The correction factor is calculated from comparison of the averaged radar observations over the whole considered area, and the analogical averaged rain gauge measurements. The mean field bias can be calculated from historical data set or dynamic time-window. The last method allows to take into consideration variability in precipitation characteristics with time, but the time-period of the dynamic window cannot be too short due to requirement of data representativeness.
- Other methods of radar precipitation correction employ the distance to radar site L as the predictor apart from rain gauge information. Correction factor C can be expressed as e.g. polynomial relationship in form proposed by Michelson et al. (2000):

$$C = aL^2 + bL + c \tag{10}$$

where a, b, and c are the empirically estimated parameters of the equation.
- More advanced methods based on multiple regression involve more predictors which play significant role in precipitation estimation. Especially in mountainous terrain the distance to radar site turned out not sufficient because of strong influence of beam blockage and shielding. Additional predictors can be height of the lowest radar beam, height above sea level, etc.

Quality index related to the adjustment with rain gauge data can be determined from magnitude of the correction.

5.6 Quality factors for precipitation accumulation

The following quality factors for precipitation accumulation can be considered:

- *Number of precipitation rate products.* Accumulated precipitation field is composed from a certain number of discrete radar measurements. The number of precipitation rate products included into the given precipitation accumulation can be used to calculate a related quality index. Lack of one or more products during the accumulation period results in a significant decrease of quality. Moreover lack of the products one after the other results in much lower quality.
- *Averaged quality index from precipitation rate products* is computed as a mean from all values of quality indices for precipitation rates (e.g. maximally seven for 10-minute time resolution and 1-hour period of accumulation) that are aggregated into the accumulation.

5.7 Combination: weather radar precipitation – rain gauges

The combination of radar precipitation and rain gauge data is treated as the next stage in precipitation field estimation: the adjustment is considered as the radar data correction, whereas the result of combination is not corrected radar data, but precipitation estimated from larger number of data sources.

The measurement techniques such as rain gauges, weather radar, and satellite are considered as independent ones, which provide rainfall information with different error characteristics. Rain gauges are assumed to measure precipitation directly with good point accuracy. However in the case of rather sparse network density, the number of rain gauges might not be sufficient to successfully reproduce spatial variability of precipitation. On the other hand weather radar is capable of reflecting the spatial pattern of rainfall with high resolution in time and space over a large area almost in the real-time. Nevertheless radar data are burdened with non-negligible errors: both non-meteorological and meteorological. Therefore, merging these two sources of information could lead to improvement in precipitation estimation. As a consequence, several methods have been developed to estimate rainfall field from radar- and raingauge-driven data.

One of them is a geostatistical approach, where spatially interpolated rain gauge data and radar field are combined employing the Cokriging technique (Krajewski, 1987). However the need for estimation of required empirical parameters might be crucial and may lead to significant errors. Velasco-Forero et al. (2004) tested different Kriging estimators (ordinary Kriging, Kriging with External Drift, Cokriging and Collocated Cokriging) to produce merged field from raingauge observations and radar data. Kriging with External Drift technique turned out to give the best final field.

In another approach (Todini, 2001) Kalman filtering is applied to optimally combine data from the two sensors (rain gauge network and weather radar) in a Bayesian sense. Radar field taken as the *a priori* estimate and the block Kriging of the raingauge observations treated as the measurement vector enable to find the *a posteriori* estimate of precipitation.

As it was pointed out, radar data is considered to be better than rain gauge network in reproduction of spatial distribution, whereas rain gauges measure precipitation accurately in their locations. This observation is a starting point in a technique proposed by Sinclair & Pegram (2005) in which the radar information is used to obtain the correct spatial structure of the precipitation field, while the field values are fitted to the raingauge observations.

5.8 Example of *QI* data

An example of the *QI* scheme application implemented in Institute of Meteorology and Water Management (IMGW) is presented below. Polish weather radar network POLRAD consists of eight C-Band Doppler radars of Gematronik with Rainbow software for basic processing of data. In Figure 3 an example of precipitation composite for selected event is presented together with quality index *QI* obtained from the aforementioned quality factors (Table 2) using additive scheme (Equation 2b).

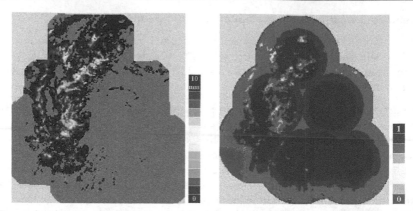

Fig. 3. Example of corrected field of precipitation rate (on the left) (composite from 5 August 2006, 03 UTC, when 7 from 8 weather radars were running) and resulting averaged quality index QI (on the right) (Szturc et. al., 2011).

The final quality index QI field depends on all quality factors included in the scheme. The most significant ones are height of the lowest radar beam, especially for places at longer distances to the nearest radar site and in mountainous areas (the zero-quality area south-west of the right map), and precipitation field variability (calculated analogically to the related 3-D algorithm) that follows the pattern of the precipitation field to some degree. It is noticeable that some quality factors are related to the precipitation field, whereas other fields are static if the set of running radars is constant, as they depend on radar locations only.

6. Conclusions

Weather radar data before being applied by the end-users must be quality controlled at all data processing stages. The main stages are generation of 3-D data (volumes) and then specialized 2-D data (products) dedicated to certain groups of the end-users. At first, the 3-D data should be corrected as they constitute the information source for generation of radar products. The corrections that are related to specific products should be made at the next stage – 2-D data processing. Due to numerous radar errors various correction techniques must be employed, moreover radar hardware limitations determine application of particular corrections. First of all dual-polarization radars, which will be a standard in the near future, open up new possibilities.

In quality control of radar data apart from the data correction, information about the data uncertainty plays also a key role. The high importance of radar data quality characterization is appreciated not only by radar people (meteorologists, hydrologists, etc.) but by end-user communities as well. Dealing with such quality information is a difficult task, however it is crucial for risk management and decision-making support.

For these reasons the quality control of radar data is becoming an essential task in weather radar data generation and processing. It has been a main subject of many international programmes, especially: the COST Action 731 ("Propagation of uncertainty in advanced meteo-hydrological forecast systems", 2005-2010), the EUMETNET OPERA ("Operational Programme for the Exchange of Weather Radar Information", from 1999), the BALTRAD

("An advanced weather radar network for the Baltic Sea Region: BALTRAD", Baltic Sea Region Programme, 2009-2014), the WMO programme RQQI ("Radar Quality Control and Quantitative Precipitation Intercomparisons", from 2011), etc. In the frame of the projects some recommendations are being developed, that will ensure harmonisation of practices in particular national meteorological services.

7. Acknowledgement

This paper contains outcomes from the BALTRAD and BALTRAD+ research projects ("An advanced weather radar network for the Baltic Sea Region: BALTRAD", Baltic Sea Region Programme) and the COST Action 731 "Propagation of uncertainty in advanced meteo-hydrological forecast systems".

8. References

Alpert, P. & Shafir, H. (1989). A physical model to complement rainfall normals over complex terrain. *Journal of Hydrology*, Vol. 110, pp. 51-62, ISSN 0022-1694

Battan, L. J. (1973). *Radar observation of the atmosphere*. University of Chicago Press, ISBN 9780226039190, Chicago – London, UK

Bebbington, D.; Rae, S.; Bech, J.; Codina, B. & Picanyol, M. (2007). Modelling of weather radar echoes from anomalous propagation using a hybrid parabolic equation method and NWP model data. *Natural Hazards and Earth System Sciences*, Vol. 7, pp. 391–398, ISSN 1561-8633

Bech, J.; Codina, B.; Lorente, J. & Bebbington, J. (2003). The sensitivity of single polarization weather radar beam blockage correction to variability in the vertical refractivity gradient. *Journal of Atmospheric and Oceanic Technology*, Vol. 20, pp. 845–855, ISSN 1520-0426

Bech, J.; Gjertsen, U. & Haase, G. (2007). Modelling weather radar beam propagation and topographical blockage at northern high latitudes. *Quarterly Journal of the Royal Meteorological Society*, Vol. 133, pp. 1191–1204, ISSN 0035-9009

Bringi, V. N. & Chandrasekar, V. (2001). *Polarimetric Doppler weather radar. Principles and applications*. Cambridge University Press, ISBN 9780521623841, Cambridge, UK

Collier, C. G. (1996). *Applications of weather radar systems. A guide to uses of radar data in meteorology and hydrology*. Wiley-Praxis, ISBN 0-7458-0510-8, Chichester, UK

Collier, C. G. (2009). On the propagation of uncertainty in weather radar estimates of rainfall through hydrological models. *Meteorological Applications*, Vol. 16, pp. 35–40, ISSN 1469-8080

Einfalt, T. & Michaelides, S. (2008). Quality control of precipitation data. In: *Precipitation: Advances in measurement, estimation and prediction*, S. Michaelides, (Ed.), 101–126, Springer-Verlag, Berlin – Heidelberg, Germany

Einfalt, T.; Szturc, J. & Ośródka, K. (2010). The quality index for radar precipitation data: a tower of Babel? *Atmospheric Science Letters*, Vol. 11, pp. 139–144, ISSN 1530-261X

Førland, E. J.; Allerup, P.; Dahlström, B.; Elomaa, E.; Jónsson, T.; Madsen, H.; Perälä, J.; Rissanen, P.; Vedin, H. & Vejen, F. (1996). *Manual for operational correction of Nordic precipitation data*. DNMI Report Nr. 24/96, Oslo, Norway, ISNN 0805-9918

Fornasiero, A.; Alberoni, P. P.; Amorati, R.; Ferraris, L. & Taramasso, A. C. (2005). Effects of propagation conditions on radar beam-ground interaction: impact on data quality. *Advances in Geosciences*, Vol. 2, pp. 201–208, ISSN 1680-7340

Franco, M.; Sempere-Torres, D.; Sánchez-Diezma, R. & Andrieu, H. (2002). A methodology to identify the vertical profile of reflectivity from radar scans and to estimate the rainrate at ground at different distances. *Proc. ERAD 2002*, 299–304.

Friedrich, K. & Hagen, M. (2004). Wind synthesis and quality control of multiple-Doppler-derived horizontal wind fields. *Journal of Applied Meteorology*, Vol. 43, pp. 38-57, ISSN 1520-0450

Friedrich, K.; Hagen, M. & Einfalt, T. (2006). A quality control concept for radar reflectivity, polarimetric parameters, and Doppler velocity. *Journal of Atmospheric and Oceanic Technology*, Vol. 23, pp. 865–887, ISSN 1520-0426

Gekat, F.; Meischner, P.; Friedrich, K.; Hagen, M.; Koistinen, J.; Michelson, D. B. & Huuskonen, A. (2004). The state of weather radar operations, networks and products. In: *Weather radar. Principles and advanced applications*, P. Meischner, (Ed.), 1–51, Springer-Verlag, ISBN 3-540-000328-2, Berlin – Heidelberg, Germany

Germann, U. & Joss, J. (2004). Operational measurement of precipitation in mountainous terrain. In: *Weather radar. Principles and advanced applications*, P. Meischner, (Ed.), 52–77, Springer-Verlag, ISBN 3-540-000328-2, Berlin – Heidelberg, Germany

Gjertsen, U.; Sálek, M. & Michelson, D.B. (2004). *Gauge-adjustment of radar-based precipitation estimates*. COST Action 717, ISBN 92-898-0000-3, Luxembourg

Goltz, C.; Einfalt, T. & Galli, G. (2006). Radar data quality control methods in VOLTAIRE. *Meteorologische Zeitschrift*, Vol. 15, pp. 497–504, ISSN 1610-1227

Gourley, J. J.; Tabary, P. & Parent-du-Châtelet, J. (2007a). Empirical estimation of attenuation from differential propagation phase measurements at C-band. *Journal of Applied Meteorology*, Vol. 46, pp. 306–317, ISSN 1520-0450

Gourley, J. J.; Tabary, P. & Parent-du-Châtelet, J. (2007b). A fuzzy logic algorithm for the separation of precipitating from nonprecipitating echoes using polarimetric radar observations. *Journal of Atmospheric and Oceanic Technology*, Vol. 24, pp. 1439–1451, ISSN 1520-0426

Holleman, I.; Michelson, D.; Galli, G.; Germann, U. & Peura, M. (2006). Quality information for radars and radar data. OPERA workpackage 1.2 (OPERA_2005_19 document).

Illingworth, A. (2004). Improved precipitation rates and data quality by using polarimetric measurements. In: *Weather radar. Principles and advanced application*, P. Meischner, Ed.), 130–166, Springer-Verlag, ISBN 3-540-000328-2, Berlin – Heidelberg, Germany

Krajewski, W. F. (1987). Cokriging radar-rainfall and rain gage data. *Journal of Geophysical Research*, Vol. 92, pp. 9571–9580, ISSN 0148–0227

Lakshmanan, V.; Fritz, A.; Smith, T.; Hondl, K. & Stumpf, G. J. (2007). An automated technique to quality control radar reflectivity data. *Journal of Applied Meteorology and Climatology*, 46, 288–305, ISSN 1558-8424

Meischner, P. (ed.) (2004). *Weather radar. Principles and advanced applications*. Springer-Verlag, ISBN 3-540-000328-2, Berlin – Heidelberg, Germany

Michelson, D. B.; Andersson, T.; Koistinen, J.; Collier, C. G.; Riedl, J.; Szturc, J.; Gjersten, U.; Nielsen, A. & Overgaard, S. (2000). *BALTEX Radar Data Centre products and their methodologies*, SMHI Reports Meteorology and Climatology, No. 90, Norrköping.

Michelson, D.; Einfalt, T.; Holleman, I.; Gjertsen, U.; Friedrich, K.; Haase, G.; Lindskog, M. & Jurczyk, A. (2005). *Weather radar data quality in Europe – quality control and characterization. Review.* COST Action 717, ISBN 92-898-0018-6, Luxembourg

Michelson, D. (2006). The Swedish weather radar production chain. *Proceedings of ERAD 2006: 4th European Conference on Radar in Meteorology and Hydrology*, pp. 382–385, ISBN 978-84-8181-227-5, Barcelona, Spain, September 18-22, 2006

Michelson, D.; Szturc, J.; Gill, R. S. & Peura, M. (2010). Community-based weather radar networking with BALTRAD. *Proceedings of ERAD 2010: 6th European Conference on Radar in Meteorology and Hydrology*, pp. 337-342, ISBN 978-973-0-09057-4, Sibiu, Romania, September 6-10, 2010

Norman, K.; Gaussiat, N.; Harrison, D.; Scovell, R. & Boscaci, M. (2010). A quality index scheme to support the exchange of volume radar reflectivity in Europe. *Proceedings of ERAD 2010: 6th European Conference on Radar in Meteorology and Hydrology*, ISBN 978-973-0-09057-4, Sibiu, Romania, September 6-10, 2010 (http://www.erad2010. org/pdf/oral/wednesday/dataex/07_ERAD2010_0259.pdf)

Ośródka, K.; Szturc, J.; Jurczyk, A.; Michelson, D.B.; Haase, G. & Peura, M. (2010). Data quality in the BALTRAD processing chain. *Proceedings of ERAD 2010: 6th European Conference on Radar in Meteorology and Hydrology, Advances in radar technology*, pp. 355–361, ISBN 978-973-0-09057-4, Sibiu, Romania, September 6-10, 2010

Ośródka, K.; Szturc, J. & Jurczyk, A. (2012). Chain of data quality algorithms for 3-D single-polarization radar reflectivity (RADVOL-QC system). *Meteorological Applications* (submitted).

Paulitsch, H.; Teschl, F. & Randeu W. L. (2009). Dual-polarization C-band weather radar algorithms for rain rate estimation and hydrometeor classification in an alpine region. *Advances in Geosciences*, Vol. 20, pp. 3-8, ISSN 1680-7340

Peura, M. (2002). Computer vision methods for anomaly removal. *Proc. ERAD 2002*, 312–317.

Peura, M.; Koistinen, J. & Hohti, H. (2006). Quality information in processing weather radar data for varying user needs. *Proceedings of ERAD 2006: 4th European Conference on Radar in Meteorology and Hydrology*, pp. 563–566, ISBN 978-84-8181-227-5, Barcelona, Spain, September 18-22, 2006

Šálek, M.; Cheze, J.-L.; Handwerker, J.; Delobbe, L. & Uijlenhoet, R. (2004). *Radar techniques for identifying precipitation type and estimating quantity of precipitation.* COST Action 717. Luxembourg.

Schuur, T.; Ryzhkov, A. & Heinselman, P. (2003). *Observations and classification of echoes with the polarimetric WSR-88D radar.* NSSL Tech. Report, Norman, OK, USA.

Sharif, H. O.; Ogden, F. L.; Krajewski, W. F. & Xue, M. (2004). Statistical analysis of radar rainfall error propagation. *Journal of Hydrometeorology*, Vol. 5, pp. 199–212, ISSN 1525-7541

Selex (2010). *Rainbow 5. Products and algorithms.* Selex SI GmbH, Neuss, Germany

Sinclair, S. & Pegram, G. (2005). Combining radar and rain gauge rainfall estimates using conditional merging. *Atmospheric Science Letters*, Vol. 6, pp. 19–22, ISSN 1530-261X

Szturc, J. & Dziewit Z. (2005). Status and perspectives on using radar data: Poland. In: *Use of radar observations in hydrological and NWP models.* COST Action 717, Final report, pp. 218–221, ISBN 92-898-0017-8, Luxembourg

Szturc, J.; Ośródka, K. & Jurczyk, A. (2006). Scheme of quality index for radar-derived estimated and nowcasted precipitation. *Proceedings of ERAD 2006: 4th European*

Conference on Radar in Meteorology and Hydrology, pp. 583–586, ISBN 978-84-8181-227-5, Barcelona, Spain, September 18-22, 2006

Szturc, J.; Ośródka, K. & Jurczyk, A. (2008a). Parameterization of QI scheme for radar-based precipitation data. *Proceedings of ERAD 2008: 5th European Conference on Radar in Meteorology and Hydrology*, Helsinki, Finland, June 30 – July 4, 2008 (CD).

Szturc, J.; Ośródka, K.; Jurczyk, A. & Jelonek, L. (2008b). Concept of dealing with uncertainty in radar-based data for hydrological purpose. *Natural Hazards and Earth System Sciences*, Vol. 8, pp. 267–279, ISSN 1561-8633

Szturc, J.; Ośródka, K. & Jurczyk, A. (2011). Quality index scheme for quantitative uncertainty characterisation of radar-based precipitation. *Meteorological Applications*, Vol. 18, pp. 407-420, ISSN 1469-8080

Tabary, P.; Le Henaff, A.; Vulpiani, G.; Parent-du-Châtelet, J.; Gourley, J.J. (2006), Melting layer characterization and identification with a C-band dual-polarization radar: a long-term analysis. *Proceedings of ERAD 2006: 4th European Conference on Radar in Meteorology and Hydrology*, pp. 17–20, ISBN 978-84-8181-227-5, Barcelona, Spain, September 18-22, 2006

Tenório, R. S.; Moraes, M. C. S. & Kwon, B. H. (2010). Raindrop distribution in the eastern coast of northeastern Brazil using disdrometer data. *Revista Brasileira de Meteorologia*, Vol. 25, pp. 415-426, ISSN 1982-4351

Testud, J.; Le Bouar, E.; Obligis, E. & Ali-Mehenni, M. (2000). The rain profiling algorithm applied to the polarimetric weather radar, *Journal of Atmospheric and Oceanic Technology*, Vol. 17, pp. 332–356, ISSN 1520-0426

Todini, E. (2001). A Bayesian technique for conditioning radar precipitation estimates to rain-gauge measurements. *Hydrology and Earth System Sciences*, Vol. 5, pp. 187-199, ISSN 1027-5606

Velasco-Forero, C. A.; Cassiraga, E. F.; Sempere-Torres, D.; Sanchez-Diezma, R. & Gomez-Hernandez, J. J. (2004). Merging radar and raingauge data to estimate rainfall fields: An improved geostatistical approach using non-parametric spatial models. *Proceedings of 6th International Symposium on Hydrological Applications of Weather Radar "Success stories in radar hydrology"*, ISBN 3-936586-29-2, Visby, Sweden, September 6-10, 2004 (CD).

Villarini, G. & Krajewski, W. F. (2010). Review of the different sources of uncertainty in single polarization radar-based estimates of rainfall. *Surveys in Geophysics*, Vol. 31, pp. 107–129, ISSN 1573-0956

Vivoni, E. R.; Entekhabi, D. & Hoffman R. N. (2007). Error propagation of radar rainfall nowcasting fields through a fully distributed flood forecasting model. *Journal of Applied Meteorology and Climatology*, Vol. 46, pp. 932–940, ISSN 1558-8424

Zawadzki, I. (2006). Sense and nonsense in radar QPE. *Proceedings of ERAD 2006: 4th European Conference on Radar in Meteorology and Hydrology*, pp. 121–124, ISBN 978-84-8181-227-5, Barcelona, Spain, September 18-22, 2006

Zejdlik, T. & Novak, P. (2010). Frequency protection of the Czech weather radar Network. *Proceedings of ERAD 2010: 6th European Conference on Radar in Meteorology and Hydrology, Advances in radar technology*, pp. 319–321, ISBN 978-973-0-09057-4, Sibiu, Romania, September 6-10, 2010

6

Doppler Weather Radars and Wind Turbines

Lars Norin and Günther Haase
Swedish Meteorological and Hydrological Institute
Sweden

1. Introduction

In many countries the number of wind turbines is growing rapidly as a response to the increasing demand for renewable energy. The cumulative capacity of wind turbines worldwide has shown a near 10-fold increase in the last decade (Global Wind Energy Council, 2011) and in the coming years many more wind turbines are expected to be built. Existing, older wind turbines are likely to be replaced by larger, next generation, turbines.

Modern wind turbines are large structures, many reach more than 150 m above the ground. Clusters of densely spaced wind turbines, so called wind farms, are being built both on- and offshore.

The continued deployment of wind turbines and wind farms is, however, not unproblematic. Radar systems, for example, are easily disturbed by wind turbines. Interference caused by wind turbines is more severe for many radar systems than interference caused by, for example, masts or towers. This is due to the rotating blades of the wind turbines. Many Doppler radars use a filter that removes echoes originating from objects with no or little radial velocity. However, these filters do not work for moving objects such as the rotating blades of wind turbines. Wind turbines located in line of sight of Doppler radars can cause clutter, blockage, and erroneous velocity measurements, affecting the performance of both military- and civilian radar systems.

Even though both radars and wind turbines have been in use for many decades it is only in the last few years that the interference problem has received substantial attention. The reason for this is simple; in recent years wind turbines have increased in number and size and at the same time radar systems have become increasingly sensitive.

In this chapter we present a brief review of some of the work made to investigate the impact of wind turbines on Doppler weather radars. Starting with a historical overview we outline the evolution of wind turbines and early studies about their impact on Doppler radars in general and Doppler weather radars in particular. Three major interference types for Doppler weather radars are identified: clutter, blockage, and erroneous velocity measurements. Observations, models, and mitigation concepts for all three interference types are discussed.

In particular, we present results from a study on average wind turbine clutter, based on long time series of data. We show that modelling wind turbine clutter using the radar cross section of a wind turbine can lead to erroneous results. We further argue that blockage due to wind turbines is difficult to analyse using operational reflectivity data. An alternative way of studying blockage is discussed and results are presented. A simple blockage model is described and its results are shown to agree with observations. Finally, examples of erroneous wind measurements are shown and mitigation measures are discussed.

2. Background

2.1 Wind turbine development

Wind power technology dates back many centuries. In the 1st century A.D. Hero of Alexandria described a simple wind wheel that could power an altar organ (Woodcroft, 1851). It is, however, not clear whether this invention was ever constructed or put to use. The first documented description of windmills that were used to perform irrigation and grinding grain comes from the region of Sistan, Persia, in the 9th century (Shepard, 1990). By the 12th century windmills were in use in Europe and in the following centuries they became increasingly important for grinding grain and pumping water. It was only after the industrial revolution their importance receded (Manwell et al., 2009).

Near the end of the 19th century the first wind turbines, used for the production of electricity, were developed. James Blyth built a 10 m high, cloth-sailed wind turbine in Scotland in 1887 (Price, 2005) and Charles Brush constructed a 25 m high wind turbine in Cleveland, Ohio, in 1887–1888 (Anon., 1890). A few years later, in the 1890s, Poul la Cour constructed over 100 wind turbines to generate electricity in Denmark (Manwell et al., 2009). In the 1970s the rising oil prices generated a renewed interest in wind power which led to serial production of wind turbines.

The increasing demand for renewable energy sources in the 21st century led to a further upswing for wind power. The pursuit of ever more powerful wind turbines lead to an increase in rotor blade diameter. A large wind turbine in the early 1980s could have a rotor diameter of 15 m and produce 55 kW whereas a large wind turbine in 2011 could have a rotor diameter larger than 150 m and produce 7 MW. Figure 1 shows the rotor diameter and the corresponding power produced by large wind turbines introduced on the market during the period 1981–2011.

Since wind turbines not only have become increasingly powerful but also grown more numerous during the last decades the global cumulative installed capacity has increased exponentially (see Fig. 2). With the exception of year 2010 the global annual installed capacity has increased monotonically since at least 1996 (cf. Fig. 2).

2.2 Wind turbine impact on Doppler radars

Wind turbines in the path of electromagnetic transmissions may cause interference by scattering parts of the transmitted signal but also by modulating the transmission's frequency. Initial studies on wind turbine interference focused on television and radio transmissions and showed that wind turbines could indeed cause interference to the reception of such signals (see, e.g., Sengupta (1984); Sengupta & Senior (1979); Senior et al. (1977); Wright & Eng (1992)).

By the end of the 20th century and beginning of the 21st century a large number of wind turbines had been installed and several investigations were conducted to analyse the impact of wind turbines on military surveillance radars and civilian air traffic control radars (see, e.g.,

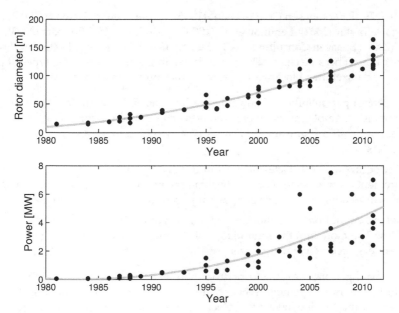

Fig. 1. Rotor diameter (top) and power (bottom) for a selection of large wind turbines introduced on the market during the period 1981–2011. Gray lines show superimposed trends. Data from The Wind Power (www.thewindpower.net) and wind turbine manufacturers' homepages.

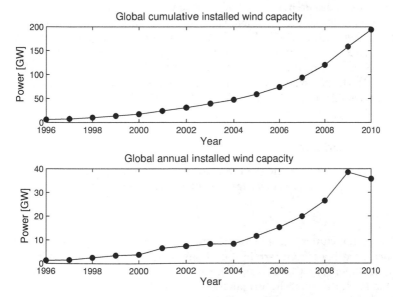

Fig. 2. Global cumulative installed wind capacity (top) and global annual installed wind capacity (bottom). Data from Global Wind Energy Council (2011).

Borely (2010); Butler & Johnson (2003); Davies (1995); Department of Defense (2006); Frye et al. (2009); Jago & Taylor (2002); Lemmon et al. (2008); Ousbäck (1999); Poupart (2003); RABC & CanWEA (2007); Sparven Consulting (2001); Summers (2001); Webster (2005a;b)). From these and other studies it became clear that wind turbines in general cause three types of problems for Doppler radars: clutter, blockage, and erroneous Doppler measurements.

- Clutter consist per definition of unwanted radar echoes. For a military surveillance radar clutter can for example consist of precipitation echoes whereas for a weather radar echoes from, e.g., aircraft are unwanted. Echoes from wind turbines are considered clutter by most radars.

- Blockage occurs when obstacles such as buildings or terrain obscure the radar line of sight. Measurements behind such obstacles become incomplete or non-existing. Wind turbines located near a radar may block a substantial part of the radar's measurement region.

- Doppler radars not only measure the echo strength of their targets but also their radial velocities. The motion of the rotor blades of a wind turbine is detected by the radar and may be interpreted as a moving target.

Clutter originating from the ground or from stationary buildings is often filtered out by built-in clutter filters that remove echoes with zero or low radial velocities. Echoes from the tower of a wind turbine have zero velocity and can therefore easily be removed but the turbine's rotating blades can have very large and variable velocities, escaping the clutter filter. Echoes from rotating wind turbine blades may therefore, for example, be mistaken for aircraft by an air traffic control radar.

Weather radar problems related to wind turbines were recognised early by Hafner et al. (2004) and Agence National des Fréquences (2005). These works have since been followed by many studies (e.g. Brenner et al. (2008); Donaldson et al. (2008); Haase et al. (2010); Hutchinson & Miles (2008); Toth et al. (2011); Tristant (2006a;b); Vogt et al. (2011; 2007a; 2009)). The increased awareness of the problems wind turbines may cause weather radars led both the World Meteorological Organization (WMO) and the Network of European Meteorological Services (EUMETNET) to issue general guidelines for the deployment of wind turbines, based on the distance from the radar (OPERA, 2010; WMO, 2010). These guidelines are summarised in Table 1 and Table 2.

3. Wind turbine interference

In the sections below we examine the three identified problems wind turbines can cause to Doppler weather radars: clutter, blockage, and erroneous wind measurements.

3.1 Clutter

For a weather radar, clutter refers to all non-meteorological radar echoes. Typical examples of clutter include echoes from terrain, buildings, and clear-air targets (e.g. insects, birds, atmospheric turbulence). Clutter can further be divided into two categories: static and dynamic. Static clutter typically originates from terrain and buildings whereas dynamic clutter is caused by moving targets such as clear-air returns. Static clutter has zero or near-zero radial velocity and can be removed by a built-in clutter filter whereas dynamic

Range	Potential impact	Guideline
0–5 km	The wind turbine may completely or partially block the radar and can result in significant loss of data that can not be recovered.	Definite Impact Zone: Wind turbines should not be installed in this zone.
5–20 km	Multiple reflection and multi-path scattering can create false echoes and multiple elevations. Doppler velocity measurements may be compromised by rotating blades.	Moderate Impact Zone: Terrain effects will be a factor. Analysis and consultation is recommended. Re-orientation or re-siting of individual turbines may reduce or mitigate the impact.
20–45 km	Generally visible on the lowest elevation scan; ground-like echoes will be observed in reflectivity; Doppler velocities may be compromised by rotating blades.	Low Impact Zone: Notification is recommended.
> 45 km	Generally not observed in the data but can be visible due to propagation conditions.	Intermittent Impact Zone: Notification is recommended.

Table 1. WMO guidance statement on weather radar/wind turbine siting. (From WMO (2010))

Range	Radar	Statement
0–5 km	C-band	No wind turbine should be deployed within this range
5–20 km	C-band	Wind farm projects should be submitted for an impact study
0–10 km	S-band	No wind turbine should be deployed within this range
10–30 km	S-band	Wind farm projects should be submitted for an impact study

Table 2. Statement of the OPERA group on the cohabitation between weather radars and wind turbines. (From OPERA (2010))

clutter originates from targets having radial velocities larger than the clutter filter limits. Dynamic clutter can therefore not be suppressed by conventional clutter filters.

Operating wind turbines generate both static and dynamic clutter. Since the static clutter from the wind turbines is suppressed by clutter filters the dynamic wind turbine clutter, mainly originating from the rotating blades, has the largest impact on weather radar measurements. Dynamic wind turbine clutter (in the following referred to as wind turbine clutter) is often difficult to separate from precipitation echoes and may therefore incorrectly be interpreted by the weather radar as precipitation.

In addition, wind turbine clutter is highly variable in time since the amplitude of the scattered signal depends sensitively on the wind turbine's yaw- and tilt angle.

3.1.1 Observations

Observations of wind turbine clutter have been presented in numerous works (e.g. Agence National des Fréquences (2005); Burgess et al. (2008); Gallardo et al. (2008); Haase et al. (2010); Isom et al. (2009); Toth et al. (2011); Tristant (2006a); Vogt et al. (2011; 2007a)). The strength

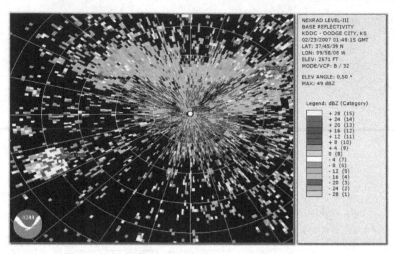

Fig. 3. Wind turbine clutter caused by two wind farms near Dodge City, Kansas, at 0149 GMT on February 23, 2007. One wind farm is located approximately 40 km to the southwest of the radar; the other near 20 km to the northeast. Range rings in white are at 10-km spacing. Adapted from Crum et al. (2008). This image was obtained from NOAA/National Climatic Data Center.

of the observed clutter can range from barely visible ($<$ 0 dBZ) to near saturation levels ($>$ 60 dBZ) (Agence National des Fréquences, 2005; Crum et al., 2008; Toth et al., 2011; Tristant, 2006a).

An example of clutter, originating from two wind farms near Dodge City, Kansas, is shown in Fig. 3. One wind farm consists of 170 wind turbines and is located approximately 40 km southwest of the radar; the other wind farm consists of 72 wind turbines, located approximately 20 km northeast of the radar. On this otherwise clear day reflectivity values close to 30 dBZ can be seen at the location of both wind farms.

Images such as Fig. 3 convincingly demonstrate that wind turbine clutter exists and that it may indeed cause problems for weather radars. However, in order to obtain a quantitative estimate of wind turbine clutter long time series of data should be studied.

In the remainder of this section we present results from a study based on long time series of wind turbine clutter. In the study operational reflectivity data from the four lowest scans of all Swedish weather radars were analysed over a period of more than three years (November 1, 2007 to March 31, 2011). In order to estimate the amount of wind turbine clutter observed by the weather radars, precipitation echoes were filtered out using a custom-designed weather filter. To further increase the quality of the wind turbine clutter, all other clutter — here referred to as background clutter — was removed from the weather-filtered reflectivity data. Finally the wind turbine clutter (z) was converted to rain rate (R) assuming the relation $z = 200R^{1.5}$ (Michelson et al., 2000).

The weather filter removed precipitation echoes from the lowest elevation angle by comparing reflectivity data cellwise to reflectivities from a higher elevation angle. If an echo from a higher

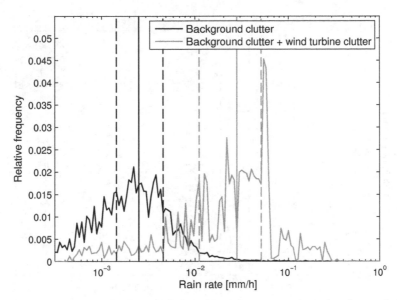

Fig. 4. Distributions of clutter before and after the construction of wind turbines. Solid, vertical lines indicate the median value and dashed, vertical lines show the first and third quartiles.

elevation angle was strong enough to indicate the presence of precipitation the corresponding value from the lowest elevation angle was filtered out.

Approximately 13 km from the weather radar in Karlskrona, Sweden, three wind turbines are located in the same radar cell (i.e., the same range bin and azimuth gate). Weather-filtered clutter distributions from this radar cell before and after the construction of the three wind turbines are shown in Fig. 4. It is seen that the two clutter distributions are easily distinguishable, having similar shapes but very different medians. It is evident that in this radar cell the existence of operational wind turbines has substantially increased the total amount of clutter.

Before the construction of the wind turbines, the reflectivity values remaining after filtering out precipitation echoes were composed of clear-air returns and other, non-identified, moving targets. In a second step of the analysis, this background clutter was removed from the weather-filtered reflectivity values recorded after the construction of the wind turbines. In this way a measure of clutter solely due to wind turbines was obtained.

The median wind turbine clutter was obtained from the difference between clutter after and before the construction of wind turbines. This analysis was carried out for all wind turbines in line-of-sight of a Swedish weather radar. The median wind turbine clutter values of 11 different radar cells, together with the first and third quartiles, are shown in Fig. 5. The median wind turbine clutter is seen to vary from close to zero to more than 0.02 mm h^{-1}. The spread of the clutter is attributed to the fact that the strength of a wind turbine echo depends on the position of the rotor blades and the yaw of the wind turbine, which in turn depends on the direction of the wind.

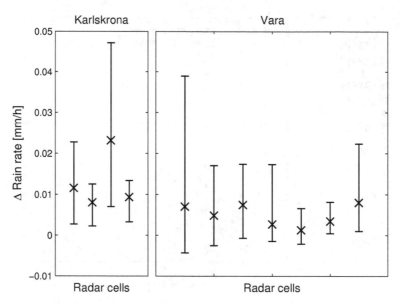

Fig. 5. Median wind turbine clutter together with first and third quartiles observed by two Swedish weather radars (Karlskrona and Vara).

Wind turbine clutter has also been observed from radar cells in which no wind turbines are located. An example of such area effects of wind turbine clutter is shown in Fig. 6. The wind turbine clutter in this figure comes from the same wind farm as in Fig. 4. All together this wind farm consists of five wind turbines with total heights of 150 m above the ground. Three of the five wind turbines are located within the same radar cell, the other two turbines each occupy a different radar cell. In Fig. 6 it is seen that not only the radar cells in which the wind turbines are located show an increase in clutter but also that several radar cells cross- and downrange of the turbines are affected.

Wind turbine clutter downrange from wind turbines (cf. Figs. 6a and b) has been observed in several other works (e.g. Crum et al. (2008); Haase et al. (2010); Isom et al. (2009); Toth et al. (2011); Vogt et al. (2011)). Such clutter tails can be visible for tens of kilometres behind wind turbines. No theoretical model has been put forward to explain this phenomenon but it has been suggested that the tails are caused by multiple scattering effects (scattering between multiple turbines and/or scattering between turbine and ground) (Crum et al., 2008; Isom et al., 2009; Toth et al., 2011). Clutter tails are not considered a problem for wind farms located further than 18 km from the weather radar (Crum & Ciardi, 2010; Vogt et al., 2009).

Cross-range clutter may also occur, as is seen in Figs. 6a and c. For the case shown in Fig. 6, the cross-range clutter is a direct result of the way the reflectivities are stored in the radar data matrix (the azimuthal resolution of the actual radar measurements is lower than the azimuthal spacing of the data matrix). However, for wind turbines generating very strong echoes it has been suggested that clutter may be seen well outside the half-power width of the radar beam, generating cross-range clutter spanning tens of degrees (Agence National des Fréquences, 2005).

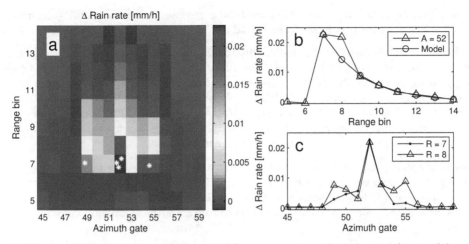

Fig. 6. Wind turbine clutter caused by a wind farm approximately 13 km northeast of the Karlskrona weather radar. a) Clutter from an area containing five wind turbines (shown by white asterisks). One wind turbine is located in radar cell [7,49], three are located in [7,52], and one in [7,55]. b) Clutter from azimuth gate 52 together with model results. c) Clutter from range bins 7 and 8.

3.1.2 Models

Most models of wind turbine clutter rely on the turbines' radar cross section (RCS) as a measure of how efficiently radar pulses are backscattered (Agence National des Fréquences, 2005; Tristant, 2006a). In order to model wind turbine clutter the RCS of a wind turbine must be converted to the equivalent radar reflectivity factor. The radar equation for point targets is given by (see, e.g. Skolnik (2008))

$$P_r = \frac{P_t G^2 \lambda^2 \sigma}{64 \pi^3 D^4} \tag{1}$$

where P_r and P_t are, respectively, the power received and transmitted by the radar, G is the antenna gain, λ is the wavelength, σ is the RCS of the target, and D is the distance from the radar to the target.

For distributed targets, such as rain, the radar equation is written as (see, e.g. Keeler & Serafin (2008))

$$P_r = \frac{P_t G^2 \theta \phi c \tau \pi^3 |K|^2 z}{1024 \ln(2) \lambda^2 D^2} \tag{2}$$

where θ and ϕ are the azimuth and elevation beamwidths, c is the speed of light, τ is the radar pulse width, $|K|^2$ is a parameter related to the complex index of refraction of the material, and z is the linear radar reflectivity factor. For a given RCS the linear radar reflectivity factor can thus be expressed as

$$z = C_1 \frac{\sigma}{D^2} \tag{3}$$

where C_1 is a constant that depends on the parameters of the radar system. For Swedish weather radars, $C_1 = 3 \times 10^{12}$ mm^{6}m^{-3}.

Let us use Eq. (3) to calculate what RCS would cause an observed rain rate of $R = 0.1$ mm h^{-1} (cf. Fig. 4) at a distance $D = 13$ km from the radar. Using $z = 200R^{1.5}$ we find that $\sigma = 4$ cm^2.

The RCS of wind turbines has been studied both experimentally and numerically (see, e.g., Greving & Malkomes (2006); Kent et al. (2008); Kong et al. (2011); Ohs et al. (2010); Poupart (2003); Zhang et al. (2011)). These studies have shown that RCSs of wind turbines display a sensitive dependence on yaw- and tilt angle. However, measurements of the RCS of large wind turbines typically range between 20 to 30 dBsm (Kent et al., 2008; Poupart, 2003) which is very far from what we obtained in the calculation. Using the RCS to calculate wind turbine clutter in this simple way may therefore lead to erroneous results.

It has been argued that the RCS is not applicable to wind turbines (Greving & Biermann, 2008; Greving et al., 2009; Greving & Malkomes, 2006; 2008). The reason is that the plane wave condition does not hold for objects on the ground. From the calculation above it is clear that more sophisticated models are needed in order to make a correct simulation of wind turbine clutter.

For downrange clutter a simple, empirical model was constructed using an exponential function to fit the limited amount of data available. The rain rate R behind a wind turbine was modelled as

$$R = R_0 \exp\left(-\frac{C_2 x}{N}\right) \tag{4}$$

where R_0 is the rain rate in mm h^{-1} from the radar cell containing the wind turbine, $C_2 = 0.7$ is an empirically determined constant, x is the distance behind the wind turbine in kilometres, and N is the number of interfering wind turbines present in the radar cell. Observations and model results are shown in Fig. 6b.

3.1.3 Mitigation concepts

Various concepts for mitigating wind turbine clutter have been suggested in different studies. Some of these concepts are listed here.

- Placing wind turbines so that they are not in line of sight of a weather radar. Under normal conditions a radar's measurements will not be affected by objects that are not in the radar line of sight. This method is therefore a certain way of limiting wind turbine clutter. It has also been suggested that wind turbines should be arranged radially from the radar. Such a formation probably does little to mitigate clutter since the blades of the different wind turbines do not move synchronously.

- Reducing the wind turbines' RCS. It has been proposed that stealth materials can be applied to wind turbines as a way of reducing the RCS (Appelton, 2005; Butler & Johnson, 2003). Studies of stealth coating wind turbine blades show that a reduction of more than 10 dB may be possible (Rashid & Brown, 2010), making it an interesting solution. An alternative way of reducing the rotor blades' RCS is to modify their shape, but this is not considered a realistic alternative as the shape of a rotor blade is optimized for efficiency.

- Adaptive clutter filters. Various filter techniques for removing or reducing effects of wind turbine clutter have been suggested. Gallardo et al. (2008) suggested using an image processing technique and Isom et al. (2009) proposed a multiquadratic interpolation technique. Other signal processing techniques have also been proposed (Bachmann et al.,

2010a;b; Gallardo-Hernando & Pérez-Martínez, 2009; Nai et al., 2011). These methods all use raw data as input, i.e., in- and quadrature phase (I/Q) data.

To speed up filtering, only radar cells containing wind turbines should ideally be processed. This may be achieved by keeping maps of all wind turbines near a weather radar or by using automatic detection schemes (Cheong et al., 2011; Gallardo-Hernando et al., 2010; Hood et al., 2009; 2010).

- Gap-filling radars. Areas contaminated by clutter may be covered by a second, nearby radar, a so-called gap-filler (Aarholt & Jackson, 2010; Department of Defense, 2006; Ohs et al., 2010). This alternative may be a convenient solution for specific cases but could also lead to even bigger problems since an introduction of additional radars introduces new sites which also must be protected.

- Adaptation of the radar scan strategy. Changing the radar scan strategy to pass over areas with wind turbines will limit the amount of clutter received. The drawback is that data will be gathered from higher altitudes which may shorten the effective range of the radar.

3.2 Blockage

For a weather radar, blockage manifests itself as a reduction of the expected precipitation echoes downrange from an obstacle. But, as we have seen in Section 3.1, obstacles in line of sight of a radar do not only cause blockage, they also cause clutter. Stationary obstacles cause static clutter which can be removed by a clutter filter. However, dynamic clutter, such as echoes from rotating blades of wind turbines, is not removed by the clutter filter. Downrange from such obstacles both clutter and blockage can appear. For wind turbines in line of sight of a weather radar the increased echo strength from the clutter can often be as large, or larger, than the reduction in echo strength due to blockage. Separating the effects of blockage and clutter is therefore often impossible using data analysis. However, large wind farms may cause substantial blockage and the effect may be visible for tens of kilometres downrange of the farm.

3.2.1 Observations

Blockage caused by wind turbines is not always visible in radar reflectivity images. As explained previously this is partly due to clutter tails but also because precipitation echoes are not always spatially homogeneous.

One example of blockage caused by a wind farm near Dodge City, Kansas, is shown in Fig. 7. In the figure a weak shadow can be seen behind a wind farm to the southwest of the weather radar. Other examples of blockage caused by wind turbines can be found in Vogt et al. (2007a) and Seltmann & Lang (2009).

As mentioned above, making a quantitative analysis of blockage behind wind farms is difficult due to clutter tails and the spatial variation of precipitation echoes. Let us therefore instead examine blockage caused by a stationary structure in line of sight of a weather radar.

The air traffic control tower of Arlanda Airport near Stockholm, Sweden, is located only 0.9 km from the Arlanda weather radar. The full width at half maximum of the radar beam is 0.9°, which at the distance of the tower corresponds to approximately 14 m. The radar beam is

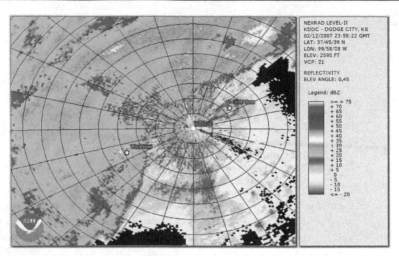

Fig. 7. Reflectivity measurement from the weather radar near Dodge City, Kansas, at 2358 GMT, February 12, 2007. Blockage caused by a wind farm to the southwest results in a reduction of precipitation echoes downrange from the wind farm (most clearly visible at a distance of 70–100 km from the radar). Range rings in black are at 10-km spacing. Adapted from Burgess et al. (2008). This image was obtained from NOAA/National Climatic Data Center.

thus wider than the width of the tower (approximately 8.5 m). For the lowest elevation angle of the radar the tower fills the entire beam height.

The average amount of precipitation per hour, for the period 1 November 2007 to 31 March 2011, is shown in Fig. 8. In this figure it is seen that some azimuth gates have considerably less measured precipitation compared to their neighbours. These gates coincide with the location of the tower.

One way to obtain a quantitative estimate of the reduction in expected precipitation due to blockage is to assume that the precipitation can be considered constant over some neighbouring azimuth gates. To validate this assumption a correlation analysis was performed. The analysis revealed that the correlation between precipitation measurements from neighbouring azimuthal radar cells depends on cross-range distance and accumulation period. The correlation decreases as the cross-range distance increases and for the same cross-range distance, shorter accumulation periods results in lower correlation. Applying the correlation analysis to the precipitation measured by the Arlanda weather radar showed that, for example, precipitation from radar cells with cross-range distances up to 5 km and an accumulation period of 24 h had a correlation over 0.9.

To obtain a measure of how much precipitation varies locally the coefficient of variation of accumulated precipitation from neighbouring azimuthal radar cells was calculated. In the analysis for the Arlanda weather radar it was shown that the coefficient of variation increased with increasing number of neighbouring azimuthal gates (i.e. window size) and decreased with accumulation period.

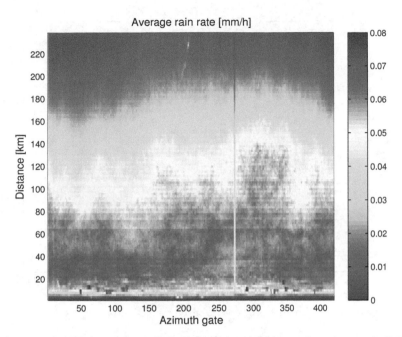

Fig. 8. Average amount of precipitation per hour from the lowest elevation angle (0.5°) for the weather radar at Arlanda airport, Sweden. Blockage caused by a nearby air traffic control tower can be seen near azimuth gate 271.

For example, for an accumulation period of 1 hour and a window size of 13 gates (corresponding to 11.7° in azimuthal angle) the coefficient of variation did not decrease lower than approximately 0.25 at any distance from the radar whereas for an accumulation period of 1 month the coefficient of variation for the same number of gates was lower than 0.05 at 10 km from the radar. The coefficient of variation can be compared with the blockage caused by an obstacle.

To find a quantitative estimate of the reduction in expected precipitation echoes caused by the Arlanda tower a 13-gate wide window was applied to the data, accumulated over the entire three-year-period. The measurements in the window were normalized over azimuth and range to the average value of unaffected gates. In Fig. 9 it is seen that the measured precipitation is reduced by close to 30% in the most severely affected gate.

Comparing the blockage of the Arlanda tower with the coefficient of variation for various accumulation periods it was found that on average between 24 hours and 1 week was needed for the coefficient of variation of the local precipitation to be lower than 30%. On individual radar images it may therefore be difficult to see the effects of the Arlanda tower blockage.

3.2.2 Models

Modelling of electromagnetic shadow effects can be done with varying accuracy and complexity. Methods and results of modelling blockage and shadow effects downrange of wind turbines can be found in, e.g., Belmonte & Fabregas (2010); Greving & Malkomes

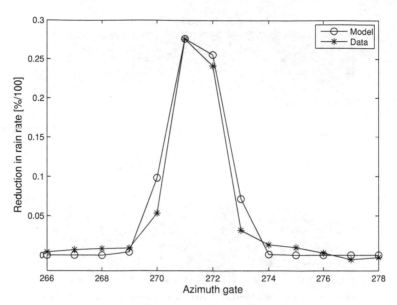

Fig. 9. Blockage caused by an air-traffic control tower at Arlanda Airport. Data and model.

(2008); Høye (2007). Here we describe a simple and computationally light method to calculate blockage caused by an obstacle.

As a first approximation of the reduction in returned power due to an obstacle we consider the obstacle's geometrical cross section. Convolving the obstacle's cross section with the radar beam's power flux and dividing by the total power from an unperturbed beam we obtain the fraction of power, P_B, blocked by the obstacle. To find the corresponding reduction in rain rate we start by noting that $1 - P_B$ is the amount of power that is unaffected by blockage. The (unaffected) power is proportional to the linear radar reflectivity factor z according to Eq. (2) and z is in turn related to rain rate R by $z \propto R^{1.5}$. Hence the reduction in rain rate, R_B, can be expressed as $R_B = 1 - (1 - P_B)^{1/1.5}$.

Applying this method to the Arlanda air-traffic control tower described in Section 3.2.1 we can estimate the reduction in rain rate it causes. The modelled reduction in rain rate is shown in Fig. 9 together with the observations. The model is seen to capture the magnitude and the cross-range shape of the blockage. This model can be used for estimating blockage caused by wind turbines, but for reasons explained in Section 3.2.1 there are no observations to compare these results with.

3.2.3 Mitigation concepts

Methods proposed to prevent or reduce blockage by wind turbines include:

- Optimising the placement of the wind turbines. Wind turbines should preferably be placed out of the line of sight of the radar. Otherwise it has been suggested that wind turbines should be arranged radially from the radar. In this way the blockage caused by the wind turbine towers may be reduced, but blockage caused by rotor blades will persist since there is no synchronisation of their movements.

- Use of a gap-filling radar. A way to remove or reduce blockage is to place an additional radar to cover areas affected by blockage.

- Adapting the radar scan strategy so that the radar beam passes over areas with wind turbines. This method ensures that measurements are not affected by blockage but in return data will be gathered from higher altitudes.

3.3 Wind measurements

A Doppler radar measures frequency shifts of the received signals and translates the shifted frequencies to radial velocities. A conventional clutter filter removes echoes with low or zero frequency shifts and thereby prevents static clutter from entering the radar products.

Signals scattered from rotating blades of a wind turbine are shifted in frequency and thereby interpreted by the radar as moving objects, escaping the clutter filter. The tip of a rotor blade can move with a velocity up to 100 m s^{-1} whereas close to the hub the blade velocity is close to zero. The scattered signals will therefore display a broad distribution in frequency space. The wind velocity is normally estimated as the strongest (non-zero) frequency component. Since echoes from wind turbines often are stronger than weather echoes this can lead the weather radar to display erroneous wind measurements.

3.3.1 Observations

There are many observations of wind turbines causing erroneous wind measurements in the literature (see, e.g., Burgess et al. (2008); Cheong et al. (2011); Crum et al. (2008); Haase et al. (2010); Isom et al. (2009); Toth et al. (2011); Vogt et al. (2007a)). One such example from the weather radar in Dodge City, Kansas, is shown in Fig. 10. From this figure it is clear that at the time of the measurements the overall wind direction was to the northwest but signals from radar cells containing a large wind farm, approximately 40 km to the southwest, show up as having close to zero velocity. In Fig. 11 is shown the spectrum width of the velocity measurements and from this figure it is clear that there is a significant broadening of the frequency spectra over the wind farm.

These observations can be understood by examining the raw I/Q data from the radar. Spectrograms of I/Q data, containing echoes from wind turbines, show highly complex and richly structured patterns. Examples of such spectrograms are given by, e.g., Bachmann et al. (2010a); Gallardo et al. (2008); Gallardo-Hernando & Pérez-Martínez (2009); Gallardo-Hernando et al. (2009); Hood et al. (2009); Isom et al. (2009); Nai et al. (2011); Poupart (2003); Vogt et al. (2007a;b). From these and other studies it is clear that echoes from wind turbine rotor blades in different positions result in broad distributions in frequency space even though the average velocity estimate is often close to zero.

As for wind turbine clutter there are observations showing tails of erroneous wind measurements behind the wind turbines (Burgess et al., 2008; Seltmann & Lang, 2009; Vogt et al., 2007a; 2009).

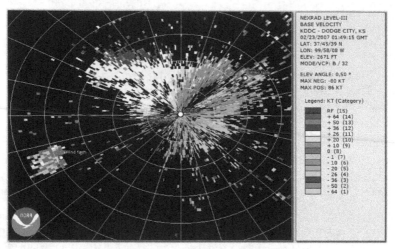

Fig. 10. Wind measurements from the weather radar near Dodge City, Kansas, at 0149 GMT, on February 23, 2007. The general wind direction is to the northwest but measurements near a wind farm to the southwest of the radar show wind velocities close to zero. Range rings in white are at 10-km spacing. Adapted from Vogt et al. (2007a). This image was obtained from NOAA/National Climatic Data Center.

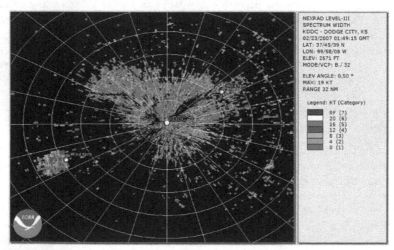

Fig. 11. Measurements of spectrum width from the weather radar near Dodge City, Kansas, at 0149 GMT, on February 23, 2007,. The spectrum widths are considerably enhanced near a wind farm to the southwest of the radar. Range rings in white are at 10-km spacing. Adapted from Vogt et al. (2007a). This image was obtained from NOAA/National Climatic Data Center.

3.3.2 Models

A few methods have been proposed to model the frequency spectra generated by wind turbines (Gallardo-Hernando et al., 2010; Hood et al., 2009; 2010; Kong et al., 2011; Zhang et al., 2011). In these models clutter from the tower, hub, and rotor blades are included. The model results show the zero-velocity returns of the tower, near-zero returns of the hub, as well as spectral broadening of the blades.

3.3.3 Mitigation concepts

Most concepts proposed to prevent or reduce clutter (see Section 3.1.3) are also valid for erroneous wind measurements.

Adaptive filters, suggested for removing wind turbine clutter, can also help mitigate erroneous wind measurements. If clutter is removed from the signal, the average wind velocity as well as the spectrum width can easily be estimated. Suggestions for adaptive wind filters can be found in, e.g., Bachmann et al. (2010a;b); Gallardo et al. (2008); Isom et al. (2009); Nai et al. (2011). As for the adaptive clutter filters, all these methods use raw I/Q data as input.

4. Conclusions

In many countries the number of wind turbines is growing rapidly as a response to the increased demand for renewable energy. As wind turbines grow larger and more numerous potential conflicts with other interests are emerging. Doppler radars, for example, are easily disturbed by wind turbines. In this chapter we have presented an overview on wind turbine-related problems experienced by Doppler weather radars. Three main wind turbine-related problems have been identified: clutter, blockage, and erroneous wind measurements.

Clutter — unwanted radar echoes — are generated by all obstacles in line of sight of a radar. Static clutter, i.e., echoes with no or low radial velocities are easily removed by the Doppler radar's built-in clutter filter. However, the moving blades of a wind turbine generate dynamic clutter which displays a wide range of radial velocities that cannot be removed by a conventional clutter filter. In this chapter it has been shown that wind turbine clutter can be problematic for weather radars since such echoes are interpreted as precipitation. Wind turbine clutter can display a large variation in strength, ranging from barely visible to near saturation levels of the radar. Behind wind turbines a tail of clutter can often be seen. This phenomenon is believed to be the result of multiple scattering effects. In an example shown in this chapter, clutter tails were seen to decrease exponentially behind the wind turbines. Cross-range clutter has also been observed. This can be caused by differences in azimuthal resolution of the actual radar measurements and azimuthal spacing of the radar data matrix but it has also been suggested that it may occur as a result of the radar sidelobes.

The magnitude of wind turbine clutter is often estimated by calculating the radar cross section of wind turbines. Such models may lead to results inconsistent with observations. Effects from the ground and terrain should be taken into account, otherwise a calibration of the model may be necessary.

Possible mitigation measures for wind turbine clutter include a) placing wind turbines out of the radar's line of sight b) reducing the wind turbines' radar cross section using stealth material c) development and application of adaptive clutter filters, d) use of gap-filling radars to cover areas contaminated by clutter and e) adapting the radar scan strategy to pass over the wind turbines.

Blockage is caused by any obstacle in line of sight of a radar and is not specific to wind turbines. For a weather radar, blockage leads to an underestimation of the precipitation behind the blocking obstacle. Blockage caused by wind turbines is difficult to analyse using reflectivity data due to spatial variation of precipitation and clutter tails that are generated behind wind turbines. Model results can, however, be compared with observations of blockage caused by stationary obstacles such as towers, masts, or wind turbines that are not in operation.

Concepts for mitigating blockage include a) placing wind turbines radially from the radar b) use of a gap-filling radar to cover affected areas and c) adapting the radar scan strategy to pass over the wind turbines.

The Doppler function of a radar detects movements in echoes such as those from the rotating blades of a wind turbine. Although such measurements are correct, the interpretation by the radar may still be wrong. For example, an air traffic control radar may interpret echoes from rotating wind turbine blades as a moving aircraft and a weather radar may interpret such measurements as an approaching thunderstorm.

A weather radar uses Doppler-shifted echoes to estimate the wind speed. Doppler-shifted echoes from wind turbine blades may therefore lead to erroneous wind measurements. Observations of wind measurements over wind farms occasionally show extremely large wind speeds but most often the wind measurements are close to zero. The non-synchronised movements of the many rotor blades of a wind farm also lead to large spectrum widths.

Mitigation measures for wind measurements are the same as those presented for wind turbine clutter.

5. References

Aarholt, E. & Jackson, C. A. (2010). Wind farm gapfiller concept solution, *European Radar Conference*, EuRAD, Paris, pp. 236–239.

Agence National des Fréquences (2005). Perturbations du fonctionnement des radars meteorologiques par les eoliennes, *Technical Report Rapport CCE5 No.1*, Commision Consultative de la Compatibilité Electromagnétique. In French.

Anon. (1890). Mr. Brush's windmill dynamo, *Scientific American* 63(25): 54.

Appelton, S. (2005). Design & manufacture of radar absorbing wind turbine blades, *Final Report W/44/00636/00/REP, DTI PUB URN 05/1409*, QinetiQ.

Bachmann, S., Al-Rashid, Y., Bronecke, P., Palmer, R. & Isom, B. (2010a). Suppression of the windfarm contribution from the atmospheric radar returns, *26th Conference on Interactive Information and Processing Systems for Meteorology, Oceanography, and Hydrology*, American Meteorological Society, pp. 81–86.

Bachmann, S., Al-Rashid, Y., Isom, B. & Palmer, R. (2010b). Radar and windfarms — mitigating negative effects through signal processing, *The Sixth European Conference on Radar in Meteorology and Hydrology*, ERAD, Sibiu, Romania, pp. 81–86.

Belmonte, A. & Fabregas, X. (2010). Analysis of wind turbines blockage on doppler weather radar beams, *IEEE Antennas and Wireless Propagation Letters* 9: 670–673.

Borely, M. (2010). Guidelines on how to assess the potential impact of wind turbines on surveillance sensors, *Technical Report EUROCONTROL-GUID-130*, EUROCONTROL.

Brenner, M., Cazares, S., Cornwall, M. J., Dyson, F., Eardley, D., Horowitz, P., Long, D., Sullivan, J., Vesecky, J. & Weinberger, P. J. (2008). Wind farms and radar, *Technical Report JSR-08-126*, JASON.

Burgess, D. W., Crum, T. D. & Vogt, R. J. (2008). Impacts of wind farms on WSR-88D radars, *24th International Conference on Interactive Information and Processing Systems for Meteorology, Oceanography, and Hydrology*, American Meteorological Society. Paper 6B.3.

Butler, M. M. & Johnson, D. A. (2003). Feasibility of mitigating the effects of windfarms on primary radar, *Department of Trade and Industry report ETSU W/14/00623/REP*, Alenia Marconi Systems Limited.

Cheong, B. L., Palmer, R. & Torres, S. (2011). Automatic wind turbine detection using level-II data, *2nd Conf. on Weather, Climate, and the New Energy Economy*, Amererican Meteorological Society, Seattle, WA. Paper 808.

Crum, T. & Ciardi, E. (2010). Wind farms and the WSR-88D: An update, *Nextrad Now* 20: 17–22.

Crum, T., Ciardi, E. & Sandifer, J. (2008). Wind farms: Coming soon to a WSR-88D near you, *Nextrad Now* 18: 1–7.

Davies, N. G. (1995). Wind farm radar study, *Technical Report ETSU-W–32/00228/49/REP*, Energy Technology Support Unit.

Department of Defense (2006). The effect of windmill farms on military readiness, *Report to the congressional defense committees*, Office of the Director of Defense Research and Engineering. [Available online at http://www.defense.gov/pubs/pdfs/windfarmreport.pdf].

Donaldson, N., Best, C. & Paterson, B. (2008). Development of wind turbine assessments for Canadian weather radars, *The Fifth European Conference on Radar in Meteorology and Hydrology*, ERAD, Helsinki, Finland.

Frye, A., Neumann, C. & Müller, A. (2009). The compatibility of wind turbines with radars, *Annual Report 54.7100.035.12*, European Aeronautic Defence and Space Company.

Gallardo, B., Pérez, F. & Aguado, F. (2008). Characterization approach of wind turbine clutter in the Spanish weather radar network, *The Fifth European Conference on Radar in Meteorology and Hydrology*, ERAD, Helsinki, Finland.

Gallardo-Hernando, B. & Pérez-Martínez, F. (2009). Wind turbine clutter, *in* G. Kouemou (ed.), *Radar Technology*, InTech, Croatia.

Gallardo-Hernando, B., Pérez-Martínez, F. & Aguado-Encabo, F. (2009). Mitigation of wind turbine clutter in C-band weather radars for different rainfall rates, *Proceedings of the 2009 International Radar Conference*, Bordeaux, France.

Gallardo-Hernando, B., Pérez-Martínez, F. & Aguado-Encabo, F. (2010). Wind turbine clutter detection in scanning weather radar tasks, *The Sixth European Conference on Radar in Meteorology and Hydrology*, ERAD, Sibiu, Romania.

Global Wind Energy Council (2011). Global wind report – annual market update 2010, *Technical report*, Global Wind Energy Council. [Available online at: http://www.gwec.net/fileadmin/images/Publications/ GWEC_annual_market_update_2010_-_2nd_ edition_ April_2011.pdf].

Greving, G. & Biermann, W.-D. (2008). Application of the radar cross section RCS for objects on the ground – example of wind turbines, *2008 International Radar Symposium*, IRS, Wroclaw, Poland.

Greving, G., Biermann, W.-D. & Mundt, R. (2009). RCS – numerical, methodological and conceptional aspects for the analysis of radar distorting objects, *11th International Radar Symposium*, IRS, Vilnius, Lithuania.

Greving, G. & Malkomes, M. (2006). On the concept of the radar cross section RCS of distorting objects like wind turbines for the weather radar, *The Fourth European Conference on Radar in Meteorology and Hydrology*, ERAD, Barcelona, Spain, pp. 333–336.

Greving, G. & Malkomes, M. (2008). Weather radar and wind turbines – theoretical and numerical analysis of the shadowing effects and mitigation concepts, *The Fifth European Conference on Radar in Meteorology and Hydrology*, ERAD, Helsinki, Finland.

Haase, G., Johnson, D. & Eriksson, K.-Å. (2010). Analyzing the impact of wind turbines on operational weather radar products, *The Sixth European Conference on Radar in Meteorology and Hydrology*, ERAD, Sibiu, Romania, pp. 276–281.

Hafner, S., Reitter, R. & Seltmann, J. (2004). Developments in the DWD radarnetwork, *The Third European Conference on Radar in Meteorology and Hydrology*, ERAD, Visby, Sweden, pp. 425–427.

Hood, K. T., Torres, S. M. & Palmer, R. D. (2009). Automatic detection of wind turbine clutter using doppler spectral features, *34th Conference on Radar Meteorology*, American Meteorological Society, Williamsburg, VA. Paper P10.1.

Hood, K., Torres, S. & Palmer, R. (2010). Automatic detection of wind turbine clutter for weather radars, *Journal of Atmospheric and Oceanic Technology* 27: 1868–1880.

Høye, G. (2007). Electromagnetic shadow effects behind wind turbines, *FFI Report FFI/ RAPPORT-2007/00842*, Norwegian Defence Research Establishment. [Avaiable online at http://rapporter.ffi.no/rapporter/2007/00842.pdf].

Hutchinson, G. & Miles, R. (2008). The protection of weather radar networks. the UK experience, *The Fifth European Conference on Radar in Meteorology and Hydrology*, ERAD, Helsinki, Finland.

Isom, B. M., Palmer, R. D., Secrest, G. S., Rhoton, R. D., L., D. S. T., Allmon, Reed, J., Crum, T. & Vogt, R. (2009). Detailed observations of wind turbine clutter with scanning weather radars, *Journal of Atmospheric and Oceanic Technology* 26: 894–910.

Jago, P. & Taylor, N. (2002). Wind turbines and aviation interests — European experience and practice, *Department of Trade and Industry report ETSU W/14/00624/REP*, STASYS Ltd.

Keeler, R. J. & Serafin, R. J. (2008). Meteorological radar, *in* M. Skolnik (ed.), *Radar Handbook*, 3 edn, McGraw-Hill.

Kent, B. M., Hill, K. C., Buterbaugh, A., Zelinski, G., Hawley, R., Cravens, L., Tri-Van, Vogel, C. & Coveyou, T. (2008). Dynamic radar cross section and radar Doppler measurements of commercial general electric windmill power turbines part 1: Predicted and measured radar signatures, *IEEE Antennas Propagation Magazine* 50: 211–219.

Kong, F., Zhang, Y., Palmer, R. & Bai, Y. (2011). Wind turbine radar signature characterization by laboratory measurements, *Radar Conference (RADAR)*, IEEE.

Lemmon, J. J., Caroll, J. E., Sanders, F. H. & Turner, D. (2008). Assessment of the effects of wind turbines on air traffic control radars, *NTIA Technical Report TR-08-454*, US Department of Commerse, National Telecommunications & Information Administration.

Manwell, J., McGowan, J. & Rogers, A. (2009). *Wind Energy Explained: Theory, Design and Application*, 2 edn, John Wiley & Sons, Ltd.

Michelson, D. B., Andersson, T., Koistinen, J., Collier, C. G., Riedl, J., Szturc, J., Gjertsen, U., Nielsen, A. & Overgaard, S. (2000). BALTEX radar data centre products and their methodologies, *RMK 90*, Swedish Meteorological and Hydrological Institute.

Nai, F., Palmer, R. & Torres, S. (2011). Wind turbine clutter mitigation using range-Doppler domain signal processing method, *27th Conf. on Interactive Information and Processing Systems*, American Meteorological Society, Seattle, WA. Paper 9.4.

Ohs, R. R., Skidmore, G. J. & Bedrosian, G. (2010). Modeling the effects of wind turbines on radar returns, *The 2010 Military Communications Conference*, MILCOM, San Jose, CA, pp. 272–276.

OPERA (2010). Statement of the OPERA group on the cohabitation between weather radars and wind turbines, [Available online at http://www.knmi.nl/opera/opera3/OPERA_2010_14_Statement_on_weather_radars_and_wind_turbines.pdf].

Ousbäck, J.-O. (1999). Försvaret och vindkraften. Huvudstudie radar, *Slutrapport 99-2936/L*, Försvarets forskningsanstalt (FOA). In Swedish.

Poupart, G. J. (2003). Wind farms impact on radar aviation interests, *Final Report FES W/14/00614/00/REP, DTI PUB URN 03/1294*, QinetiQ.

Price, T. J. (2005). James Blyth—Britains first modern wind power pioneer, *Wind Engineering* 29(3): 191–200.

RABC & CanWEA (2007). Technical information and guidelines of the assessment of the potential impact of wind turbines on radiocommunication, radar and seismoacoustic systems, *Technical report*, Radio Advisory Board and Canadian Wind Energy Association.

Rashid, L. & Brown, A. (2010). Partial treatment of wind turbine blades with radar absorbing materials (RAM) for RCS reduction, *Proceedings of the Fourth European Conference on Antennas and Propagation*, EuCAP, Barcelona, Spain.

Seltmann, J. & Lang, P. (2009). Impact of wind turbines on radar measurements and tracking processes. Internal study commisioned by TI.

Sengupta, D. L. (1984). Electromagnetic interference effects of wind turbines, *The Working Committee on EMI*, International Energy Association, Copenhagen, Denmark.

Sengupta, D. L. & Senior, T. B. A. (1979). Electromagnetic interference to television reception caused by horizontal axis windmills, *Proceedings of the IEEE* 67: 1133–1142.

Senior, T. B. L., Sengupta, D. L. & Ferris, J. E. (1977). TV and FM interference by windmills, *Technical Report E(11-1)-2846*, Energy Research and Development Administration.

Shepard, D. G. (1990). Historical development of the windmill, *Contractor Report 4337 DOE/NASA/5266-1*, NASA.

Skolnik, M. (2008). An introduction and overview of radar, *in* M. Skolnik (ed.), *Radar Handbook*, 3 edn, McGraw-Hill.

Sparven Consulting (2001). Wind turbines and radar: Operational experience and mitigation measures, *Technical report*, Sparven Consulting.

Summers, E. (2001). The operational effects of wind farm developments on ATC procedures for Glasgow Prestwick international airport, *Technical report*, Glasgow Prestwick International Airport.

Toth, M., Jones, E., Pittman, D. & Solomon, D. (2011). DOW radar observations of wind farms, *Bulletin of the American Meteorological Society* 92(11): 987–995.

Tristant, P. (2006a).   Impact of wind turbines on weather radars band, *Report to WMO, Commission for Basic Systems Steering Group on Radio Frequency Coordination CBS/SG-RFC 2006/Doc. 3.1(6)*, Météo France.   [Available online at http://www.wmo.int/ pages/prog/www/TEM/SG-RFC06/Wind turbine vs weather radars.doc].

Tristant, P. (2006b). Radio frequency threats on meteorological radar operations, *The Fourth European Conference on Radar in Meteorology and Hydrology*, ERAD, Barcelona, Spain.

Vogt, R. J., Crum, T. D., Greenwood, W., Ciardi, E. J. & Guenther, R. G. (2011). New criteria for evaluating wind turbine impacts on NEXRAD radars, *WINDPOWER 2011*, American Wind Energy Association Conference and Exhibition, Anaheim, CA.

Vogt, R. J., Crum, T. D., Reed, J. R., Ray, C. A., Chrisman, J. N., Palmer, R. D., Isom, B., Snow, J. T., Burgess, D. W. & Paese, M. S. (2007a).   Weather radars and wind farms — working together for mutual benefit, *WINDPOWER 2007*, American Wind Energy Association Conference and Exhibition, Los Angeles, CA.

Vogt, R. J., Crum, T. D., Sandifer, J. B., Ciardi, E. J. & Guenther, R. G. (2009).   A way forward wind farm — weather radar coexistence, *WINDPOWER 2009*, American Wind Energy Association Conference and Exhibition, Seattle, WA.

Vogt, R. J., Reed, J., Crum, T., Snow, J. T., Palmer, R., Isom, B. & Burgess, D. W. (2007b).   Impacts of wind farms on WSR-88D operations and policy considerations, *23rd International Conference on Interactive Information and Processing Systems for Meteorology, Oceanography, and Hydrology*, American Meteorological Society, San Antonio, TX. Paper 5B.7.

Webster, D. M. (2005a). The effects of wind turbine farms on air defence radars, *Technical Report AWC/WAD/72/652/TRIALS*, Air Warfare Centre, Waddington, United Kingdom.

Webster, D. M. (2005b).   The effects of wind turbine farms on ATC radar, *Technical Report AWC/WAD/72/665/TRIALS*, Air Warfare Centre, Waddington, United Kingdom.

WMO (2010).   Commission for instruments and methods of observation, *Fifteenth session WMO-No.1046*, World Meteorological Organization.   [Available online at   http://www.wmo.int/pages/prog/www/CIMO/CIMO15-WMO1064/1064 _en.pdf].

Woodcroft, B. (1851). *The Pneumatics of Hero of Alexandria from the original Greek*, Taylor Walton and Maberly, London.   [Available online at http://www.history.rochester.edu/ steam/hero].

Wright, D. T. & Eng, C. (1992). Effects of wind turbines on UHF television reception, *Fifteenth session BBC RD 1992/7*, BBC Research Department.

Zhang, Y., Huston, A., Palmer, R. D., Albertson, R., Kong, F. & Wang, S. (2011).   Using scaled models for wind turbine EM scattering characterization: Techniques and experiments., *IEEE T. Instrumentation and Measurement* 60: 1298–1306.

7

Effects of Anomalous Propagation Conditions on Weather Radar Observations

Joan Bech[1], Adolfo Magaldi[2], Bernat Codina[1] and Jeroni Lorente[1]
[1]Dep. Astronomy and Meteorology, University of Barcelona
[2]Institute of Space Sciences, Spanish National Research Council (CSIC), Bellatera
Spain

1. Introduction

The effect of atmospheric propagation on radar observations is an important topic both for radar application developers and end-users of radar products, particularly of weather radar systems. An excellent review of this subject is given by Patterson (2008), and most general books about weather radars have a chapter on the topic –see for example Battan (1973), Collier (1996), Doviak and Zrnic (2006), Rinehart (2001) or Sauvageot (1991).

In this chapter our objective is to provide an overview of the effects of anomalous propagation conditions on weather radar observations, based mostly on studies performed by the authors during the last decade, summarizing results from recent publications, presentations, or unpublished material. We believe this chapter may be useful as an introductory text for graduate students, or researchers and practitioners dealing with this topic. Throughout the text a spherical symmetric atmosphere is assumed and the focus is on the occurrence of ground and sea clutter and subsequent problems for weather radar applications. Other related topics such as long-path, over-the-horizon propagation and detection of radar targets (either clutter or weather systems) at long ranges is not considered here; however readers should be aware of the potential problems these phenomena may have as range aliasing may cause these echoes appear nearer than they are – for more details see the discussion about second trip echoes by Zrnic, this volume.

Despite the motivation and results shown here are focused on ground-based weather radar systems (typically X, C or S band radars, i.e. cm-radars), a large part of these results are applicable to other types of radar, in fact also to micro-wave links or, in general terms, for propagation of electromagnetic waves in the atmosphere. As discussed in detail below, the main effect of anomalous propagation on weather radar observation is a lower height of the observed echoes than expected in normal conditions. This may imply an increase of ground clutter or, for radars operating near the coast, an increase of sea clutter, which will be hardly corrected by the standard Doppler filtering, affecting inevitably precipitation estimates.

This chapter is organized as follows. Section 2 introduces the fundamental concepts of refractivity and modified refractivity and the various propagation conditions associated with refractivity profiles. Section 3 presents some results on propagation condition variability, and Section 4 focuses specifically upon the impact of that variability on radar beam blockage corrections and subsequent precipitation estimates. Section 5 deals with the topic of propagation conditions forecasting and Section 6 presents a method to correct the effects of intense anomalous propagation conditions on weather radar precipitation estimates using satellite observations. Finally Section 7 provides a summary and concluding remarks.

2. Weather radar beam propagation conditions

This section presents qualitatively the different propagation regimes affecting the radar beam refraction. By radar beam we mean the energy emitted (and received) by the radar, limited by the half-power (3 dB) antenna main lobe (see Zrnic, this volume, for more details). In the vacuum, as in any media with constant index of refraction, a radar beam follows a straight trajectory. But in the atmosphere the index of refraction changes and therefore the variation of the air refractive index plays a key role when characterizing the propagation conditions of a radar beam in the troposphere, i.e. the lowest part of the atmosphere. In particular, the vertical profiles of the air temperature, moisture and pressure are mostly responsible for the way the radar energy will propagate in a given air layer. A number of assumptions on these vertical profiles are usually made, assuming the so-called "standard" or normal propagation conditions which are associated with the average state of the atmosphere accepted as the most representative, as discussed below. Under those conditions, the radar beam bends downward with a radius of curvature greater than that of the Earth surface. Consequently, the net effect is an increase of the height of the centre of the beam with respect to the ground as the distance from the radar increases (in Section 4 the equation for the radar beam height is given).

However, due to the inherent variability of the atmosphere, it is a well-known fact that propagation conditions may differ, sometimes significantly, from those considered standard resulting in anomalous propagation (AP). As illustrated schematically in Fig. 1, subrefraction causes the radar beam to bend less than usual, and therefore follows a higher trajectory than in normal conditions. Super refraction of a weather radar beam produces more bending towards the ground surface than expected for standard conditions and therefore increases and intensifies ground clutter echoes (AP or anaprop echoes). An extreme case of superrefraction, known as ducting, occurs when the beam has a curvature smaller than that of the Earth surface.

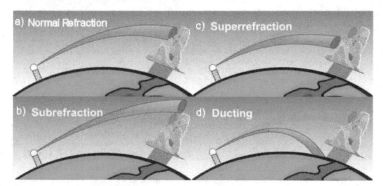

Fig. 1. Radar beam propagation conditions (adapted from US NOAA National Weather Service, introductory radar tutorial, "Doppler radar beams", http://www.srh.noaa.gov/jetstream/doppler/beam_max.htm).

Note that the term AP literally means "anomalous propagation" but AP echoes are associated with superrefraction and ducting, not to subrefraction. The occurrence of AP echoes may be particularly negative for automated quantitative precipitation estimates (QPE) such as those required for operational weather surveillance and hydrological flood warning. On the other hand, it should be noted that ducting may occur not only in the lowest air layer (surface ducting) as represented in Fig. 1d, but also on an elevated layer above which there is normal refraction. In that case, the duct (known as elevated duct), may trap the radar energy for a long distance without producing evident signs – AP echoes. ·

Figure 2 illustrates the effect of AP echoes on weather radar observations. It shows two radar reflectivity Plan Position Indicator (PPI) images recorded by the weather radar of the Meteorological Service of Catalonia located in Vallirana (41°22′N, 1°52′E, about 20 km west of Barcelona). The PPIs were obtained in two different days, one with normal propagation conditions, and the other under superrefraction conditions; none on those images show real precipitation, only ground and sea clutter. To see more clearly the change in AP echoes no Doppler filtering was applied to these images. In Fig. 2b arrows indicate some of the new or intensified AP echoes, either ground clutter (southernmost arrow pointing to the coast, or easternmost arrow pointing to the small island of Minorca), or sea clutter (around the centre of the image). PPI images corresponding to Fig. 2b where Doppler filtering was applied reduced largely AP ground clutter but not sea clutter, or other moving targets such as wind turbines, which may yield spurious hourly accumulations exceeding 50 mm.

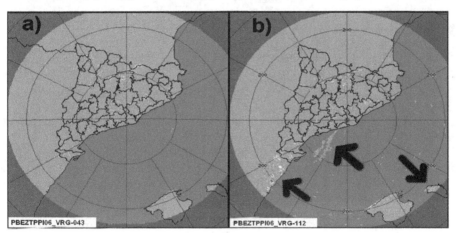

Fig. 2. Radar reflectivity base PPI images (0.6°) with no Doppler filtering showing ground and sea clutter on a normal propagation day (a) and a superrefractive day (b). Arrows indicate new or more intense AP echoes.

Despite the fact that AP echoes may be detected and cleaned with several techniques, this does not prevent that radar observations may be affected because of the difference between their real height and that expected assuming standard conditions. If this difference is important enough for a given application, any procedure which requires a precise knowledge of the echo altitude may be potentially affected by AP. For example, if radar data (either echo intensity or Doppler winds) are to be assimilated in a NWP model or if the radar echo intensity is corrected for beam blockage due to mountain sheltering (Bech et al., 2003), the effect may be relevant.

2.1 Refractivity N

As anomalous propagation is due to relatively small variations of the air refractive index n, the magnitude known as refractivity N, defined as one million times n-1, is commonly used in anaprop studies. As shown by Bean and Dutton (1968), or more recently in ITU (2003), N can be written as:

$$N = (n-1)10^6 = \frac{77.6}{T}\left(p + \frac{4810 \cdot e}{T}\right), \tag{1}$$

where T is the air temperature (K), p atmospheric pressure (hPa), and e is the water vapour pressure (hPa). According to ITU (2003), this expression may be used for all radio frequencies; for frequencies up to 100 GHz, the error is less than 0.5%. This formula takes into account only air gases and does not consider liquid water content (usually with negligible effects), or free electron density (important for high atmospheric altitudes, typically above 60 km).

Note that N is a dimensionless magnitude, though quite often the term "N units" is employed. N is sometimes considered the sum of two different terms of (1): the dry term, N_d, which depends only on p and T, and the wet term, N_w, which is also function of e, i.e. is related to moisture content. Typical values of N of air at ground level are within the range 250 to 450.

2.2 Modified refractivity M

A magnitude related to N is the modified refractivity M, which is defined as:

$$M = N + \frac{z}{10^{-6}r}, \tag{2}$$

where z is altitude and r is the radius of the Earth, expressed in meters (m). Modified refractivity is very useful to characterize propagation conditions as for constant M the curvature of the ray path is that of the Earth's surface and, therefore, when there are negative M vertical gradients the ray path may be bent towards the surface and then radio waves get trapped like in a wave guide (ducting). Based on M gradients, Johnson et al. (1999) suggested the use of a ducting index, with positive values proportional to the probability of occurrence of ducting.

2.3 Propagation conditions

Propagation characteristics may vary largely, depending for instance on the type of air mass (Gossard, 1977). When characterizing the radio propagation environment it is usual to consider the vertical refractivity gradient (VRG) of the air of the first kilometre above ground level to estimate propagation effects such as ducting, surface reflection and multipath on terrestrial line-of-sight links. However, the effect on weather radar beam refraction not only depends on the refractivity gradient of a layer but also on the angle of incidence between the beam and the trapping layer considered or the frequency of the electromagnetic wave (ITU, 2003). In the following paragraph, specific VRG values are given for the propagation conditions described earlier qualitatively.

For weather radar applications, if the vertical refractivity gradient of the first kilometre (VRG) of the atmosphere is around $-1/4r$ (i.e. -39 N units km^{-1} or 118 M units km^{-1}, where r is the Earth's radius) then standard propagation will occur for any angle of incidence (Doviak and Zrnic, 2006). An increase in VRG bends the radar beam more slowly than normal (subrefraction) and reduces the microwave radar horizon. With regard to ground clutter echoes, subrefraction implies a decrease in their frequency and intensity. On the other hand, a decrease in VRG generates the opposite effect, bending the beam faster than normal (super refraction) for the interval between (typically) -78.7 km^{-1} and -157 km^{-1} (the threshold to distinguish between standard propagation and superrefraction varies in the literature around 80 km^{-1}). Trapping, or ducting, the most extreme case of anomalous propagation, occurs for values lower than -157 km^{-1}, and in this case the microwave energy may travel for long distances before intercepting ground targets producing anomalous propagation (i.e., anaprop or AP) echoes. In fact the exact threshold for ducting depends on the precise local value of the Earth radius, which means that it is not a constant value (for example varies with latitude) – see Table 1 for a summary of ranges of refractivity and modified refractivity gradients for different propagation conditions. As a reference, the two examples of radar images shown in Fig. 2 were recorded with VRGs of -43 and -112 km^{-1}.

Characteristic	dN/dZ (km^{-1})	dM/dZ (km^{-1})
Subrefraction	$(0,+\infty)$	$[157, +\infty)$
Normal	$(-79,0]$	$(157, 79)$
Superrefraction	$[-79,-157)$	$[79,0)$
Ducting	$[-157, -\infty)$	$[0,-\infty)$

Table 1. Effects upon propagation under different ranges of dN/dZ and dM/dZ (adapted from Bech et al. 2007a).

On the other hand, a careful analysis of the fluctuation of target reflectivity may be a way to monitor variations in atmospheric conditions (changes in moisture content, etc.) as shown by Fabry et al. (1997). Subsequent research from that work triggered new interest in the analysis and characterization of refractivity profiles near ground level – see for example Park & Fabry (2011).

Superrefraction and ducting in particular, is usually associated with temperature inversions or sharp water vapour vertical gradients. During cloudless nights, radiation cooling over land favours the formation of ducts which disappear as soon as the sun heats the soil surface destroying the temperature inversion. This process may be sometimes clearly observed in the daily evolution of clutter echoes, as reported by Moszkowicz et al. (1994) and others.

3. Propapagation condition variability

As radiosoundings have been traditionally the only source of upper air information available on a routine basis, they have been used for years to calculate long term averages of propagation conditions –see, for example, Gossard (1977) or Low and Huddak (1997)–. Since 1997, radiosonde observations have been made in Barcelona to support the operations of the regional government's Subdirectorate of Air Quality and Meteorology, which later became the Meteorological Service of Catalonia.

Results presented below were derived from observations collected from Vaisala RS-80 sondes (from 41.38°N, 2.12°E and 98 m asl) which sampled every 10 s providing much higher vertical resolution than the usual standard operational radiosounding observations. This allowed better characterization of the air refractive index variability and the detection of thinner super refractive layers that may not be detected by standard radiosounding observations but may have significant effects in the propagation of the radar beam. Most results presented in this and the next section, are based on data collected between 1997 and 2002, at 00 and 12 UTC in Barcelona (Bech et al., 1998, 2000, 2002). From the original 2485 radiosoundings available, 86% passed the quality control process (based both in data format and content analysis, adapted from Météo-France, 1997).

3.1 Surface refractivity

Surface refractivity is an important factor in radiometeorology; it appears in the refractivity exponential model and is one of the terms used in the standard computation of the VRG (ITU, 2003). Table 2 shows Barcelona N_s statistics.

00Z MONTH	Mean	St_dev	Min	P25	P50	P75	Max
J	315	8	291	310	315	320	335
F	317	10	293	309	318	325	334
M	316	10	296	310	319	323	334
A	320	10	292	314	322	329	335
M	329	13	294	319	332	339	351
J	341	13	297	334	343	350	366
J	347	15	302	336	351	357	372
A	354	15	303	346	355	364	382
S	344	13	309	338	345	354	371
O	336	14	305	326	335	347	367
N	316	13	286	309	314	321	367
D	313	11	284	305	312	318	339
Total 00Z	330	19	284	315	328	346	382
12Z MONTH	Mean	St_dev	Min	P25	P50	P75	Max
J	312	10	286	305	311	317	340
F	309	11	284	302	310	317	331
M	316	12	292	306	316	324	342
A	313	13	268	306	315	324	336
M	326	13	300	316	328	336	352
J	335	13	285	326	338	344	367
J	341	16	265	332	341	352	388
A	344	16	298	331	345	356	369
S	337	17	300	322	340	350	368
O	328	15	299	316	327	340	359
N	312	12	283	305	311	319	348
D	311	11	278	303	310	318	338
Total 12Z	325	18	265	312	324	339	388
Total 00Z & 12Z	327	19	265	313	325	341	388

Table 2. N_s statistics for Barcelona calculated from 00Z and 12Z data.

It may be noted that nocturnal N_s values were lower than noon values (about 5 N units in the monthly means) and also the existence of a marked seasonal pattern with a peak in August and a minimum in December. This yearly cycle may be explained by examining the behaviour of the magnitudes considered in the computation of refractivity and also by considering separately the dry and wet terms (Fig. 3).

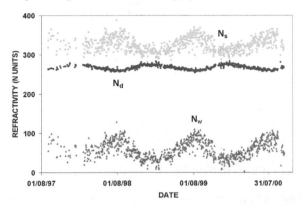

Fig. 3. Evolution of surface refractivity (N_s) and the wet (N_w) and dry terms (N_d) over Barcelona (Bech, 2003).

Monthly variations of these magnitudes show different behaviours. While the temperature follows a very clear seasonal pattern (highs in summer and lows in winter, as expected), in the case of the pressure it is much weaker (approximately winter maxima and summer minima). The humidity, changing constantly throughout the year, exhibits no apparent pattern. These behaviours are reflected in the evolution of N_d and N_w. The first one, proportional to pT^{-1}, is nearly constant with maxima in summer and minima in winter; the second, proportional to eT^{-2}, is much more variable (because of e) but maxima and minima are swapped with respect to N_d (because of T^{-2}). Therefore, N_w, which represents about 30% of N, contributes mostly to its variation: at short scale, it adds variability and also, at monthly scale, modulates the summer maximum and winter minimum cycle which is slightly compensated by the opposite cycle shown by N_d.

Surface refractivity distributions in Barcelona are shown in Fig. 4, exhibiting larger variations at 12 UTC (aprox. 265 – 385) than at 00 UTC.

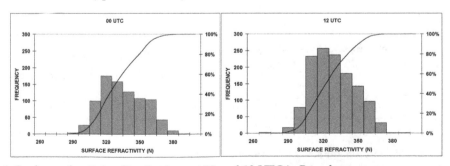

Fig. 4. Surface refractivity distributions at 00 and 12 UTC in Barcelona.

3.2 Vertical refractivity gradient

Vertical refractivity gradient in the first 1000 m (VRG) exhibits, like N_s, lower values for night conditions and a similar seasonal pattern both in the 00 Z & 12 Z data (Fig. 5).

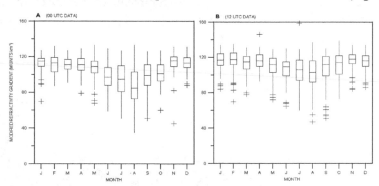

Fig. 5. Box-whisker plots of VRG in Barcelona for 00 Z and 12 Z data.

These box plots show that in summer not only there is a minimum monthly median value (August), but also that the interquartile range (IQR) is increased compared to cold months. Another significant feature is that outliers seldom represent subrefractive events but are quite common for superrefraction; besides, they appear almost at any month, in particular for 12Z data. A similar behaviour is observed using 2 years of radiosonde data recorded at several northern latitude observatories (Fig. 6).

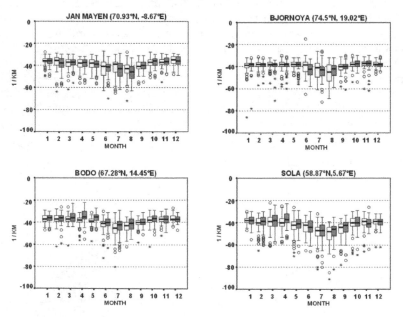

Fig. 6. Box-whisker plots for several Norwegian radiosonde sites showing 00 Z (clear boxes) and 12 Z (dark boxes) data. Adapted from Bech et al. (2007b).

The yearly minima of VRG, below –80 km⁻¹ sometimes reaching –120 km⁻¹ (maximum superrefraction), at the end of the warm season is also appreciated in the VRG time series plot of Barcelona shown in Fig. 7.

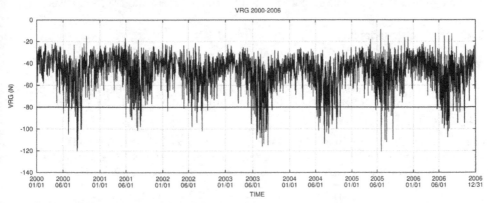

Fig. 7. Time series plot of VRG (N units km⁻¹) for the period 2000-2006 in Barcelona.

The seasonal pattern noted in Barcelona is already indicated in the VRG World Wide maps prepared by the International Telecommunications Union (ITU, 2003). In particular, in August, an area of maximum superrefraction affects the Western Mediterranean region, comparable in intensity to the maximum above the SW Pacific coast of N. America, and somewhat weaker than the Arabian Peninsula –where the world maximum is located for that month–. Using the Historical Electromagnetic Propagation Condition Data Base from the US Naval Systems Ocean Center (Patterson, 1987) a comparison with ten radiosonde stations located in the area was performed. Median monthly values allowed to check similar patterns both in Ns and VRG. A related study was carried out recently by Lopez (2009) using global analysis data from the European Centre for Medium-range Weather Forecasts (ECMWF) to assess the occurrence of superrefraction, or with a similar approach, but at a local scale, by Mentes and Kaymaz (2007) in Turkey, or Mesnard and Sauvageot (2010) in France.

The frequency and cumulative probability distributions for Barcelona VRG are shown in Fig 8. A similar unimodal left skewed pattern, with stepper slopes for higher VRG values (tending to super refraction) is shown for both 00 and 12 Z data. However, modal values are very near the nominal standard propagation value of -40 N units/km (-49 N units/km at night and -42 N/km units at noon).

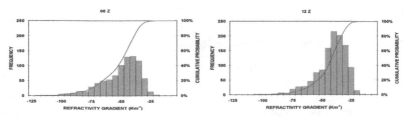

Fig. 8. Frequency and cumulative probability distributions for the Barcelona VRG.

The relationship between surface refractivity and the vertical refractivity gradient for the first kilometre was investigated during the sixties for data collected in the UK (Lane, 1961) and the US (Bean and Dutton, 1968). In both cases a high correlation was found for monthly averages of both magnitudes. For the data set collected in Barcelona, a correlation of 0.9745 was found.

3.3 Anaprop echo variability

Quality control procedures for QPE have traditionally dealt with anaprop and, in general, clutter echoes (see, for example, Anderson et al., 1997; Archibald, 2000; da Silveira and Holt, 1997; Fulton et al., 1998; Joss and Lee, 1995; Kitchen et al., 1994; Sánchez-Diezma et al., 2001, Steiner and Smith, 2002; Szturc et al., in this volume; and Villarini and Krajewski, 2010, among others).

Fornasiero et al. (2006a, 2006b), studied AP echoes occurrence in two radars in the Po Valley, Italy, with a methodology developed by Alberoni et al. (2001). With a three year dataset, they examined the seasonal variability of AP echoes in the diurnal cycle (Fig. 9).

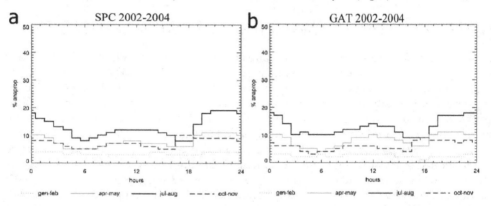

Fig. 9. Mean percentage of anaprop clutter detected. The average is calculated for each hour during the time range 1 January 2002–31 December 2004 for San Pietro Capofiume (a) and for Gattatico radar (b) in the Po Valley, Italy (adapted from Fornasiero et al. 2006a).

They found that in the warm season there were more AP echoes (reaching nearly 20% of the time) with a maximum in the late evening and a secondary maximum at noon, probably associated with local circulations such as sea breeze. In winter the variability was much lower and AP echoes were generally below 5%. These results were helpful to characterize the incidence of AP in precipitation estimates and to design an adequate quality control procedure.

4. Radar beam blockage and propagation conditions

In this section the effect of propagation conditions on beam blockage corrections is described. This type of correction is a classical post-processing step applied to radar reflectivity measurements in order to obtain quantitative precipitation estimates in hilly terrain. A particular implementation of this correction developed during the COST 717 action (Rossa 2000) is described.

4.1 Radar beam blockage

Weather radars installed in complex orographic areas may suffer from partial or total beam blockage caused by surrounding mountains. This effect can restrict seriously the use of the lowest antenna elevation angles which typically provide the most useful information for precipitation estimation at ground level – see for example Joss and Waldvogel (1990), Sauvageot (1994), Collier (1996), or Smith (1998) among others. Therefore, in hilly terrain, beam blockage correction schemes are needed to minimize the effect of topography if quantitative precipitation estimations (QPE) are required. Such corrections are usually included in operational QPE procedures as can be seen in, for example, Crochet (2009), Harrold et al. (1974), Kitchen et al. (1994), Joss and Lee (1995), or Fulton et al. (1998) and may be combined with correction techniques based in the analysis of the 3-D echo structure (Krajewski and Vignal, 2001; or Steiner and Smith, 2002).

The idea that assuming normal propagation conditions for radar observations may not always be a good choice and the use of local climatological refractive data for a specific radar site was already proposed, for example, in the COST 73 Project (Newsome, 1992) and, in a different context, evaluated by Pittman (1999) to improve radar height measurements. In this section the effect of changing the radar beam propagation conditions upon an ordinary single polarization reflectivity blockage correction is described – note that polarimetric radars allow other type of corrections (Giangrande and Ryzhkov 2005; Lang et al. 2009). A simplified interception function is proposed to simulate beam blockage and particular results for the Vallirana weather radar, located at 650 m above sea level near Barcelona (NE Spain) in a complex orography zone are obtained considering real atmospheric propagation conditions.

4.2 Beam blockage simulation

To describe in full detail the interception of the energy transmitted by the radar with the surrounding topography, a precise description of the antenna radiation pattern is required. As this pattern is rather complex, it is common to assume the usual geometric-optics approach and consider that the radar energy is concentrated in the main lobe of the radar antenna pattern (Skolnik, 1980). Then, when a radar beam intercepts a mountain, two situations are possible: 1) only part of the beam cross section illuminates the intercepted topography (partial blockage) or 2) the radar beam is completely blocked (total blockage). The percentage area of the radar beam cross section blocked by topography may be expressed as a function of the radius of the beam cross section, a, and the difference of the average height of the terrain and the centre of the radar beam, y (Fig. 10)

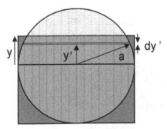

Fig. 10. Elements considered in the radar beam blockage function: a, radius of the radar beam cross section, y, difference between the centre of the radar beam and the topography, dy' differential part of blocked beam section and y' the distance from the center to dy'.

$$PBB = \frac{y\sqrt{a^2 - y^2} + a^2 \arcsin \frac{y}{a} + \frac{\pi}{2} a^2}{\pi \, a^2}. \tag{3}$$

Depending on the relative position of the beam height respect to topography, y may be either positive or negative. According to these definitions, partial beam blockage occurs when $-a < y < a$, total beam blockage means that $y \geq a$, and finally, $y \leq -a$ implies there is no blockage at all. Using the notation introduced above, it can be seen that integrating dy' partial beam blockage, PBB, may be written as an analytical expression (Bech et al. 2003):

On the other hand, the height of the centre of the radar beam, h, is given at a distance r by the expression (see, for example, Doviak and Zrnic, 2006):

$$h = \sqrt{r^2 + \left(k_e R\right)^2 + 2 r \, k_e \, R \sin \theta} - k_e R + H_0 \qquad , \tag{4}$$

where R is the Earth's radius, k_e is the ratio between R and the equivalent Earth's radius, θ the antenna elevation angle and H_0 the antenna height. Information about atmospheric propagation conditions is contained in k_e, which may be written in terms of the refractivity gradient as:

$$k_e = \frac{1}{1 + R\left(\dfrac{dN}{dh}\right)} \tag{5}$$

The usual value for k_e in the first kilometre of the troposphere, assuming the normal VRG value of 40 km[-1], is approximately $4/3$. Substituting (5) and (4) in (3), an expression of the beam blockage in terms of the propagation conditions is obtained (Bech et al. 2003).

Three clutter targets (MNT, LML and MNY), which presented partial beam blockage under normal propagation conditions, were chosen to examine the effects of changing the VRG. The Vallirana radar (41 22' 28" N, 1 52' 52" E) is a C band Doppler system with a 1.3 ° beam width antenna at 3 dB. The targets chosen are normally used to check the radar antenna alignment on a routine basis and are located within the region of interest of radar QPE.

The targets were located at different ranges, had different heights and showed different degrees of blockage, in order to be representative of the topography surrounding the radar. They are located in the so called Pre-coastal Range sharing a similar propagation environment and comparable to that obtained by the Barcelona radiosonde. For example the area considered is usually influenced by a marked sea-breeze circulation pattern, just like the city of Barcelona (Redaño et al., 1991).

4.3 Beam blockage correction

To evaluate the effects of anomalous propagation, the partial beam blocking correction scheme used in the NEXRAD Precipitation Processing System has been considered. This scheme (Fulton et. al, 1998) is applied to radar beams partially shielded. In particular, this type of beam blockage correction is applied to radar pixels (or radar bins) whose shielding ranges between 10% and 60% and it consists of modifying radar equivalent reflectivity factor measurements by adding 1 to 4 dB depending on the degree of occultation. The correction is also applied to all pixels further out in range of the same blocked radar ray, neglecting diffraction below shadow boundary. The correction depends only on the percentage of beam cross section shielded and, in the description provided by Fulton et al. (1998), no specific mention is made about which part of the beam is shielded. This approach allows consideration of a simple interception function, as the one proposed in the previous section, assuming that the correction additive factors contain considerations about interception details such as the beam power distribution. This beam blockage procedure is used with other corrections such as a test on the vertical echo continuity and a sectorized hybrid scan (Shedd et al., 1991). Other approaches to this question with different degrees of sophistication have been used in the past (see for example Delrieu et al. 1995, Gabella and Perona 1998, Michelson et al. 2000, Park et al. 2009). All of them have in common the assumption of standard propagation conditions of the radar beam.

4.4 Refractivity gradient vs beam blockage

The radar beam blockage under a particular VRG can be simulated considering both the observed propagation conditions and the interception function described in the previous sections. This may be achieved by assuming an homogeneous VRG for the whole radar beam and calculating the associated beam blockage for each selected target for a given initial antenna elevation angle.

In Fig. 11 a set of beam blockages vs VRG plots is shown for different antenna elevation angles. The refractivity gradient values considered contain the observed extreme VRG values (-119 km^{-1} and -15 km^{-1}) and are also extended to include pure subrefraction (0 km^{-1}) and almost ducting conditions (-156 km^{-1}) to illustrate their effects. These extreme cases seem realistic taking into account the presence of thin ducting layers that may have high VRG embedded in others with lower VRG and considering the fact that the bending of the ray path is an additive process throughout the whole layer crossed by the radar beam.

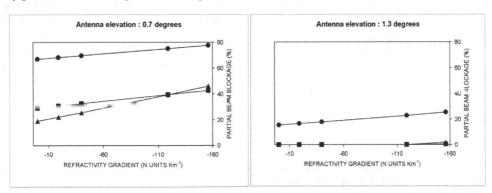

Fig. 11. Simulated beam blockage vs vertical refractivity gradient for targets MNT, (circle), LML (square) and MNY (triangle) at different antenna elevation angles.

As expected, as the antenna angle increases, beam blockage is reduced. For example, for an antenna elevation of 0.7 ° a relatively high beam blockage rate is expected as the lowest part of the main lobe in a 1.3° beam width antenna is pointing to the surrounding hills, producing values of blockage ranging mostly between 30% and 80%. On the other hand, the 1.3° elevation beam blockage values are mostly below 20% and for some targets are always null (no blockage at all) except for the most super refractive situations.

In Fig. 12, target MNT, shows moderate (around 40%) to low (10%) rate of beam blockage, respectively (similar results were obtained for LML). On the other hand, we found that the most distant target, MNY, intercepted the radar beam mostly between 8% and 14%. The range of variations in the beam blockage observed in the above mentioned histograms oscillates from 8% (LML) and 10% (MNT) to 18% (MNY). From the cumulative probability plots obtained it may be noted that MNT and LML show single classes representing more than 50% while a more smoothed distribution is found for MNY.

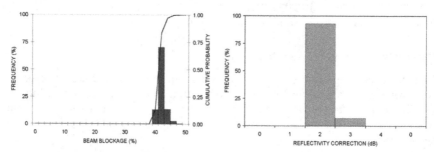

Fig. 12. Simulated beam blockage frequency and cumulative probability distributions (left) and the corresponding correction histograms (right) for 1° antenna elevation for target MNT.

The corresponding correction histogram is also shown. Should the beam blockage correction have been a continuous function, where for a particular value of blockage a different correction factor would be applied, then the spread of the beam blockage histograms would have been reflected in the spread of the correction histograms. However, this is not the case for the particular type of correction considered where only four different correction values are possible depending on the beam blockage. Therefore, a big variability in the beam blockage occurrence does not necessarily produce the same variability in the blockage correction. An additional conclusion of this analysis (Bech et al. 2003) was that errors in beam blockage corrections derived from propagation variability were comparable to antenna pointing errors of 0.1°, which is a typical value for operational systems. This confirms the need for hardware calibration control and monitoring, particularly if quantitative precipitation estimates are required.

4.5 Improved quantitative precipitation estimates

The methodology proposed in the previous section to simulate the radar beam blockage by topography has been implemented to derive correction factors which were applied to improve precipitation estimates. For example Fornasiero et al. (2006b) performed corrections in different events, calculating specific corrections assuming both standard and non-standard propagation conditions and finding some improvement with the corrections. In Bech et al. (2007b, 2010a) results reported were carried out in the framework of the COST-731 action (Rossa et al. 2010) using the so-called BPM model (which implements the blockage function presented above. Larger data sets were considered for blockage corrections under standard conditions and individual ducting events were examined in detail. Here we illustrate some of the results obtained assuming standard propagation conditions.

Figure 13 shows details of Bømlo radar (59.5°N, 5.1°E) from the Norwegian Meteorological Service (met.no). A panorama from the radar site shows some of the hills which block the radar coverage (three of them are numbered). One year of precipitation, illustrating the blocked areas is also shown, as well as the correction factors computed with the BPM model. The improvement in the bias, defined here as 10 times the decimal logarithm of the ratio of gauge to radar derived precipitation amounts, is shown in Table 3. At all ranges the correction reduced the bias.

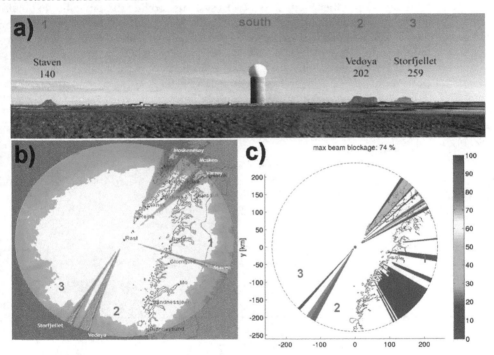

Fig. 13. a). Southern view from the Bømlo radar in Norway; three of the surrounding hills are numbered and indicated on the other panels. b). One year of radar precipitation estimates, illustrating clearly the blocked sectors with less (or no) precipitation. c). Modelled blockage with the BPM system. Figure courtesy of Dr. Uta Gjertsen (met.no).

Blockage	Range (km)							
(%)	40-100			100–160			160–240	
0	2.3		(12)	5.2		(19)	11.2	(16)
1–50	4.2	**3.1**	(16)	9.3	**8.5**	(26)	15.0	**14.1**	(70)
50–70	8.6	**6.0**	(5)	14.4	**11.8**	(15)	21.4	**18.2**	(29)

Table 3. Bias (dB) of uncorrected and blockage-corrected (bold) radar estimates from the Bømlo radar for 2004 grouped in different ranges. Sample size is in parentheses. Adapted from Bech et al. (2007b).

5. Radar propagation condition forecasting

This section deals with anomalous propagation forecasting using mesoscale numerical weather prediction models. It is illustrated with several examples, discussing capabilities and limitations found in this application.

5.1 VRG forecasts

Anticipating the occurrence of AP may be an advantage for monitoring purposes of radar quality control or to obtain a deeper understanding of processes related to anomalous propagation. Numerical Weather Prediction (NWP) systems provide the capability to obtain forecasts of propagation conditions from temperature and humidity forecast profiles in a similar way as they are obtained from radiosonde observations. Despite NWP systems allow to study anomalous propagation events with more spatial detail than that given by the synoptic radiosonde network, they have a number of accuracy limitations that may hamper the operational production of AP forecasts. For example Bech et al. (2007a) compared 4 months of vertical refractivity gradient forecasts over Barcelona retrieved from numerical model output of the MASS system (Codina et al. 1997a, 1997b; Koch et al. 1985) with actual radiosonde observations and found a systematic bias of the model towards subrefraction (Fig. 14).

In order to reduce the bias, a simple heuristic approach was suggested combining linearly model output and previous radiosonde observations. As illustrated in the Taylor diagram (Taylor, 2001) shown in Fig. 15, the modified forecasts, labelled here as H2b, H4b, H6b and H8b, produced better results in terms of RMS and correlation compared to the original forecasts (MASS).

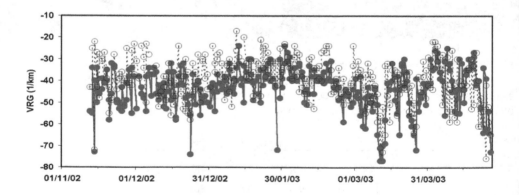

Fig. 14. Time series of Vertical Refractivity Gradient (VRG) over Barcelona from NWP–derived forecasts (dashed line) and radiosonde–based diagnostics (solid line).

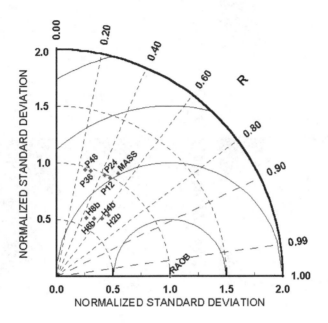

Fig. 15. Taylor Diagram of Vertical Refractivity Gradient VRG radiosonde observations (RAOB), original MASS forecasts, persistence of the observations and modified forecasts.

5.2 AP case studies

A number of anomalous propagation case studies examined with an electromagnetic propagation model with different degrees of sophistication and NWP data or simply with a radiosonde profile can be found in the literature, covering different geographic areas, such as Burk and Thompson (1997) in California, Atkinson et al (2001) over the Persian Gulf, or Bebbington et al. (2007) in the Mediterranean. Applications of this type of modelling tool include radar coverage computation (Haase et al. 2006), or even correction of improvement of radar data in NWP assimilation systems (Haase et al. 2007).

Fig. 16 shows an example of AP case study for the Røst radar (met.no), where NWP data provided by the HIRLAM system provided better results, even 24 h forecasts, than actual radiosonde data, which in this case was not representative of the radar coverage environment. In Bech et al. (2007b) this and two other case studies were discussed, highlighting the quality of HIRLAM forecasts for examining and anticipating AP cases with the BPM model.

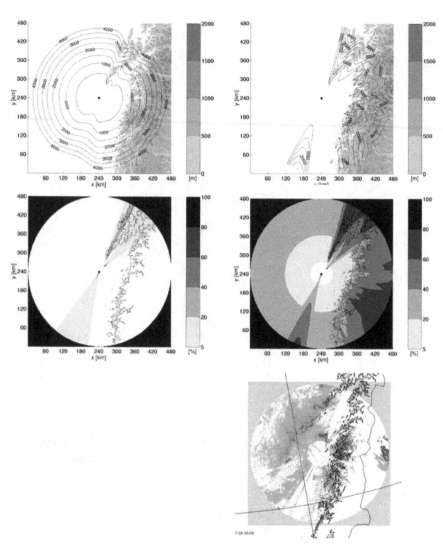

Fig. 16. Lowest unblocked radar coverage (top row) and beam blockage (middle) computed with radiosonde data (left column) and NWP-derived profiles (right column). The bottom panel shows actual radar observations, 6 July 2005 00 UTC (Røst radar, met.no). Adapted from Bech et al. (2007a).

6. Detection and correction of AP echoes with satellite data

Several studies have been reported regarding the use of satellite images to detect AP echoes, based on the simple approach of removing echoes in cloudless conditions. However, in practice this procedure is not as straight forward as might seem and requires substantial fine tuning to obtain a reasonable balance between false alarms and detection, particularly in cloudy, and most importantly, rainy conditions. Some correction procedures to remove non-precipitating echoes rely only on radar data (e.g. Berenguer et al. 2006, Sánchez-Diezma et al. 2001, Steiner & Smith, 2002) but others consider as well the use of satellite observations – see for example Michelson and Sunhede (2004), Bøvith et al. (2006) or Magaldi et al. (2009). In any case, quantitative applications of radar data such as thunderstorm tracking (Rigo et al., 2010), precipitation estimates (Trapero et al. 2009), or radar-based precipitation forecasts (Atencia et al., 2010), or even qualitative use of radar images by a non-specialized audience (as discussed in Bech et al. 2010b), clearly require the use of proper clutter filtering, particularly considering anomalous propagation.

6.1 Methodology

We summarize in this section the methodology proposed by Magaldi et al. (2009) to detect and remove AP echoes in radar images using satellite observations and NWP model data. They took advantage of the improved temporal and spatial resolution of the Meteosat Second Generation (MSG) satellite to update the procedure developed by Michelson and Sunhede (2004), based on the first generation of Meteosat satellites, and incorporated the use of enhanced precipitating cloud masks. Fig. 17 illustrates the basic idea behind the proposed methodology, showing a radar reflectivity image with real precipitation and clutter (in this case sea clutter, near the coast), the precipitating cloud mask associated, and the new image where clutter has been removed.

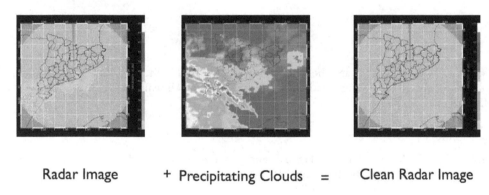

Radar Image   + Precipitating Clouds   =   Clean Radar Image

Fig. 17. Illustration of the correction procedure of radar reflectivity echoes affected by clutter (left panel) with a precipitating cloud mask (centre panel) and the resulting cleaned radar image (Vallirana radar, 1 January 2004 14 UTC).

The basic algorithm is shown on Fig. 18, where a data flow diagram showing the different processes involved is displayed. Analysis of radio propagation conditions with radiosonde (RAOB) data (vertical refractivity gradients below -80 km^{-1} or ducting index above 20) was used to select AP events. For those events, MSG satellite and NWP MASS model data were used to build precipitating cloud masks based on the SAF (SAF 2004, 2007; hereafter S) and Michelson and Sunhede (2004) algorithms (hereafter M). These masks were compared pixel by pixel with radar data, and non-precipitating pixels were removed in the final corrected radar data.

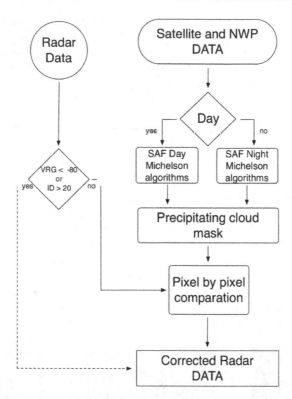

Fig. 18. Flow diagram showing the main processes involved in the algorithm to detect radar AP echoes with precipitating cloud masks derived from satellite images. Adapted from Magaldi et al. (2009).

6.2 Results

Using the SMC Vallirana radar (Fig. 17) and a network of 155 raingauges and manually edited radar data as verification data sets, Magaldi et al. (2009) tested the performance of this procedure for several case studies, considering the original uncorrected data (UC), and data corrected with the M and S algorithms, all compared against manually corrected data. They obtained statistics considering Percentage Correct (PC), False Alarm Rate (FAR), and Hanssen-Kuipers skill (HKS) scores - see Wilks (1995) for details. The HKS suggested that S performed better, despite for strong echoes M yielded lower false alarms (Table 4).

Echo class	Mean sample	FAR			PC			HKS		
		UC	M	S	UC	M	S	UC	M	S
Weak	883713	0.34	0.28	0.21	78.12	92.65	96.59	0.80	0.74	0.96
Strong	769162	0.25	0.07	0.37	73.12	94.21	95.81	0.88	0.87	0.80
All	929055	0.30	0.22	0.27	75.55	93.16	96.35	0.83	0.78	0.90

Table 4. Verification scores for different echo intensities (strong echoes are higher than 15 dBZ; weak, the rest).

A larger data set of six months (January to July 2007) using the SMC Vallirana radar provided additional insight to the performance of this correction technique of AP echoes. It also allowed to evaluate the performance of the technique applied by Bøvith et al. (2006) which made use of cloud type information (SAF, 2011) as precipitating echo mask. A parallax correction (Vicente et al., 2002) was introduced in the mask in order to improve the matching between the two data types and, as seen in Table 5, substantial changes were found for some of the cloud types.

Uncorrected   Corrected

Cloud type	Precipitation frequency(%)
01.- Cloud free land	0.64
02.- Cloud free sea	0.62
03.- Snow/ice land	0.41
06.-Very low Cu.	0.77
08.- Low St.	3.60
10.- Medium St.	15.30
12.-High & opaque St.	29.18
14.-Very High & opaque St.	31.90
15.- Thin Ci.	0.79
16.-Moderate thick Ci.	1.53
17. Ci. above lower cloud	6.80
19.- Fractional cloud	1.40
20.-Holes	0.00

Cloud type	Precipitation frequency(%)
01.- Cloud free land	0.30
02.- Cloud free sea	0.40
03.- Snow/ice land	0.40
06.-Very low Cu.	0.60
08.- Low St.	3.48
10.- Medium St.	16.38
12.-High & opaque St.	30.85
14.-Very High & opaque St.	32.76
15.- Thin Ci.	0.99
16.-Moderate thick Ci.	1.68
17. Ci. above lower cloud	6.62
19.- Fractional cloud	0.63
20.-Holes	0.85

Table 5. Precipitation frequency for Cloud type product using the parallax corrected (left) and uncorrected products (right), both generated using six months of SMC radar data (January-July 2007).

7. Final remarks

In this chapter, an overview of the effects of radio propagation conditions upon radar observations has been given. Though we have focused in ground-based weather radar systems, many of the concepts presented apply as well to other types of radar and applications. Particular emphasis has been given to aspects with potential impact on radar quantitative precipitation estimates, considering beam blockage corrections or anomalous propagation echoes detection and removal. These items should be considered in quality control for weather radars, particularly those operating in complex topography environments and located near the coast where anomalous propagation may affect dramatically radar observations.

8. Acknowledgments

Part of the results presented here were obtained by the authors in research projects made in collaboration with a number of individuals including Uta Gjertsen (formerly at met.no and now at the Norwegian Defence Research Establishment, Norway), Günther Haase and Anke

Toss (Swedish Meteorological and Hydrological Institute, SMHI, Sweden), Pier Paolo Alberoni and Anna Fornasiero (ARPA, Italy), and David Bebbington (Essex University, UK). Weather radar observations, and radiosonde and surface automated observations data were provided by ARPA, SMHI, met.no and the Meteorological Service of Catalonia (SMC). This chapter was produced in the framework of the ProFEWS project (CGL2010-15892) and the Hymex project (Hymex.es CGL2010-11757-E).

9. References

Alberoni, P. P.; Anderson, T.; Mezzasalma, P.; Michelson, D. B. & Nanni, S. (2001). Use of the vertical reflectivity profile for identification of anomalous propagation, *Meteorological Applications*, Vol. 8, pp. 257–266

Anderson, T.; Alberoni, P. P.; Mezzalsama, P.; Michelson, D. & Nanni, S. (1997). Anomalous propagation identification from terrain and sea waves using vertical reflectivity profile analysis, *Proceedings of 28th International Conference On Radar Meteorology*, pp. 93-94, American Meteorologial Society, Austin, Texas, USA

Archibald, E. (2000). Enhanced clutter processing for the UK weather radar network. *Physics and Chemistry of the Earth. Part B-Hydrology, Oceans and Atmosphere*, Vol. 25, No.10-12, pp. 823-828

Atencia, A.; Rigo, T.; Sairouni, A.; Moré, J.; Bech, J.; Vilaclara, E.; Cunillera, J.; Llasat, M.C. & Garrote, L. (2010). Improving QPF by blending techniques at the Meteorological Service of Catalonia. *Natural Hazards and Earth System Science*, Vol.10, No.7, pp. 1443-1455

Atkinson, B.W.; Li, J.G. & Plant, R.S. (2001). Numerical modelling of the propagation environment in the atmospheric boundary layer over the Persian Gulf. *Journal of Applied Meteorology*, Vol.40, pp. 586–603

Battan, L. J. (1973). *Radar Observation of the atmosphere*. University of Chicago Press, Chicago, USA, 324 pp.

Bean, B. R. & Dutton, E. J. (1968). *Radio meteorology*. 435 pp., Dover Publications, New York USA

Bebbington, D.; Rae, S.; Bech, J.; Codina B. & Picanyol, M. (2007). Modelling of weather radar echoes from anomalous propagation using a hybrid parabolic equation method and NWP model data. *Natural Hazards and Earth System Science*, Vol.7, No.3, pp. 391-398

Bech, J. (2003). *Observational analysis and numerical modelling of weather radar anomalous propagation echoes*. PhD Thesis, ISBN 8468854506, University of Barcelona, Barcelona, Spain

Bech, J.; Bebbington, D.; Codina, B., Sairouni, A. & Lorente, J. (1998). Evaluation of atmospheric anomalous propagation conditions: an application for weather radars. *EUROPTO Conference on Remote Sensing for Agriculture, Ecosystems, and Hydrology.* SPIE Vol. 3499, pp. 111-115, Barcelona, Spain.

Bech, J.; Codina, B. & Lorente, J. (2007a). Forecasting weather radar propagation conditions. *Meteorology and Atmospheric Physics*, Vol.96, No.3-4, pp. 229-243

Bech, J.; Codina, B.; Lorente, J. & Bebbington, D. (2002). Monthly and daily variations of radar anomalous propagation conditions: How "normal" is normal propagation?". *Proceedings 2nd European Conference on Radar Meteorology*, pp. 35–39. Copernicus GmbH © 2002, Delft, Netherlands

Bech, J.; Codina, B.; Lorente, J. & Bebbington, D. (2003). The sensitivity of single polarization weather radar beam blockage correction to variability in the vertical refractivity gradient. *Journal of Atmospheric and Oceanic Technology*, Vol. 20, No.6, pp. 845-855

Bech, J.; Gjertsen, U. & Haase, G. (2007b). Modelling weather radar beam propagation and topographical blockage at northern high latitudes. *Quarterly Journal of the Royal Meteorological Society*, Vol.133, No.626, pp. 1191–1204

Bech, J.; Gjertsen, U. & Haase, G. (2010a). Reply to comment of J.I. Dahl on DOI: 10.1002/qj.98. *Quarterly Journal of the Royal Meteorological Society*, Vol.136, No.648, pp. 817–818

Bech, J.; Molina, T.; Vilaclara, E. & Lorente, J. (2010b). Improving TV weather broadcasts with technological advancements. *Meteorological Applications*, Vol.17, No.2, pp. 142-148

Bech, J.; Sairouni, A.; Codina, B.; Lorente, J. & Bebbington, D. (2000). Weather radar anaprop conditions at a Mediterranean coastal site. *Physics and Chemistry of the Earth. Part B-Hydrology, Oceans and Atmosphere*, Vol.25, No.10-12, pp. 829-832

Berenguer, M.; Sempere-Torres, D.; Corral, D. & Sánchez-Diezma, R. (2006). A Fuzzy Logic Technique for Identifying Nonprecipitating Echoes in Radar Scans. *Journal of Atmospheric and Oceanic Technology*, Vol.23, No.9, pp. 1157-1180

Bøvith, T; Gill, R.S.; Overgaard, S; Nielsen, A.A. & Hansen, L.K. (2006). Detecting weather radar clutter using satellite-based nowcasting products. Proc fourth European conference on radar meteorology (ERAD), pp 153–156, Barcelona, Spain

Burk, S.D. & Thompson, W.T. (1997). Mesoscale modelling of summertime refractive conditions in the southern California bight. *Journal of Applied Meteorology*, Vol. 36, pp. 22-31

Codina, B.; Aran, M.; Young, S. & Redaño, A. (1997a). Prediction of a mesoscale convective system over Catalonia (Northeastern Spain) with a nested numerical model. *Meteorology and Atmospheric Physics*, Vol.62, pp. 9-22

Codina, B.; Sairouni, A.; Bech, J. & Redaño, A. (1997b). Operational application of a nested mesoscale numerical model in Catalonia (Meteo'96 Project). *Proceedings of the INM/WMO International Symposium of Cyclones and Hazardous Weather in the Mediterranean*, pp. 657-667, ISBN 84-7632-329-8, Palma de Mallorca, Spain

Collier, C.G. (1996). *Applications of Weather Radar Systems. A Guide to Uses of Radar Data in Meteorology and Hydrology*. pp 390, 2d ed. John Wiley & Sons., ISBN 0471960136, Chichester, UK

Crochet, P. (2009) Enhancing radar estimates of precipitation over complex terrain using information derived from an orographic precipitation model. *Journal of Hydrology*, 377:3-4, 417-433

da Silveira, R. B. & Holt, A.R. (1997). A neural network application to discriminate between clutter and precipitation using polarisation information as feature space. *Proceedings of the 28th Internat. Conf. on Radar Meteor.*, pp. 57-58, Amer. Meteor. Soc., Austin, Texas

Delrieu, G.; Creutin, J.D. & Andrieu, H. (1995). Simulation of radar mountain returns using a digitized terrain model. *Journal of Atmospheric and Oceanic Technology*, Vol.12, pp.

Doviak, R.J. & Zrnic, D.S. (2006). *Doppler radar and weather observations*, 2nd edition. Academic Press, ISBN 0122214226, London, UK

Fabry, F.; Frush, C.; Zawadki, I. & Kilambi, A. (1997). On the extraction of near-surface index of refraction using radar phase measurements from ground targets. *Journal of Atmospheric and Oceanic Technology*, Vol.14, No.4, pp. 978-897

Fornasiero, A.; Alberoni, P.P. & Bech, J. (2006a). Statistical analysis and modelling of weather radar beam propagation conditions in the Po Valley (Italy). *Natural Hazards and Earth System Science*, Vol.6, No.2, pp. 303-314

Fornasiero, A.; Bech, J. & Alberoni, P.P. (2006b). Enhanced radar precipitation estimates using a combined clutter and beam blockage correction technique. *Natural Hazards and Earth System Science*, Vol.6, No.5, pp. 697-710

Fulton, R.A.; Breidenbach, J.P.; Seo, D.; Miller, D. & O'Bannon, T. (1998). The WSR-88D Rainfall Algorithm. *Weather and Forecasting*, Vol.13, No.2, 377-395

Gabella, M. & Perona, G. (1998). Simulation of the orographic influence on weather radar using a geometric-optics approach. *Journal of Atmospheric and Oceanic Technology*, Vol.15, No.6, pp. 1486-1495

Giangrande, S.E. & Ryzhkov, A.V. (2005). Calibration of Dual-Polarization Radar in the Presence of Partial Beam Blockage. *Journal of Atmospheric and Oceanic Technology*, Vol.22, No.8, pp. 1156-1166

Gossard, E.E. (1977). Refractive index variance and its height distribution in different air masses. *Radio Science*, Vol. 12, No.1, pp. 89-105

Haase, G.; Bech, J.; Wattrelot, E.; Gjertsen, U.; Jurasek, M. (2007). *Towards the assimilation of radar reflectivities: improving the observation operator by applying beam blockage information*, 33rd Conference on Radar Meteorology, AMS, Cairns, Australia.

Haase, G.; Gjertsen, U.; Bech, J.; Granström, Å. (2006). *Assessment of potential radar locations using a beam propagation model.* 4th European Conf. Radar Meteorol. Hydrol., Barcelona, Spain

Harrold, T.; English, E. & Nicholass, C. (1974). The accuracy of radar-derived rainfall measurements in hilly terrain. *Quarterly Journal of the Royal Meteorological Society*, Vol. 100, No.425, pp. 331-350

ITU (2003). ITU-R P.453-9 Recommendation, *The Radio Refractive Index: Its Formula and Refractivity data*, ITU Radiocommunication Assembly, ITU-R P-Series, 2003)

Johnson, C.; Harrison, D. & Golding, B. (1999). *Use of atmospheric profile information in the identification of anaprop in weather radar images.* Observation Based Products Technical Report No. 17, Forecasting Systems, UK Meteorological Office, 30 pp. [Available from the National Meteorological Library, London Road, Bracknell, RG12, 2SZ, UK]

Joss, J. & Lee, R. (1995). The application of radar-gauge comparisons to operational precipitation profile corrections. *Journal of Applied Meteorology*, Vol.4, No.12, pp. 2612-2630

Joss, J. & Waldvogel, A. (1990). *Precipitation measurement and hydrology, a review.* In: Radar in Meteorology, D. Atlas (Ed.), Chapter 29a, pp. 577-606. American Meteorological Society, Boston, USA

Kitchen, M.; Brown, R. & Davies, A.G. (1994). Real-time correction of weather radar data for the effects of bright band, range and orographic growth in widespread precipitation. *Quarterly Journal of the Royal Meteorological Society*, Vol.120, No.519, pp. 1231-1254

Koch, S.E.; Skillman, W.C.; Kocin, P.J.; Wetzel, P.J.; Brill, K.F.; Keyser, D.A. & McCumber, M.C. (1985). Synoptic scale forecast skill and systematic errors in the MASS 2.0 model. *Monthly Weather Review*, Vol.113, No. 10, pp. 1714-1737

Krajewski, W. F. & Vignal, B. (2001). Evaluation of anomalous propagation echo detection in WSR-88D Data: a large sample case study. *Journal of Atmospheric and Oceanic Technology*, Vol. 18, No.5, pp. 807-814

Lane, J.A (1961). The radio refractive index gradient over the British Isles. *Journal of Atmospheric and Terrestrial Physics*, Vol.21, No.2-3, pp. 157-166

Lang, T.J.; Nesbitt, S.W. & Carey, L.D. (2009). On the Correction of Partial Beam Blockage in Polarimetric Radar Data. *Journal of Atmospheric and Oceanic Technology*, Vol.26, No.5, pp. 943-957

Lopez, P. (2009). A 5-yr 40-km-Resolution Global Climatology of Superrefraction for Ground-Based Weather Radars. *Journal of Applied Meteorology and Climatology*, Vol. 48, pp. 89–110

Low, T.B. & Hudak, D.R. (1997). Development of Air Mass Climatology Analysis for the Determination of Characteristic Marine Atmospheres. Part I: North Atlantic. *Theoretical and Applied Climatology*, Vol.57, No.3-4, pp. 135-153

Magaldi, A.V.; Bech, J. & Lorente, J. (2009). A multisource scheme based on NWP and MSG data to correct non-precipitating weather radar echoes. *Meteorology and Atmospheric Physics*, Vol.105, No.3-4, pp. 121-132

Mentes, Ş., & Kaymaz, Z. (2007). Investigation of Surface Duct Conditions over Istanbul, Turkey. *Journal of Applied Meteorology and Climatology*, Vol. 46, pp. 318–337

Mesnard, F., Sauvageot, H. (2010). Climatology of Anomalous Propagation Radar Echoes in a Coastal Area. *Journal of Applied Meteorology and Climatology*, Vol. 49, pp. 2285–230

Météo-France (1997). *Quality control on GTS data at Météo-France*. Météo-France, Service Centrale d'Exploitation de la Météorologie. 42, Av. Coriolis, 31057 Toulouse Cedex 1, France

Michelson, D. B.; Andersson, T.; Koistinnen, J.; Collier, C. G.; Riedl, J.; Szturc, J. ; Gjertsen, U.; Nielsen, A. & Overgaard, S. (2000). *BALTEX radar data centre products and their methodologies*. RMK 90. Swedish Meteorological and Hydrological Institute, Norrköpping, Sweden

Michelson, D.B. & Sunhede, D. (2004). Spurious weather radar echo identification and removal using multisource temperature information. *Meteorological Applications*, Vol.11, pp. 1–14

Moszkowicz, S.; Ciach, G.J. & Krajewski, W.F. (1994). Statistical detection of anomalous propagation in radar reflectivity patterns. *Journal of Atmospheric and Oceanic Technology*, Vol.11, No.4, pp. 1026-1034

Newsome, D.H. (1992). *Weather Radar Networking COST Project 73 Final Report*. Kluwer Academic Publishers, Dordrecht, Netherlands, 254 pp., ISBN 0792319397

Park, S. & Fabry, F. (2011). Estimation of Near-Ground Propagation Conditions Using Radar Ground Echo Coverage. *Journal of Atmospheric and Oceanic Technology*, Vol.28, No.2, pp. 165-180

Park, S.; Jung, S.; Lee, J. & Kim, K. (2009) Correction of Radar Reflectivity over Beam Blocking Area by Accumulated Radar Reflectivity. *Journal of Korea Water Resources Association*, Vol.42, No.8, pp. 607-617

Patterson, W.L. (1987). *Historical Electromagnetic Propagation Condition Database Description*. Technical Document 1149, ADA-A189 157, NOSC 1149, 71 pp., US NAVY, USA

Patterson, W.L. (2008). The Propagation Factor, F_p, in the radar equation, In: *Radar Handbook*, 3rd edition, M. Skolnik (Ed.), Ch. 26, ISBN 0071485473. McGraw Hill, New York, USA.

Pittman, T.S. (1999). *A climatology-based model for long-term prediction of radar beam refraction*. Master's Thesis, 184 pp., US Air Force Institute of Technology, USA

Redaño, A.; Cruz, J. & Lorente, J. (1991). Main features of sea breeze in Barcelona. *Meteorology and Atmospheric Physics*, Vol.46, No.3-4, pp. 175-179

Rigo, T.; Pineda, N. & Bech, J. (2010). Analysis of warm season thunderstorms using an object-oriented tracking method based on radar and total lightning data. *Natural Hazards and Earth System Science*, Vol.10, No.9, pp. 1881-1893

Rinehart, R. (2001). *Radar for Meteorologists*. 3rd edition. Rinehart Publications, ISBN 0-9658002-0-2, 428 pp., P.O. Box 30800, MO, USA

Rossa, A.; Haase, G; Keil, C.; Alberoni,P.P; Ballard, S.; Bech, J.; Germann, U.; Pfeifer, M. & Salonen, K. (2010). Propagation of uncertainty from observing systems into NWP. *Atmospheric Science Letters*, Vol.11, No.2, pp. 145-152

Rossa, A.M. (2000). The COST 717 action: use of radar observations in hydrological and NWP models. *Physics and Chemistry of the Earth. Part B-Hydrology, Oceans and Atmosphere*, Vol.25, No.10-12, pp. 1221–1224

SAF (2004) Software user manual for PGE04 of the NWCSAF/MSG: scientific part. EUMETSAT Satellite Application Facility to Nowcasting and Very Short Range Forecasting, 31 pp

SAF (2007) Validation report for "Precipitating Clouds" (PC-PGE04 v1.4). EUMETSAT Satellite Application Facility to Nowcasting and Very Short Range Forecasting, 29 pp

SAF (2011) Algorithm Theoretical Basis Document for "Cloud Products" (CMa-PGE01v3.1, CT-PGE02 v2.1 & CTTH-PGE03 v2.2). EUMETSAT Satellite Application Facility to Nowcasting and Very Short Range Forecasting, 87 pp

Sánchez-Diezma, R.; Sempere-Torres, D.; Delrieu, G. & Zawadki, I. (2001). An improved methodology for ground clutter substitution based on a pre-classification of precipitation types. *Proceedings of the 30th Internat. Conf. on Radar Meteor.*, pp. 271-273, Amer. Meteor. Soc., Münich, Germany

Sauvageot, H. (1991). *Radar Meteorology*, Artech House, ISBN 978-0-89006-318-7, 366 pp., London, UK

Sauvageot, H. (1994). Rainfall measurement by radar: a review. *Atmospheric Research*, Vol.35, No.1 , pp. 27-54

Shedd, R.; Smith, J. & Walton, M. (1991). Sectorized hybrid scan strategy of the NEXRAD precipitation-processing system. In *Hydrological Applications of Weather Radar*, I. Cluckie & C. Collier, Eds., pp. 151-159, ISBN 0134414780, Ellis Horwood Limited, New York, USA

Skolnik, M. (1980). *Introduction to radar systems*. 581 pp., ISBN 0070665729, McGraw-Hill, New York, USA

Smith, P.L., Jr. (1998). On the minimum useful elevation angle for weather surveillance radar scans. *Journal of Atmospheric Oceanic Technology*, Vol.15, No.3 , pp. 841-843

Steiner, M. & Smith, J.A. (2002). Use of three-dimensional reflectivity structure for automated detection and removal of non-precipitating echoes in radar data. *Journal of Atmospheric and Oceanic Technology*, Vol.19, No.5, pp. 673-686

Taylor, K.E. (2001). Summarizing multiple aspects of model performance in single diagram. *Journal of Geophysical Research*, Vol. 106, D7, pp. 7183-7192.

Trapero, L.; Bech, J.; Rigo, T.; Pineda, N. & Forcadell, D. (2009). Uncertainty of precipitation estimates in convective events by the Meteorological Service of Catalonia radar network. *Atmospheric Research*, Vol. 93, No.1-3, pp. 408-418

Vicente, G.; Davenport, J C. & Scofield, R. A. (2002). The role of orographic and parallax corrections on real time high resolution satellite rainfall rate distribution. *International Journal of Remote Sensing*, Vol.23, pp.221–230

Villarini, G. & Krajewski, W.F. (2010). Review of the Different Sources of Uncertainty in Single Polarization Radar-Based Estimates of Rainfall. *Surveys in Geophysics*, Vol.31, No.1, pp. 107-129, ISBN 0169-3298

Wilks, D.S. (1995). *Statistical Methods in the Atmospheric Sciences*. Academic Press, ISBN , 467 pp., London, UK

Part 3

Advanced Techniques for Probing the Ionosphere

8

Incoherent Scatter Radar — Spectral Signal Model and Ionospheric Applications

Erhan Kudeki[1] and Marco Milla[2]
[1]*University of Illinois at Urbana-Champaign*
[2]*Jicamarca Radio Observatory, Lima*
Peru

1. Introduction

Doppler radars find a widespread use in the estimation of the velocity of discrete *hard-targets* as described elsewhere in this volume. In case of *soft-targets* — collections of vast numbers of weakly scattering elements filling the radar beam — the emphasis typically shifts to collecting the *statistics* of random motions of the scattering elements — i.e., Doppler spectral estimation — from which thermal or turbulent state of the target can be inferred, as appropriate. For instance, in case of a plasma in thermal equilibrium, e.g., the quiescent *ionosphere*, a Doppler radar of sufficient power-aperture-product can detect, in addition to the plasma drift velocities, the densities, temperatures, and even current densities of charged particle populations of the probed plasma — such Doppler radars used in ionospheric research are known as *incoherent scatter radars* (ISR). In this chapter we will provide a simplified description of ISR spectral theories (e.g., Kudeki & Milla, 2011) and also discuss magnetoionic propagation effects pertinent to ionospheric applications of ISR's at low latitudes. A second chapter in this volume focusing on in-beam imaging of soft-targets by Hysell & Chau (2012) is pertinent to non-equilibrium plasmas and complements the topics covered in this article.

The chapter is organized as follows: The working principles of ISR's and the general theory of incoherent scatter spectrum are described in Sections 2 and 3. ISR spectral features in unmagnetized and magnetized plasmas are examined in Sections 4 and 5, respectively. Coulomb collision process operating in magnetized ionosphere is described in Section 6. Effects of Coulomb collisions on particle trajectories and ISR spectra are discussed in Sections 7 and 8. Finally, Section 9 discusses the magnetoionic propagation effects on incoherent scattered radar signals. The chapter ends with a brief summary in Section 10.

2. Working principles of ISR's

The basic physical mechanism underlying the operation of ISR's is *Thomson scattering* of elecromagnetic waves by ionospheric free electrons. Thomson scattering refers to the fact that free electrons brought into oscillatory motions by incident radar pulses will re-radiate like Hertzian dipoles at the frequency of the incident field. The total power of scattered fields in an ISR experiment is a resultant of interference effects between re-radiated field components arriving from free electrons occupying the radar field of view. Furthermore the frequency spectrum of incoherent scatter signal is shaped by the same interference effects in addition to the distribution of random velocities of the electrons in the radar frame of reference in accordance with a two-way Doppler effect.

The "incoherent scatter" concept refers, in essence, to a scattering scenario where each of the Thomson scattering electrons would have statistically independent random motions. The total scattered power would then be reduced to a simple sum (see below) of the return power of individual electrons in the radar field of view treated as hard targets in terms of a standard *radar equation*, i.e.,

$$P_r = \frac{P_t G_t A_r}{(4\pi r^2)^2} \sigma_e,$$ (1)

with *transmitted power* and *gain* P_t and G_t, respectively, effective area A_r of the receiving antenna, radar range r, and backscatter *radar-cross-section* (RCS) of an individual electron, $\sigma_e \equiv 4\pi r_e^2$, where $r_e = e^2(4\pi\epsilon_o mc^2)^{-1} \approx 2.181 \times 10^{-15}$ m is the *classical electron radius*.

Ionospheric electron motions are not fully independent — i.e., particle trajectories are partially correlated — however, and, as a consequence, the scattered radar power from the ionosphere deviates form such a simple sum in a manner that depends on several factors including the radar frequency, electron and ion temperatures, as well as ambient magnetic field of the ionospheric plasma. This deviation is just one of many manifestations of the correlations — also known as "collective effects" — between ionospheric charge carriers, including the deviation of the Doppler frequency spectrum of the scattered fields from a simple Gaussian shape (of thermal velocity distribution of electrons) implied by the ideal incoherent scatter scenario. It turns out that the "complications" introduced by the collective effects in the Doppler spectrum of this "not-exactly-incoherent-scatter" from the ionosphere amount to a wealth of information that can be extracted from the ISR spectrum given its proper forward model. This model will be described in the following sections.

Historical note: When ISR's were first proposed (Gordon, 1958), it was expected that ionospheric scattering from free electrons would be fully incoherent. First ISR measurements (Bowles, 1958) showed that not to be the case. Realistic spectral models compatible with the measurements and correlated particle motions were developed subsequently. Rapid theoretical progress took place in the 1960's, but issues related to ISR response at small magnetic aspect angles were resolved only very recently (e.g., Milla & Kudeki, 2011) as explained in Section 8.

3. From Thomson scatter to the general formulation of ISR spectrum

Since oscillating free electrons radiate like Hertzian dipoles, it can be shown, using elementary antenna theory, that the backscattered field amplitude[1] from an electron at a distance r to a radar antenna is (using phasor notation)

$$E_s = -\frac{r_e}{r}e^{-jk_o r}E_i = -\frac{r_e}{r}E_o e^{-j2k_o r},$$ (2)

where $E_i = E_o e^{-jk_o r}$ is the incident field phasor and $k_o = \omega_0/c$ is the wavenumber of the incident wave with a carrier frequency ω_0. It follows that a collection of scattering electrons

---

[1] Since transmitted and scattered fields are co-polarized we can avoid using a vector notation here.

filling a small radar volume ΔV will produce a scattered field[2]

$$E_s = -\sum_{p=1}^{N_0\Delta V} \frac{r_e}{r_p} E_{op} e^{-j2k_o r_p} \approx -\frac{r_e}{r} E_0 \sum_{p=1}^{N_0\Delta V} e^{j\mathbf{k}\cdot\mathbf{r}_p}. \tag{3}$$

Here N_0 is the mean density of free electrons within ΔV and the rightmost expression amounts to invoking a plane wave approximation[3] of the incident and scattered fields in terms of scatterer position vector $\mathbf{r}_p$ and a Bragg wave vector $\mathbf{k} = -2k_o\hat{r}$ pointing from the center of subvolume ΔV to the location of the radar antenna (assuming a mono-static backscatter radar geometry).

With electrons in (non-relativistic) motion, scattered field phasor (3) turns into

$$E_s(t) = -\frac{r_e}{r} E_0 \sum_{p=1}^{N_0\Delta V} e^{j\mathbf{k}\cdot\mathbf{r}_p(t-\frac{r}{c})} \tag{4}$$

including a propagation time delay r/c of the scattered field from the center of volume[4] ΔV. It then follows that the auto-correlation function (ACF) of the scattered field is

$$\langle E_s^*(t)E_s(t+\tau)\rangle = \frac{r_e^2}{r^2}|E_i|^2 \sum_{p=1}^{N_0\Delta V}\sum_{q=1}^{N_0\Delta V} \langle e^{j\mathbf{k}\cdot[\mathbf{r}_q(t+\tau-\frac{r}{c})-\mathbf{r}_p(t-\frac{r}{c})]}\rangle, \tag{5}$$

where angular brackets denote an expected value (ensemble average) operation. Using $\langle e^{j\mathbf{k}\cdot[\mathbf{r}_q-\mathbf{r}_{p\neq q}]}\rangle = \langle e^{j\mathbf{k}\cdot\mathbf{r}_q}\rangle\langle e^{-j\mathbf{k}\cdot\mathbf{r}_p}\rangle = 0$ for statistically independent electrons ($p \neq q$), this reduces to

$$\langle E_s^*(t)E_s(t+\tau)\rangle = \frac{r_e^2}{r^2}|E_i|^2 N_0\Delta V \langle e^{j\mathbf{k}\cdot[\mathbf{r}_q(t+\tau-\frac{r}{c})-\mathbf{r}_q(t-\frac{r}{c})]}\rangle = \frac{r_e^2}{r^2}|E_i|^2 N_0\Delta V \langle e^{j\mathbf{k}\cdot\Delta\mathbf{r}}\rangle \tag{6}$$

with $\Delta\mathbf{r} \equiv \mathbf{r}_q(t+\tau-\frac{r}{c}) - \mathbf{r}_q(t-\frac{r}{c})$ denoting particle displacements over time intervals τ. Only with (6), i.e., only under a strict incoherent scatter scenario, we can obtain

$$\langle |E_s(t)|^2\rangle = \frac{r_e^2}{r^2}|E_i|^2 N_0\Delta V, \tag{7}$$

a result that implies a total scattered power which is a simple sum over all scatterers individually described by (1).

Collective effects in general invalidate the results (6) and (7) from being directly applicable. Nevertheless the desired spectral model for ionospheric incoherent scatter can be expressed in terms of (6) and (7) after suitable corrections and transformations. To obtain the model let us first re-express (4) as

$$E_s(t) = -\frac{r_e}{r} E_0 n_e\left(\mathbf{k}, t-\frac{r}{c}\right) \tag{8}$$

---

[2] We assume here that ω_o is sufficiently large so that dispersion effects due to plasma density N_0 can be neglected (or treated as perturbation effects). Also, multiple scattering is neglected.

[3] Justified for $r > 2k_o\Delta V^{2/3}/\pi$, the far-field condition for an antenna of size $\Delta V^{1/3}$.

[4] ΔV is sufficiently small for electrons to move only an insignificant fraction of the radar wavelength during an interval for light to propagate across ΔV.

in terms of 3D spatial Fourier transform

$$n_e(\mathbf{k}, t) \equiv \sum_{p=1}^{N_o \Delta V} e^{-j\mathbf{k} \cdot \mathbf{r}_p t} \tag{9}$$

of the *microscopic* density function $n_e(\mathbf{r}, t) = \sum_p \delta(\mathbf{r} - \mathbf{r}_p(t))$ of the electrons[5] in volume ΔV. The scattered field spectrum for volume ΔV can then be expressed[6] as

$$\langle |E_s(\omega)|^2 \rangle = \frac{r_e^2}{r^2} |E_i|^2 \langle |n_e(\mathbf{k}, \omega)|^2 \rangle \Delta V \tag{11}$$

in terms of the electron density frequency spectrum

$$\langle |n_e(\mathbf{k}, \omega)|^2 \rangle \equiv \int_{-\infty}^{\infty} d\tau e^{-j\omega\tau} \frac{1}{\Delta V} \langle n_e^*(\mathbf{k}, t - \frac{r}{c}) n_e(\mathbf{k}, t - \frac{r}{c} + \tau) \rangle \tag{12}$$

which simplifies as

$$\langle |n_e(\mathbf{k}, \omega)|^2 \rangle = N_o \int_{-\infty}^{\infty} d\tau e^{-j\omega\tau} \langle e^{j\mathbf{k} \cdot \Delta \mathbf{r}} \rangle \equiv \langle |n_{te}(\mathbf{k}, \omega)|^2 \rangle \tag{13}$$

for independent electrons. We also have an identical expression $\langle |n_{ti}(\mathbf{k}, \omega)|^2 \rangle$ describing the density spectrum independent ions in the same volume in terms of ion displacements $\Delta \mathbf{r}$.

While neither $\langle |n_{te}(\mathbf{k}, \omega)|^2 \rangle$ nor $\langle |n_{ti}(\mathbf{k}, \omega)|^2 \rangle$ are accurate representations of the density spectra of electrons and ions in a real ionosphere (because of the neglect of collective effects), it turns out that an accurate model for $\langle |n_e(\mathbf{k}, \omega)|^2 \rangle$ can be expressed as a linear combination of $\langle |n_{te}(\mathbf{k}, \omega)|^2 \rangle$ and $\langle |n_{ti}(\mathbf{k}, \omega)|^2 \rangle$ given by

$$\langle |n_e(\mathbf{k}, \omega)|^2 \rangle = \frac{|j\omega\epsilon_o + \sigma_i|^2 \langle |n_{te}(\mathbf{k}, \omega)|^2 \rangle}{|j\omega\epsilon_o + \sigma_e + \sigma_i|^2} + \frac{|\sigma_e|^2 \langle |n_{ti}(\mathbf{k}, \omega)|^2 \rangle}{|j\omega\epsilon_o + \sigma_e + \sigma_i|^2}, \tag{14}$$

where $\sigma_{e,i}$ denote the AC conductivities of electrons and ions in the medium. This result can be derived (e.g., Kudeki & Milla, 2011) by enforcing charge conservation (i.e., continuity equation) in a plasma carrying quasi-static *macroscopic* currents $\sigma_{e,i}E$ forced by longitudinal polarization fields[7] E produced by the mismatch of thermally driven electron and ion density fluctuations $n_{te}(\mathbf{k}, t)$ and $n_{ti}(\mathbf{k}, t)$. Furthermore, Nyquist noise theorem (e.g., Callen & Welton, 1951) stipulates that the required conductivities are related to the thermal density spectra via relations

$$\frac{\omega^2}{k^2} e^2 \langle |n_{te,i}(\mathbf{k}, \omega)|^2 \rangle = 2KT_{e,i} \mathrm{Re}\{\sigma_{e,i}(\mathbf{k}, \omega)\}. \tag{15}$$

---

[5] Here $\delta(\cdot)$'s denote Dirac's deltas utilized to highlight the trajectories $\mathbf{r}_p(t)$ of individual electrons.

[6] This expression can be generalized as a soft-target radar equation

$$P_r = \int \int \frac{|E_i|^2 / 2\eta_o}{4\pi r^2} A_r 4\pi r_e^2 \langle |n_e(\mathbf{k}, \omega)|^2 \rangle \frac{d\omega}{2\pi} dV \tag{10}$$

for backscatter ISR's having a scattering volume defined by the beam pattern associated with the effective area function $A_r(\mathbf{r})$.

[7] Note that it is the response of individual particles to the quasi-static E that produces the mutual correlations in their motions.

And since $\sigma_{e,i}(\mathbf{k},\omega)$ can be uniquely obtained from $\text{Re}\{\sigma_{e,i}(\mathbf{k},\omega)\}$ using Kramer-Kronig relations (e.g., Yeh & Liu, 1972), a full blown solution of the modeling problem can be formulated in terms of "single particle correlations" $\langle e^{j\mathbf{k}\cdot\Delta\mathbf{r}}\rangle$ underlying the thermal density spectra $\langle |n_{te}(\mathbf{k},\omega)|^2\rangle$ and $\langle |n_{ti}(\mathbf{k},\omega)|^2\rangle$.

This general formulation is as follows (see Appendix 2 in Kudeki & Milla, 2011, for a detailed derivation): In terms of a one-sided integral transformation

$$J_s(\omega) \equiv \int_0^\infty d\tau\, e^{-j\omega\tau}\langle e^{j\mathbf{k}\cdot\Delta\mathbf{r}_s}\rangle, \tag{16}$$

known as *Gordeyev integral* for species s (e or i for the single-ion case), we have

$$\frac{\langle |n_{ts}(\mathbf{k},\omega)|^2\rangle}{N_0} = 2\text{Re}\{J_s(\omega_s)\} \quad \text{and} \quad \frac{\sigma_s(\mathbf{k},\omega)}{j\omega\epsilon_0} = \frac{1 - j\omega_s J_s(\omega_s)}{k^2 h_s^2}, \tag{17}$$

where $\omega_s \equiv \omega - \mathbf{k}\cdot\mathbf{V}_s$ is a Doppler-shifted frequency in the radar frame due to mean velocity $\mathbf{V}_s$ of species s, $h_s \equiv \sqrt{\epsilon_0 K T_s/N_0 e^2}$ is the corresponding Debye length, and the $\mathbf{k}$-ω spectrum of electron density fluctuations in the equilibrium plasma is given by (14) or its multi-ion generalizations.

The "general framework" of ISR spectral models represented by (16)-(17) and (14) (as well as (10)) takes care of the macrophysics of the incoherent scatter process due to collective effects, while microphysics details of the process remain to be addressed in the specification of single particle ACF's $\langle e^{j\mathbf{k}\cdot\Delta\mathbf{r}}\rangle$.

4. Single particle ACF's $\langle e^{j\mathbf{k}\cdot\Delta\mathbf{r}}\rangle$ for un-magnetized plasmas

We have just seen that ISR spectrum of ionospheric plasmas in thermal equilibrium can be specified in terms of single particle ACF's $\langle e^{j\mathbf{k}\cdot\Delta\mathbf{r}}\rangle$. In general, an ACF $\langle e^{j\mathbf{k}\cdot\Delta\mathbf{r}}\rangle$ can be explicitly computed if the probability distribution function (pdf) $f(\Delta r)$, where Δr is the component of $\Delta\mathbf{r}$ along $\mathbf{k}$, is known. Alternatively, $\langle e^{j\mathbf{k}\cdot\Delta\mathbf{r}}\rangle$ can also be computed directly given an ensemble of realizations of Δr for a given time delay τ. In either case, pdf's $f(\Delta r)$ or pertinent sets of Δr data will reflect the dynamics of random particle motions taking place in ionospheric plasmas.

When Δr is a Gaussian random variable with a pdf

$$f(\Delta r) = \frac{e^{-\frac{\Delta r^2}{2\langle\Delta r^2\rangle}}}{\sqrt{2\pi\langle\Delta r^2\rangle}}, \tag{18}$$

the single-particle ACF

$$\langle e^{j\mathbf{k}\cdot\Delta\mathbf{r}}\rangle = \int e^{jk\Delta r} f(\Delta r)d(\Delta r) = e^{-\frac{1}{2}k^2\langle\Delta r^2\rangle} \tag{19}$$

depends on the mean-square displacement $\langle\Delta r^2\rangle$ of the particles. In such cases incoherent scatter modeling problem reduces to finding the appropriate variance expressions $\langle\Delta r^2\rangle$.

In a non-magnetized and collisionless plasma the charge carriers will move along straight line (unperturbed) trajectories with random velocities $\mathbf{v}$. In that case the displacement vectors will be

$$\Delta\mathbf{r} = \mathbf{v}\tau \tag{20}$$

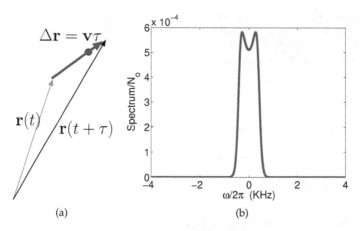

(a)   (b)

Fig. 1. (a) A cartoon depicting particle displacements Δr in a plasma with straight line charge carrier trajectories, and (b) a sample ISR spectrum for a non-magnetized and collisionless plasma in thermal equilibrium.

over intervals τ. Assuming Maxwellian distributed velocity components v along wavevector $\mathbf{k}$, we then have Gaussian distributed displacements $\Delta r = v\tau$ with variances

$$\langle \Delta r^2 \rangle = \langle v^2 \rangle \tau^2 = C^2 \tau^2, \tag{21}$$

where $C = \sqrt{KT/m}$ is the thermal speed of the charge carrier. The corresponding single particle ACF is in that case

$$\langle e^{j\mathbf{k}\cdot\Delta\mathbf{r}} \rangle = e^{-\frac{1}{2}k^2 C^2 \tau^2}, \tag{22}$$

which leads (via the general framework equations) to the most basic incoherent scatter spectral model exhibiting double humped shapes as depicted in Figure 6b when (22) is applied to both electrons and ions (with $C = C_e$ and C_i, respectively).

The ACF (22) is also applicable in collisional plasmas so long as the relevant "collision frequency" v is small compared to the product kC, i.e., $v \ll kC$, so that an average particle moves a distance of many wavelengths $\frac{2\pi}{k}$ in between successive collisions. Otherwise, (22) will only be valid until the "first collisions" take place at $\tau \sim v^{-1}$. At larger τ, the mean-square displacement $\langle \Delta r^2 \rangle$ as well as the pdf $f(\Delta r)$ will in general depend on the details of the dominant collision process.

Long range Coulomb collisions between charged particles (e.g., electrons and ions) are frequently modeled as a "Brownian motion" process[8], a procedure which leads (e.g., Kudeki & Milla, 2011) to a Gaussian $f(\Delta r)$ with a variance

$$\langle \Delta r^2 \rangle = \frac{2C^2}{v^2}(v\tau - 1 + e^{-v\tau}). \tag{23}$$

---

[8] As discussed in Kudeki & Milla (2011) and here in Section 7, in Brownian motion the position and velocity increments are Gaussian random variables and correspond to stochastic solutions of a first-order Langevin update equation with constant coefficients.

The corresponding single particle ACF is

$$\langle e^{j\mathbf{k}\cdot\Delta\mathbf{r}}\rangle = e^{-\frac{k^2 C^2}{\nu^2}(\nu\tau - 1 + e^{-\nu\tau})}, \tag{24}$$

having the asymptotic limits (22) as well as

$$\langle e^{j\mathbf{k}\cdot\Delta\mathbf{r}}\rangle = e^{-\frac{k^2 C^2}{\nu}\tau} \tag{25}$$

for $\nu \ll kC$ and $\nu \gg kC$, respectively. Note that when (25) is applicable, with $\nu \gg kC$, an average particle moves across only a small fraction of a wavelength $\frac{2\pi}{k}$ in between successive collisions. In Coulomb interactions, the time ν^{-1} between "effective collisions" (an accumulated effect of interactions with many collision partners via their microscopic Coulomb fields) can be interpreted as the time interval over which the particle velocity vector rotates by about $90°$.

Binary collisions of charge carriers with neutral atoms and molecules — dominant in the lower ionosphere — can be modeled as a Poisson process (Milla & Kudeki, 2009) and treated kinetically using the BGK collision operator (e.g., Dougherty & Farley, 1963). As shown in Milla & Kudeki (2009), in binary collisions with neutrals the mean-squared displacement of charge carriers is still given by (23), but the relevant pdf $f(\Delta r)$ is a Gaussian only for short and long delays τ satisfying $\nu\tau \ll 1$ and $\nu\tau \gg 1$, respectively. At intermediate τ's the ACF of a collisional plasma dominated by binary collisions will then deviate from (24) and as a result collisional spectra will in general exhibit minor differences between binary and Coulomb collisions except in $\nu \ll kC$ and $\nu \gg kC$ limits (Hagfors & Brockelman, 1971; Milla & Kudeki, 2009).

As the above discussion implies, the single particle ACF in the high collision limit ($\nu \gg kC$) is insensitive to the distinctions between Coulomb and binary collisions and obeys a simple relation (25). In that limit it is fairly straightforward to evaluate the corresponding Gordeyev integrals analytically, and obtain (via the general framework equations) a Lorentzian shaped electron density spectrum (mainly the "ion-line"),

$$\frac{\langle |n_e(\mathbf{k},\omega)|^2\rangle}{N_0} \approx \frac{2k^2 D_i}{\omega^2 + (2k^2 D_i)^2}, \tag{26}$$

valid for $kh \ll 1$ (wavelength larger than Debye length), where $D_i \equiv C_i^2/\nu_i = KT_i/m_i\nu_i$ denotes the ion diffusion coefficient in the collisional plasma. This result is pertinent to D-region incoherent scatter observations (see Figure 2) neglecting possible complications due to the presence of negative ions (e.g., Mathews, 1984). Also, from (26) it follows that

$$\langle |n_e(\mathbf{k})|^2\rangle \equiv \int_{-\infty}^{\infty} \frac{d\omega}{2\pi}\langle |n_e(\mathbf{k},\omega)|^2\rangle = \frac{N_0}{2}, \tag{27}$$

which is in fact true in general — i.e., for all types of plasmas with or without collisions and/or DC magnetic field — so long as $T_e = T_i$ and $kh \ll 1$. In view of radar equation (10), this result leads to a well-known volumetric radar cross-section (RCS) formula

$$4\pi r_e^2 \langle |n_e(\mathbf{k})|^2\rangle = 2\pi r_e^2 N_0 \tag{28}$$

for ISR's that is valid under the same conditions as (27). Hence, RCS measurements with ISR's can provide us with ionospheric mean densities N_0.

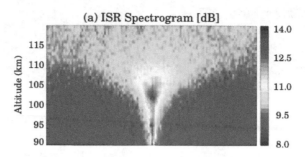

Fig. 2. Collisional D-region spectrograms from Jicamarca Radio Observatory (from Chau & Kudeki, 2006).

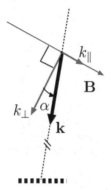

Fig. 3. Backscattering geometry in a magnetized ionosphere parametrized by wavevector components $k_\parallel$ and $k_\perp$ and aspect angle $\alpha = \tan^{-1}(k_\parallel/k_\perp)$.

5. Incoherent scatter from a magnetized ionosphere

In a magnetized ionosphere with an ambient magnetic field **B**, it is convenient to express the scattered wavevector as $\mathbf{k} = \hat{b}k_\parallel + \hat{p}k_\perp$, where $\hat{b}$ and $\hat{p}$ are orthogonal unit vectors on **k**-**B** plane which are parallel and perpendicular to **B**, respectively, as depicted in Figure 3. We can then express the single particle ACF as

$$\langle e^{j\mathbf{k}\cdot\Delta\mathbf{r}} \rangle = \langle e^{j(k_\parallel \Delta r + k_\perp \Delta p)} \rangle = \langle e^{jk_\parallel \Delta r} \times e^{jk_\perp \Delta p} \rangle, \tag{29}$$

where Δr and Δp are particle displacements along unit vectors $\hat{b}$ and $\hat{p}$. Assuming independent Gaussian random variables Δr and Δp, we can then write

$$\langle e^{j\mathbf{k}\cdot\Delta\mathbf{r}} \rangle = e^{-\frac{1}{2}k_\parallel^2 \langle \Delta r^2 \rangle} \times e^{-\frac{1}{2}k_\perp^2 \langle \Delta p^2 \rangle} \tag{30}$$

in analogy with the non-magnetized case. The assumptions are clearly justified in case of a collisionless ionosphere (or for intervals τ such that $\tau\nu \ll 1$), in which case

$$\langle \Delta r^2 \rangle = C^2\tau^2 \tag{31}$$

and, as shown in Kudeki & Milla (2011),

$$\langle \Delta p^2 \rangle = \frac{4C^2}{\Omega^2}\sin^2(\Omega\tau/2), \tag{32}$$

where $\Omega \equiv \frac{qB}{m}$ is the particle gyrofrequency. The mean-square displacement (32) which is periodic in τ is can be derived by invoking circular particle orbits with periods $2\pi/\Omega$ and mean radii $\sqrt{2}C/\Omega$ on the plane perpendicular to $\mathbf{B}$. As a consequence of (31) and (32), ISR spectra in a magnetized but collisionless ionosphere can be derived from the single particle ACF

$$\langle e^{j\mathbf{k}\cdot\Delta\mathbf{r}} \rangle = e^{-\frac{1}{2}k_\parallel^2 C^2 \tau^2} \times e^{-\frac{2k_\perp^2 C^2}{\Omega^2}\sin^2(\Omega\tau/2)} \tag{33}$$

for electrons and ions.

Note that the ACF (33) becomes periodic and the associated Gordeyev integrals and spectra become singular (expressed in terms of Dirac's deltas) in $k_\parallel \to 0$ limit. Spectral singularities are of course not observed in practice since collisions in a real ionosphere end up limiting the width of single particle ACF's in τ in the limit of small "aspect angles" $\alpha = \tan^{-1}(k_\parallel/k_\perp)$.

Despite the singularities in (33), it turns out that for finite aspect angles α larger than a few degrees, the collisionless result (33) leads us to the most frequently used ISR spectral model at F-region heights. This is true because given a finite $k_\parallel$, the term $e^{-\frac{1}{2}k_\parallel^2 C^2 \tau^2}$ in (33) restricts the width of the ACF to a finite value of $\sim (k_\parallel C)^{-1}$ even in the absence of collisions (or when collision frequencies are smaller than $k_\parallel C$). It can then be shown that for $\tau \ll (k_\parallel C)^{-1}$, as well as $\Omega\tau \ll 2\pi$ (easily satisfied by massive ions), the ACF (33) for ions recombines to a simplified form $e^{-\frac{1}{2}k^2 C^2 \tau^2}$ as if the plasma were non-magnetized. Also with finite $k_\parallel$, the ACF (33) for electrons simplifies to $e^{-\frac{1}{2}k^2 \sin^2\alpha C^2 \tau^2}$, since for the light electrons a condition $k_\perp C \ll \Omega$ can be easily invoked to ignore the rightmost exponential in (33) (or even more accurately, replace it with its average value over τ, namely, $1 - \frac{k_\perp^2 C^2}{\Omega^2}$). These ion and electron ACF's exhibit similar τ dependencies and lead to similar shaped Gordeyev integrals. The resulting ISR spectra are of the "double humped" type shown in Figure 1b.

6. Modeling the Coulomb collision effects in magnetized plasmas

As we have noted, the form (33) of the single particle ACF indicates that magnetic field effects in ISR response are confined to small aspect angles, which is also the regime where collision effects cannot be neglected (e.g., Farley, 1964; Sulzer & González, 1999; Woodman, 1967) given the non-physical behavior of ACF (33) in $\alpha \to 0°$ limit.

Historical note: The need to account for the effects of collisions in incoherent scatter theory of ionospheric F-region returns was first pointed out by Farley (1964). Based on a qualitative analysis, Farley recognized that ion Coulomb collisions would be responsible for the lack of O^+ gyroresonance signatures on incoherent scatter observations carried out at 50 MHz at the Jicamarca Radio Observatory located near Lima, Peru. This analysis was later verified by the theoretical work of Woodman (1967) which was based on the simplified Fokker-Planck collision model of Dougherty (1964). Many years later, after the application of modern radar and signal processing techniques to the measurement and analysis of ISR signals (e.g., Kudeki et al., 1999), Sulzer & González (1999) noted that, in addition to ion collisions, electron Coulomb collisions also have an influence on the shape of the ISR spectra at small magnetic aspect angles. Based on a more complex Fokker-Planck Coulomb collision model, Sulzer & González found that the collisional spectrum is narrower (just like the observations of Kudeki et al., 1999) than what the collisionless theory predicts and that the effect of electron collisions extends up to relatively large magnetic aspect angles. Recently, this work has been refined and extended by Milla & Kudeki (2011). The new procedure allows the calculation of collisional IS spectra at all magnetic aspect angles including the perpendicular-to-**B** direction ($\alpha = 0°$) as needed for IS radar applications. In this section, we present the procedure developed by Milla & Kudeki (2011) to model the effects of Coulomb collisions on the incoherent scatter spectrum.

The single-particle ACF $\langle e^{j\mathbf{k}\cdot\Delta\mathbf{r}} \rangle$ in a collisional plasma including a magnetic field can in principle be calculated by taking the spatial Fourier transform of the probability distribution $f(\Delta\mathbf{r}, \tau)$ of the particle displacement $\Delta\mathbf{r}$ appropriate for such plasmas, and $f(\Delta\mathbf{r}, \tau)$ in turn can be derived from the solution $f(\mathbf{r}, t)$ of the Boltzmann kinetic equation with a collision operator, e.g., the Fokker-Planck kinetic equation of Rosenbluth et al. (1957). Although, analytical solutions of simplified versions of the Fokker-Planck kinetic equation are available (e.g., Chandrasekhar, 1943; Dougherty, 1964), determining $f(\Delta\mathbf{r}, \tau)$ would be a daunting task when the full Fokker-Planck equation is considered.

We will discuss here an alternative and more practicable approach that involves Monte Carlo simulations of sample paths $\mathbf{r}(t)$ of particles undergoing Coulomb collisions. A sufficiently large set of samples of trajectories $\mathbf{r}(t)$ can then be used to compute $\langle e^{j\mathbf{k}\cdot\Delta\mathbf{r}} \rangle$ as well as any other statistical function of $\Delta\mathbf{r}$ assuming the random process $\mathbf{r}(t)$ to be ergodic. This alternate procedure requires the availability of a stochastic equation describing how the particle velocities

$$\mathbf{v}(t) \equiv \frac{d\mathbf{r}}{dt} \tag{34}$$

may evolve under the influence of Coulomb collisions.

Assuming that under Coulomb collisions the velocities $\mathbf{v}(t)$ constitute a Markovian random process — meaning that past values of $\mathbf{v}$ would be of no help in predicting its future values if the present value is available — the stochastic evolution equation of $\mathbf{v}(t)$ will be constrained by very strict self-consistency conditions discussed by Gillespie (1996a;b) to acquire the form of a Langevin equation

$$\frac{d\mathbf{v}(t)}{dt} = \mathbf{A}(\mathbf{v}, t) + \bar{\mathbf{C}}(\mathbf{v}, t)\mathbf{W}(t) \tag{35}$$

where vector $\mathbf{A}(\mathbf{v}, t)$ and matrix $\bar{\mathbf{C}}(\mathbf{v}, t)$ consist of arbitrary smooth functions of arguments $\mathbf{v}$ and t, and $\mathbf{W}(t)$ is a random vector having statistically independent Gaussian white noise components

$$W_i(t) = \lim_{\Delta t \to 0} \mathcal{N}(0, 1/\Delta t), \tag{36}$$

compatible with the requirement that $\langle W_i(t + \tau)W_i(t) \rangle = \delta(\tau)$. Here and elsewhere $\mathcal{N}(\mu, \sigma^2)$ denotes the normal random variable with mean μ and variance σ^2.

A more natural way of expressing the Langevin equation (35) is to cast it in an update form, namely

$$\mathbf{v}(t + \Delta t) = \mathbf{v}(t) + \mathbf{A}(\mathbf{v}, t)\,\Delta t + \bar{\mathbf{C}}(\mathbf{v}, t)\,\Delta t^{1/2}\mathbf{U}(t), \tag{37}$$

where Δt is an infinitesimal update interval and $\mathbf{U}(t)$ is a vector composed of independent zero-mean Gaussian random variables with unity variance, i.e., $U_i(t) = \mathcal{N}(0, 1)$.

Note that the Langevin equation describing a Markovian process has the form of Newton's second law of motion, with the terms on the right representing forces per unit mass exerted on plasma particles. Considering the Lorentz force on a charged particle in a magnetized plasma with a constant magnetic field **B**, and not violating the strict format of (35), we can modify the equation by adding a term $q\mathbf{v}(t) \times \mathbf{B}/m$ to its right hand side.

Another relevant fact is that a special type of Markov process characterized by a linear $\mathbf{A}(\mathbf{v}, t) = -\beta\mathbf{v}$ and a constant matrix $\bar{\mathbf{C}} = D^{1/2}\bar{\mathbf{I}}$, independent of $\mathbf{v}$ and t, is known as Brownian motion process (e.g., Chandrasekhar, 1942; Uhlenbeck & Ornstein, 1930), which is often invoked in simplified models of collisional plasmas (e.g., Dougherty, 1964; Holod et al., 2005; Woodman, 1967) including our earlier result (24) with $\nu = \beta$. In these models, friction and diffusion coefficients, β and D, are constrained to be related by

$$D = \frac{2KT}{m}\beta \tag{38}$$

for a plasma in thermal equilibrium.

In return for having restricted $\mathbf{v}(t)$ to the space of Markovian processes, we have gained a stochastic evolution equation (35) with a plausible Newtonian interpretation and with the potential of taking us beyond Brownian motion based collision models. Furthermore, the evolution of probability density $f(\mathbf{v}, t)$ of a random variable $\mathbf{v}(t)$ is known to be governed, when $\mathbf{v}(t)$ is Markovian, by the Fokker-Planck kinetic equation having a "friction vector" and "diffusion tensor"

$$\left\langle \frac{\Delta\mathbf{v}}{\Delta t} \right\rangle_c = \mathbf{A}(\mathbf{v}, t), \tag{39}$$

and

$$\left\langle \frac{\Delta\mathbf{v}\Delta\mathbf{v}^{\mathsf{T}}}{\Delta t} \right\rangle_c = \bar{\mathbf{C}}(\mathbf{v}, t)\bar{\mathbf{C}}^{\mathsf{T}}(\mathbf{v}, t), \tag{40}$$

respectively, specified in terms of the input functions of the Langevin equation. This intimate link between the Langevin and Fokker-Planck equations — in describing Markovian processes from two different but mutually compatible perspectives — was first pointed out by Chandrasekhar (1943) and discussed in detail by Gillespie (1996b).

Since the Fokker-Planck friction vector and diffusion tensor for equilibrium plasmas with Coulomb interactions have already been worked out by Rosenbluth et al. (1957) as

$$\left\langle \frac{\Delta\mathbf{v}}{\Delta t} \right\rangle_c = -\beta(v)\mathbf{v} \tag{41}$$

and

$$\left\langle \frac{\Delta\mathbf{v}\Delta\mathbf{v}^{\mathsf{T}}}{\Delta t} \right\rangle_c = \frac{D_\perp(v)}{2}\bar{I} + \left(D_\parallel(v) - \frac{D_\perp(v)}{2}\right)\frac{\mathbf{v}\mathbf{v}^{\mathsf{T}}}{v^2}, \tag{42}$$

in terms of scalar functions $\beta(v)$, $D_\parallel(v)$, $D_\perp(v)$, it follows that the Langevin update equation, magnetized version of (37), can be written as

$$\mathbf{v}(t + \Delta t) = \mathbf{v}(t) + \frac{q}{m}\mathbf{v}(t) \times \mathbf{B}\,\Delta t$$

$$- \beta(v)\Delta t\,\mathbf{v}(t) + \sqrt{D_\parallel(v)\Delta t}\,U_1\,\hat{v}_\parallel + \sqrt{D_\perp(v)\frac{\Delta t}{2}}\,(U_2\,\hat{v}_{\perp 1} + U_3\,\hat{v}_{\perp 2}), \tag{43}$$

where $\hat{v}_\parallel(t)$, $\hat{v}_{\perp 1}(t)$, and $\hat{v}_{\perp 2}(t)$ denote an orthogonal set of unit vectors parallel and perpendicular to the particle trajectory and $U_i(t) = \mathcal{N}(0, 1)$ are independent random numbers. For weakly magnetized plasmas of interest here, where Debye lengths are smaller than the mean gyro radii, the "friction coefficient" $\beta(v)$ and velocity-space diffusion

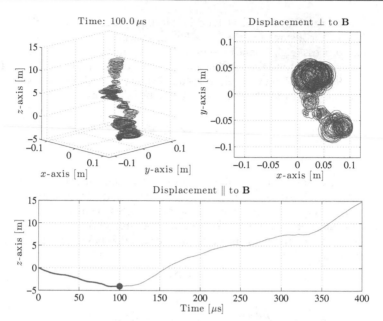

Fig. 4. Sample trajectory of an electron moving in an O+ plasma with density $N_e = 10^{12}\,\mathrm{m}^{-3}$, temperatures $T_e = T_i = 1000\,\mathrm{K}$, and an ambient magnetic field $\mathbf{B} = \hat{z}25000\,\mathrm{nT}$. Top left panel depicts the trajectory in 3D space; projection on the x-y plane (the plane perpendicular to $\mathbf{B}$) is shown on the right; displacements parallel to $\mathbf{B}$ are depicted in the bottom plot (from Milla & Kudeki, 2011).

coefficients $D_{\parallel}(v)$ and $D_{\perp}(v)$ needed in (43) take the forms derived by Rosenbluth et al. (1957) which, for Maxwellian plasmas, have the Spitzer forms given in Milla & Kudeki (2011).

The velocity update equation (43) just described, along with its position counterpart

$$\mathbf{r}(t + \Delta t) = \mathbf{r}(t) + \mathbf{v}(t)\,\Delta t, \tag{44}$$

constitute our model equations for examining the effects of Coulomb collisions on incoherent scatter response from magnetized plasmas. These equations are used to simulate particle trajectories such as one shown in Figure 4 from which particle displacement statistics needed in ISR spectral models are estimated as explained in Sections 7 and 8.

7. Coulomb collision effects on ion and electron trajectories

7.1 Statistics of ion displacements

First we use the update equations (43) and (44) to simulate sample trajectories $\mathbf{r}(t)$ of an ion, e.g., an oxygen ion O^{+}, moving in an ionospheric plasma with suppressed collective interactions but experiencing Coulomb collisions. Using the trajectory data, we can build up the probability distributions of the displacements Δr in directions perpendicular and parallel to the magnetic field for different time delays. Analyzing both distributions (parallel and perpendicular), we notice that their shapes are in essence Gaussian for time delays smaller than the inverse of the corresponding collision frequency. In Figure 5, we show examples

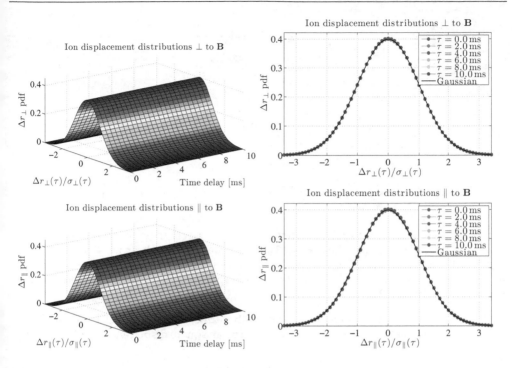

Fig. 5. Probability distributions of the displacements of a test ion in the directions perpendicular (top panels) and parallel (bottom panels) to the magnetic field. On the left, the displacement pdf's are displayed as functions of time delay τ. On the right, sample cuts of the pdf's are compared to a Gaussian distribution. Note that all distributions at all time delays are normalized to unit variance. The displacement axis of each distribution at every delay τ is scaled with the corresponding standard deviation of the simulated displacements (from Milla & Kudeki, 2011).

of the distributions of the ion displacements in the directions perpendicular and parallel to the magnetic field. In this case, we have considered an oxygen ion moving in a plasma with density $N_e = 10^{12}\,\mathrm{m}^{-3}$, temperatures $T_e = T_i = 1000\,\mathrm{K}$ and magnetic field $B_o = 25000\,\mathrm{nT}$. Note that, at every delay τ, the distributions have been normalized to unit variance by scaling the displacement axis of each distribution with the corresponding standard deviation of the particle displacements. On the left panels, the distributions are displayed as functions of τ, while, on the right panels, sample cuts of these distributions are compared to a Gaussian pdf showing good agreement. In addition, we can verify that the components of the vector displacement (i.e., Δr_x, Δr_y, and Δr_z) are mutually uncorrelated.

This analysis implies that ion particle displacements can be represented as jointly Gaussian $\Delta \mathbf{r}$ components, therefore the single-particle ACF takes the form (e.g., Kudeki & Milla, 2011)

$$\langle e^{j\mathbf{k}\cdot\Delta\mathbf{r}} \rangle = e^{-\frac{1}{2}k^2 \sin^2\alpha \langle \Delta r_\parallel^2 \rangle} \times e^{-\frac{1}{2}k^2 \cos^2\alpha \langle \Delta r_\perp^2 \rangle}, \tag{45}$$

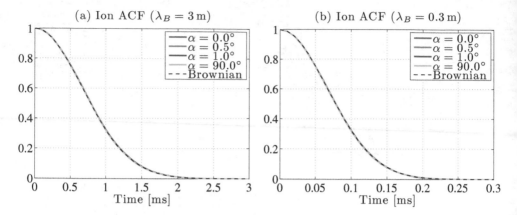

Fig. 6. Simulated single-ion ACF's at different magnetic aspect angles α for two radar Bragg wavelengths: (a) $\lambda_B = 3\,\mathrm{m}$ and (b) $\lambda_B = 0.3\,\mathrm{m}$. The simulation results (color lines) are compared to theoretical ACF's computed using expression (51) of the Brownian-motion approximation (dashed lines). Note that there is effectively no dependence on aspect angle α (from Milla & Kudeki, 2011).

where, assuming a Brownian-motion process with distinct friction coefficients $\nu_\parallel$ and $\nu_\perp$ in the directions parallel and perpendicular to **B**, the mean square displacements will vary as

$$\langle \Delta r_\parallel^2 \rangle = \frac{2C^2}{\nu_\parallel^2} \left(\nu_\parallel \tau - 1 + e^{-\nu_\parallel \tau} \right) \tag{46}$$

and

$$\langle \Delta r_\perp^2 \rangle = \frac{2C^2}{\nu_\perp^2 + \Omega^2} \left(\cos(2\gamma) + \nu_\perp \tau - e^{-\nu_\perp \tau} \cos(\Omega \tau - 2\gamma) \right) \tag{47}$$

in which $\gamma \equiv \tan^{-1}(\nu_\perp / \Omega)$, and $C \equiv \sqrt{KT/m}$ and $\Omega \equiv qB/m$ are, respectively, the thermal speed and gyrofrequency of the particles. Furthermore, simulated $\langle \Delta r_{\parallel,\perp}^2 \rangle$ match (46) and (47) with

$$\nu_\perp \approx \nu_\parallel \approx \nu_{i/i}, \tag{48}$$

where

$$\nu_{i/i} = \frac{N_e \, e^4 \, \ln \Lambda_i}{12 \, \pi^{3/2} \, \epsilon_0^2 \, m_i^2 \, C_i^3} \tag{49}$$

is the Spitzer ion-ion collision frequency given by Callen (2006) and Milla & Kudeki (2011).

The simulations also indicate, in the case of oxygen ions,

$$\langle \Delta r_\parallel^2 \rangle \approx \langle \Delta r_\perp^2 \rangle \approx C_i^2 \tau^2 \tag{50}$$

for short time delays $\nu_\parallel \tau \ll 1$ and $\nu_\perp \tau < \Omega_i \tau \ll 1$, in consistency with (46) and (47). Hence (45) simplifies to

$$\langle e^{j\mathbf{k} \cdot \Delta \mathbf{r}_i} \rangle \approx e^{-\frac{1}{2} k^2 C_i^2 \tau^2}. \tag{51}$$

Evidently, the single-oxygen-ion ACF's are essentially the same as in collisionless and non-magnetized plasmas because (a) the ions move by many Bragg wavelengths $\lambda_B = 2\pi/k$

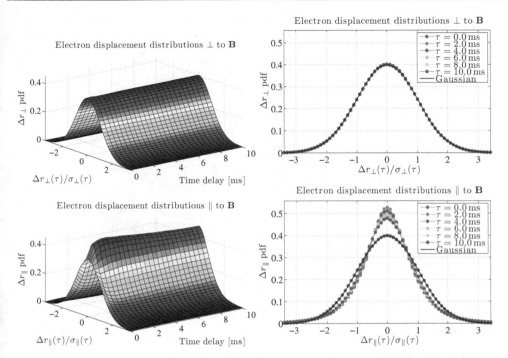

Fig. 7. Same as Figure 5 but for the case of a test electron. All distributions at all time delays are normalized to unit variance. Note that the distributions of the displacements parallel to **B** become narrower than a Gaussian distribution (from Milla & Kudeki, 2011).

between successive Spitzer collisions, and (b) the ions are unable to return to within $\lambda_B/2\pi$ of their starting positions after a gyro-period as a consequence of Coulomb collisions. As an upshot, we will be able to handle the ion terms analytically in spectral calculations.

7.2 Statistics of electron displacements

Next, we study the effects of Coulomb collisions on electron trajectories using procedures similar to those applied to ions. In Figure 7, the displacement distributions resulting from an electron moving in an O+ plasma are presented. The top and bottom panels in Figure 7 correspond, respectively, to displacement distributions in perpendicular and parallel directions. On the left, the distributions are displayed as functions of τ, while on the the right, sample cuts of the distributions are compared to a Gaussian pdf. As in the ion case, we note that the normalized distributions for perpendicular direction to be invariant with τ and closely match a Gaussian. However, the distributions of parallel displacements change with τ, and the shapes are distinctly non-Gaussian for intermediate values of τ. More specifically, at very small time delays (lower than the inverse of a collision frequency), the distributions are Gaussian, but then, in a few "collision" times, the distribution curves become more "spiky" (positive kurtosis) than a Gaussian. Although, at even longer delays τ the distributions once again relax to a Gaussian shape, it is clear that the electron displacement in the direction parallel to **B** is not a Gaussian random variable at all time delays τ.

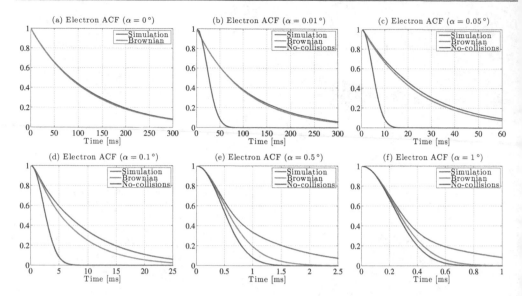

Fig. 8. Electron ACF's for $\lambda_B = 3\,\text{m}$ at different magnetic aspect angles: (a) $\alpha = 0°$, (b) $\alpha = 0.01°$, (c) $\alpha = 0.05°$, (d) $\alpha = 0.1°$, (e) $\alpha = 0.5°$, and (f) $\alpha = 1°$. Note the different time scales used in each plot (from Milla & Kudeki, 2011).

Fitting the simulated $\langle \Delta r_{\|,\perp}^2 \rangle$ to match (46) and (47) we find

$$\nu_\| \approx \nu_{e/i} \tag{52}$$

and

$$\nu_\perp \approx \nu_{e/i} + \nu_{e/e}, \tag{53}$$

where

$$\nu_{e/e} = \frac{N_e\, e^4\, \ln \Lambda_e}{12\, \pi^{3/2}\, \epsilon_0^2\, m_e^2\, C_e^3} \tag{54}$$

and

$$\nu_{e/i} = \sqrt{2}\nu_{e/e} = \frac{\sqrt{2}\, N_e\, e^4\, \ln \Lambda_e}{12\, \pi^{3/2}\, \epsilon_0^2\, m_e^2\, C_e^3} \tag{55}$$

are the Spitzer electron-electron and electron-ion collision frequencies. However, the Brownian ACF model (45) fails to fit the electron ACF's $\langle e^{j\mathbf{k}\cdot\Delta\mathbf{r}_e} \rangle$ computed with simulated trajectories as shown in Figure 8 for a range of magnetic aspect angles and $\lambda_B = 3\,\text{m}$. The blue curves correspond to the ACF's calculated with the Fokker-Planck model (simulations), while the green curves are the electron ACF's calculated using expression (45) together with our approximations for $\nu_\|$ and $\nu_\perp$. Additionally, the electron ACF's for a collisionless magnetized plasma are also plotted (red curves). We can see that the Fokker-Planck and the Brownian ACF's matched almost perfectly at $\alpha = 0°$, and also that the agreement is still good at very small magnetic aspect angles (see panels a, b, and c). However, substantial differences between the Brownian and estimated ACF's become evident as the magnetic aspect angle increases (see panels d, e, and f).

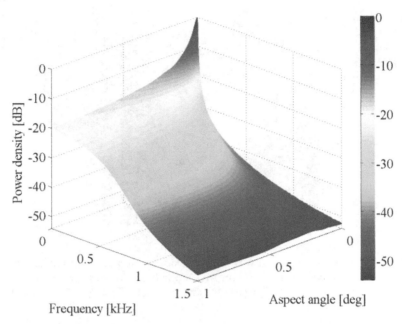

Fig. 9. Collisional incoherent scatter spectra as a function of magnetic aspect angle and Doppler frequency for $\lambda_B = 3\,\mathrm{m}$ (e.g., Jicamarca radar Bragg wavelength). An O+ plasma is considered (from Milla & Kudeki, 2011).

In summary, the single-electron ACF's needed for ISR spectral calculations cannot be obtained from Brownian motion model (45) at small aspect angles. This necessitates the construction of a numerical "library" compiled from Monte Carlo simulations based on the Langevin equation. The fundamental reason for this is the deviation of the electron displacements parallel to **B** from Gaussian statistics, despite the fact that displacement variances are well modeled by the Brownian model. Certainly, a non-Gaussian process cannot be fully characterized by a model that specifies its first and second moments only; this is particularly true for the estimation of the characteristic function of the process $\langle e^{j\mathbf{k}\cdot\Delta\mathbf{r}_e}\rangle$ that depends on all the moments of the process distribution.

8. ISR spectrum for the magnetized ionosphere including Coulomb collision effects

The general framework of incoherent scatter theory formulates the spectrum in terms of the Gordoyev integrals or the corresponding single-particle ACFs for each plasma species. As discussed above, in the case of Coulomb collisions, the single-ion ACF can be approximated using the analytical expression (45). However, in the case of the electrons, the approximation of the electron motion as a Brownian process is not accurate, and thus, Monte Carlo calculations were needed to model single-electron ACFs and Gordeyev integrals for different sets of plasma parameters.

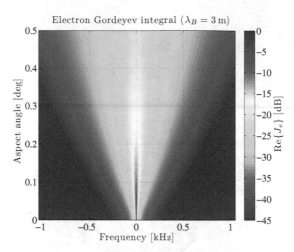

Fig. 10. Electron Gordeyev integral as functions of Doppler frequency and magnetic aspect angle for radar Bragg wavelength $\lambda_B = 3$ m. An O+ plasma with electron density $N_e = 10^{12}$ m^{-3}, temperatures $T_e = T_i = 1000$ K, and magnetic field $B_o = 25\,$ T is considered (from Milla & Kudeki, 2011).

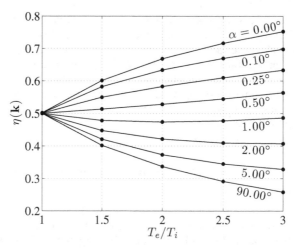

Fig. 11. Electron scattering efficiency factor $\eta(k)$ resulting from the frequency integration of the collisional incoherent scatter spectra as a function of electron-to-ion temperature ratio T_e/T_i and magnetic aspect angle α. An O+ plasma with $N_e = 10^{12}$ m^{-3}, $T_i = 1000$ K, and $B_o = 25\,$ T is considered (from Milla & Kudeki, 2011).

Figure 9 shows a surface plot constructed from full IS spectrum calculations for $\lambda_B = 3$ m (e.g., for the 50 MHz Jicamarca ISR system located near Lima, Peru) using the ACF library constructed with the Monte Carlo procedure for electrons. The underlying electron Gordeyev integral $J_e(\omega)$ is presented in Figure 10 where only $\mathrm{Re}\{J_e(\omega)\} \propto \langle |n_{te}(\mathbf{k}_B, \omega)|^2 \rangle$ is displayed. The plots are displayed as a functions of aspect angle α and Doppler frequency $\omega/2\pi$. In both figures it can be observed how these spectral functions sharpen significantly at small aspect angles. In particular in the case of the IS spectrum, we can see that, just in the range between $0.1°$ and $0°$, the amplitude of the spectrum becomes ten times larger while its bandwidth is reduced by the same factor.

Some interesting features of the IS spectrum caused by collisions can be pointed out. As discussed by Milla & Kudeki (2011), in the absence of collisions, the magnetic field restricts the motion of electrons in the plane perpendicular to $\mathbf{B}$, forcing them to gyrate perpetually around the same magnetic field lines — this would generate infinite correlation time of the IS signal. With collisions, the electrons manage to diffuse *across* the field lines, and consequently the correlation time of the IS signal becomes finite. As a result, in the limit of $\alpha \to 0°$, the width of the spectrum becomes proportional to the collision frequency. However, at other magnetic aspect angles, the effects are slightly different. In a few hundredths of a degree from perpendicular to $\mathbf{B}$ ($\alpha > 0.01°$), the shape of the IS spectrum is dominated by electron diffusion *along* the magnetic field lines. As collisions impede the motion of particles, electrons diffuse slower in a collisional plasma than in a collisionless one (where electrons move freely), which implies that the electrons stay closer to their original locations for longer periods of time. As a result, the correlation time of the signal scattered by the electrons also becomes longer, causing the broadening of the IS signal ACF and the associated narrowing of the signal spectrum in this aspect angle regime, as first explained by Sulzer & González (1999).

Spectrum dependence on electron density N_e and temperatures T_e and T_i has been studied by Milla & Kudeki (2011). Since at very small aspect angles the electron Gordeyev integral dominates the shape of the overall incoherent scatter spectrum, Milla & Kudeki (2011) found that in the limit of $\alpha \to 0°$ the bandwidth of $\mathrm{Re}\{J_e(\omega)\}$, and therefore the IS spectrum, varies according to

$$k^2 C_e^2 \frac{\nu_\perp}{\nu_\perp^2 + \Omega_e^2}. \tag{56}$$

Furthermore, using $\nu_\perp \approx \nu_{e/i} + \nu_{e/e}$ from the last section and taking $\Omega_e \gg \nu_\perp$, we can verify that the bandwidth dependence (56) is proportional to

$$\frac{N_e}{\sqrt{T_e}}. \tag{57}$$

However, as α increases, in a few hundredths of a degree, the dependance of the IS spectral width on N_e and T_e is exchanged, i.e., the bandwidth increases as either the density decreases or the temperature increases. The reason for this is the exchange of roles between particle diffusion in the directions across and along the magnetic field lines. It should be mentioned that collision effects become less significant at even larger aspect angles where the spectrum is shaped by ion dynamics. In that regime, the spectral shapes become independent of N_e as long as $k h_e \ll 1$.

The volumetric radar cross section (RCS) pertinent in ISR applications is given by (e.g., Farley, 1966; Milla & Kudeki, 2006)

$$\sigma_v \equiv 4\pi r_e^2 N_e \eta(\mathbf{k}) \tag{58}$$

where

$$\eta(\mathbf{k}) \equiv \int \frac{d\omega}{2\pi} \frac{\langle |n_e(\mathbf{k}, \omega)|^2 \rangle}{N_e}, \tag{59}$$

is an electron scattering efficiency factor (see Milla & Kudeki, 2006) and depends on the temperature ratio T_e/T_i and magnetic aspect angle α. A plot of this factor obtained from our collisional IS model is shown in Figure 11. As we can observe, if the plasma is in thermal equilibrium (i.e., if $T_e = T_i$), this factor is $1/2$ at all angles α and compatible with (28). We can also see that $\eta(\mathbf{k})$ at $\alpha - 0°$ increases in proportion to T_e/T_i. However, at large magnetic aspect angles, the efficiency factor shows a decrease with increasing T_e/T_i. In particular, note that our calculations for $\alpha = 90°$ match the well-known formula $(1 + T_e/T_i)^{-1}$, as expected for moderate values of T_e/T_i and negligible Debye length (e.g., Farley, 1966). Note that for $\alpha \approx 1°$ the factor is approximately independent of T_e/T_i, but otherwise it increases and decreases with the temperature ratio at small and large aspect angles, respectively.

9. Magnetoionic propagation effects on IS spectrum

A radiowave propagating through the ionosphere experiences changes in its polarization caused by the presence of the Earth's magnetic field. In this section, a model for incoherent scatter spectrum and cross-spectrum measurements that takes into account magnetoionic propagation effects is developed.

A mathematical description of radiowave propagation in an inhomogeneous magnetoplasma based on the Appleton-Hartree solution is presented. The resultant wave propagation model is used to formulate a soft-target radar equation in order to account for magnetoionic propagation effects on incoherent scatter spectrum and cross-spectrum models.

9.1 Propagation of electromagnetic waves in a homogeneous magnetoplasma

In the presence of an ambient magnetic field $\mathbf{B}_o$, there are two possible and orthogonal modes of electromagnetic wave propagation in a plasma, and, therefore, any propagating field can be represented as the weighted superposition of these characteristic modes. Labeling the modes as ordinary (O) and extraordinary (X), the transverse component of an outgoing (transmitted) electric wave field, at a distance r from the origin, can be written in phasor form as

$$\mathbf{E}^t = A_O \left(\hat{\theta} - j\hat{\phi}\frac{F_O}{Y_L} \right) e^{-jk_o n_O r} + A_X \left(\hat{\theta} - j\hat{\phi}\frac{F_X}{Y_L} \right) e^{-jk_o n_X r}, \tag{60}$$

where A_O and A_X are the amplitudes of the O- and X-mode waves with refractive indices

$$n_{O/X}^2 = 1 - \frac{X}{1 - F_{O/X}} \tag{61}$$

specified by Appleton-Hartree equations (e.g., Budden, 1961), in which

$$F_{O/X} = \frac{Y_T^2 \mp \sqrt{Y_T^4 + 4Y_L^2 (1 - X)^2}}{2(1 - X)}, \tag{62}$$

$$X \equiv \frac{\omega_p^2}{\omega^2}, \qquad Y_L \equiv \frac{\Omega_e}{\omega} \cos\theta, \qquad \text{and} \quad Y_T \equiv \frac{\Omega_e}{\omega} \sin\theta. \tag{63}$$

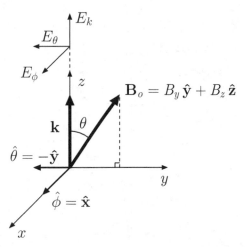

Fig. 12. Coordinate system used for analyzing wave propagation in a magnetized plasma. The magnetic field $\mathbf{B}_o$ is on the yz-plane and angle θ is measured from $\mathbf{B}_o$ to the propagation vector $\mathbf{k}$ which is parallel to the $\hat{\mathbf{z}}$-direction. The wave field $\mathbf{E}$ has three mutually orthogonal components E_k, E_θ, and E_ϕ in directions $\hat{\mathbf{k}} = \hat{\mathbf{z}}$, $\hat{\boldsymbol{\theta}} = -\hat{\mathbf{y}}$, and $\hat{\boldsymbol{\phi}} = \hat{\mathbf{x}}$, respectively. $\hat{\boldsymbol{\theta}}$ is the direction of increasing θ and $\hat{\boldsymbol{\phi}} \equiv \hat{\mathbf{k}} \times \hat{\boldsymbol{\theta}}$.

Above $k_o = \omega/c$ is the free-space wavenumber, $\omega_p \equiv \sqrt{N_e e^2 / \epsilon_o m_e}$ and $\Omega_e = e B_o / m_e$ are the plasma- and electron gyro-frequencies, respectively, and θ is the angle measured from the magnetic field vector to the propagation direction $\hat{\mathbf{k}}$. Also, $\hat{\boldsymbol{\theta}}$ and $\hat{\boldsymbol{\phi}}$ are orthogonal unit vectors normal to $\hat{\mathbf{k}}$ as shown in Figure 12.

Note that $F_O F_X = -Y_L^2$ as demanded by the orthogonality of O- and X-mode terms in (60). Thus, $a \equiv \frac{F_O}{Y_L} = -\frac{Y_L}{F_X}$ denotes the axial ratio of elliptically polarized modes in (60), which in turn can be expressed in matrix notation as

$$
\begin{bmatrix} E_\theta \\ E_\phi \end{bmatrix} = \begin{bmatrix} e^{-jk_o n_O r} & e^{-jk_o n_X r} \\ -jae^{-jk_o n_O r} & ja^{-1} e^{-jk_o n_X r} \end{bmatrix} \begin{bmatrix} A_O \\ A_X \end{bmatrix},
\tag{64}
$$

where E_θ and E_ϕ are the transverse field components in $\hat{\boldsymbol{\theta}}$ and $\hat{\boldsymbol{\phi}}$ directions. Note that a can take values within the range $0 \le |a| < 1$ and that the limits 0 and 1 correspond to the cases of linearly and circularly polarized propagation modes. Defining $\bar{n} \equiv \frac{n_O + n_X}{2}$ and $\Delta n \equiv \frac{n_O - n_X}{2}$, and considering $E_{\theta,o}$ and $E_{\phi,o}$ as the field components at the origin, the propagating electric field (64) can be recast as

$$
\begin{bmatrix} E_\theta \\ E_\phi \end{bmatrix} = \frac{e^{-jk_o \bar{n} r}}{1 + a^2} \underbrace{\begin{bmatrix} e^{-jk_o \Delta n r} + a^2 e^{jk_o \Delta n r} & 2a \sin(k_o \Delta n r) \\ -2a \sin(k_o \Delta n r) & a^2 e^{-jk_o \Delta n r} + e^{jk_o \Delta n r} \end{bmatrix}}_{\bar{\mathbf{T}}} \begin{bmatrix} E_{\theta,o} \\ E_{\phi,o} \end{bmatrix},
\tag{65}
$$

where $\bar{\mathbf{T}}$ is a propagator matrix that maps the fields at the origin into the fields at a distance r. Note that in the case of waves traveling in $-\hat{\mathbf{k}}$ direction, the same matrix $\bar{\mathbf{T}}$ can be used to propagate the fields from a distance r to the origin.

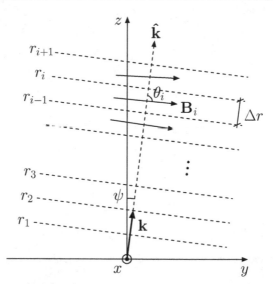

Fig. 13. Geometry of wave propagation in an inhomogeneous magnetized ionosphere.

Using the components of (65), we can re-express the outgoing electric field phasor $E_\theta\, \hat{\theta} + E_\phi\, \hat{\phi}$ as

$$\mathbf{E}^t = e^{-jk_o \bar{n} r}\left[e^{-jk_o \Delta n r}\hat{\mathbf{p}}_O\hat{\mathbf{p}}_O^H + e^{jk_o \Delta n r}\hat{\mathbf{p}}_X\hat{\mathbf{p}}_X^H\right]\mathbf{E}_o^t, \tag{66}$$

where $\mathbf{E}_o^t$ is the wave field at the origin,

$$\hat{\mathbf{p}}_O = \frac{\hat{\theta} - ja\,\hat{\phi}}{\sqrt{1 + a^2}} \qquad \text{and} \qquad \hat{\mathbf{p}}_X = \frac{-ja\,\hat{\theta} + \hat{\phi}}{\sqrt{1 + a^2}} \tag{67}$$

are the orthonormal polarization vectors of the O- and X-mode waves, while $\hat{\mathbf{p}}_O^H$ and $\hat{\mathbf{p}}_X^H$ refer to their conjugate transpose counterparts.

9.2 Model for radiowave propagation in an inhomogeneous ionosphere

A radiowave propagating through an inhomogeneous magnetoplasma will experience refraction and polarization effects. At VHF frequencies, however, ionospheric refraction effects can be considered negligible for most propagation directions because the wave frequency ω exceeds the ionospheric plasma frequency ω_p by a wide margin (i.e., $X \ll 1$). But for the same set of frequencies, polarization changes are still significant despite the fact that the electron gyrofrequency Ω_e is much smaller than the wave frequency ω (i.e., $Y \ll 1$). The reason for this is that the distances traveled by the propagating fields are long enough (hundreds of kilometers) so that phase differences between wave components propagating in distinct modes accumulate to significant and detectable levels. Taking these elements into consideration and noting that, at VHF frequencies, the longitudinal components of the wave fields are negligibly small (as $X \ll 1$ and $Y \ll 1$), waves propagating through the ionosphere can be represented as TEM (transverse electromagnetic) waves.

Consider plane wave propagation in an inhomogeneous magnetized ionosphere in an arbitrary direction $\hat{\mathbf{k}}$. To model the electric field of the propagating wave, we can divide the ionosphere in slabs of equal width (see Figure 13) perpendicular to the propagation direction such that within each slab the physical plasma parameters (as electron density, electron and ion temperatures, and magnetic field) can be considered constants.[9] The transverse component of the wave electric field propagates from the bottom to the top of the i-th slab according to (66), that is

$$\mathbf{E}_i = \underbrace{e^{-jk_o\bar{n}_i\Delta r}\left[e^{-jk_o\Delta n_i\Delta r}\hat{\mathbf{p}}_O\hat{\mathbf{p}}_O^H + e^{jk_o\Delta n_i\Delta r}\hat{\mathbf{p}}_X\hat{\mathbf{p}}_X^H\right]}_{\bar{\mathbf{T}}_i}\mathbf{E}_{i-1},\tag{68}$$

which is the superposition of the O- and X-modes of magnetoionic propagation detailed in the previous section. Above, $\bar{\mathbf{T}}_i$ denotes the i-th propagator matrix (expressed in cartesian coordinates), where $k_o \equiv 2\pi/\lambda_o$ is the free-space wavenumber, Δr is the width of the slab, and where $\bar{n}_i \equiv \frac{n_{O,i}+n_{X,i}}{2}$ and $\Delta n_i \equiv \frac{n_{O,i}-n_{X,i}}{2}$ are the mean and half difference between the refractive indices of the propagation modes in the i-th layer. The polarization vectors of the O- and X-modes are

$$\hat{\mathbf{p}}_O = \frac{\hat{\theta} - ja_i\,\hat{\phi}}{\sqrt{1+a_i^2}} \qquad\text{and}\qquad \hat{\mathbf{p}}_X = \frac{-ja_i\,\hat{\theta} + \hat{\phi}}{\sqrt{1+a_i^2}}\tag{69}$$

where $a_i \equiv \frac{F_{O,i}}{Y_{L,i}} = -\frac{Y_{L,i}}{F_{X,i}}$ is the polarization parameter, and $\hat{\theta}_i$ and $\hat{\phi}_i$ are a pair of mutually orthogonal unit vectors perpendicular to $\hat{\mathbf{k}}$ whose directions depend on the relative orientation of the propagation vector $\mathbf{k}$ and the magnetic field $\mathbf{B}_i$ (see Figure 12). Neglecting reflection from the interfaces between slabs, the field components of an upgoing plane wave propagating in the $+\hat{\mathbf{k}}$ direction (at a distance $r_i = i\Delta r$ from the origin) can be computed by the successive application of the propagator matrices; that is,

$$\mathbf{E}_i^u = \bar{\mathbf{T}}_i \cdots \bar{\mathbf{T}}_2\bar{\mathbf{T}}_1\mathbf{E}_o^u,\tag{70}$$

where $\mathbf{E}_o^u$ is the wave field at the origin (perpendicular to $\hat{\mathbf{k}}$), and $\bar{\mathbf{T}}_1 \ldots \bar{\mathbf{T}}_i$ are the propagator matrices from the bottom layer to the i-th layer. Similarly, taking advantage of the bidirectionality of the propagator matrices, the field components of a downgoing plane wave propagating in the $-\hat{\mathbf{k}}$ direction (from the i-th layer to the ground) can be written as

$$\mathbf{E}_o^d = \bar{\mathbf{T}}_1\bar{\mathbf{T}}_2 \cdots \bar{\mathbf{T}}_i\mathbf{E}_i^d,\tag{71}$$

where $\mathbf{E}_i^d$ is the field at the top of the i-th layer.

In radar experiments, the transverse field component of the signal backscattered from a radar range $r_i = i\Delta r$ can be modeled as

$$\mathbf{E}_o^r \propto \kappa_i\,\underbrace{\bar{\mathbf{T}}_1\bar{\mathbf{T}}_2 \cdots \bar{\mathbf{T}}_i\bar{\mathbf{T}}_i \cdots \bar{\mathbf{T}}_2\bar{\mathbf{T}}_1}_{\bar{\Pi}_i}\,\mathbf{E}_o^t,\tag{72}$$

where $\mathbf{E}_o^t$ and $\mathbf{E}_o^r$ are the fields transmitted and received by the radar antenna in the $\hat{\mathbf{k}}$ direction. Above, $\bar{\Pi}_i$ denotes a two-way propagator matrix that accounts for the polarization effects on

---

[9] In the ionosphere, electron density and plasma temperatures can be considered to be functions of altitude $f(z)$. Thus, the values of these physical parameters at any position $\mathbf{r}$ from a radar placed at the origin are given by $f(r\cos\psi)$ where r is the radar range and ψ is the zenith angle.

the waves incident on and backscattered from the radar range r_i (upgoing and downgoing waves, respectively). In addition, κ_i is a random variable related to the radar cross section (RCS) of the scatterers at the range r_i (e.g., randomly moving ionospheric electrons).

We now consider an $\hat{x}$ polarized radar antenna transmitting

$$\hat{p}_1 = \frac{\hat{k} \times \hat{k} \times \hat{x}}{|\hat{k} \times \hat{k} \times \hat{x}|} \tag{73}$$

polarized waves field in $\hat{k}$ direction. On reception, the same antenna would be co-polarized with incoming fields of identical polarization direction $\hat{p}_1$. For an orthogonal $\hat{y}$ polarized antenna

$$\hat{p}_2 = \frac{\hat{k} \times \hat{k} \times \hat{y}}{|\hat{k} \times \hat{k} \times \hat{y}|} \tag{74}$$

would be the polarization direction of co-polarized fields. Let's assume that these two antennas, located at the geomagnetic equator, scan the ionosphere from north to south to construct power maps of the backscattered signals. In every pointing direction, narrow pulses are transmitted so that range filtering effects (due to the convolution of the pulse shape with the response of the ionosphere) can be ignored. In transmission, only the first antenna ($\hat{x}$ polarized) is excited, while, in reception, both antennas are used to collect the backscattered signals. The two antennas then provide us with co- and cross-polarized output voltages

$$v_1(\hat{k}) \propto \kappa_i \, \hat{p}_1^{\mathsf{T}} \, \bar{\Pi}_i \, \hat{p}_1 \quad \text{and} \quad v_2(\hat{k}) \propto \kappa_i \, \hat{p}_2^{\mathsf{T}} \, \bar{\Pi}_i \, \hat{p}_1, \tag{75}$$

sampled at each range r_i, where the two-way propagator matrix $\bar{\Pi}_i$ (defined above) is dependent on the electron density and magnetic field values along $\hat{k}$ up to the radar range r_i. As κ_i is a random variable, the statistics of voltages (75) would be needed to characterize the scattering targets. For instance, the mean square values of v_1 and v_2 can be modeled as

$$\langle |v_1|^2 \rangle \propto \sigma_v \Gamma_1 \quad \text{and} \quad \langle |v_2|^2 \rangle \propto \sigma_v \Gamma_2, \tag{76}$$

where $\sigma_v = \langle |\kappa_i|^2 \rangle$ is the volumetric RCS of the medium, which is dependent on the electron density, temperature ratio, and magnetic aspect angle at any given range. In addition, Γ_1 and Γ_2 are polarization coefficients defined as

$$\Gamma_1 = \left| \hat{p}_1^{\mathsf{T}} \, \bar{\Pi}_i \, \hat{p}_1 \right|^2 \quad \text{and} \quad \Gamma_2 = \left| \hat{p}_2^{\mathsf{T}} \, \bar{\Pi}_i \, \hat{p}_1 \right|^2. \tag{77}$$

To simulate radar voltages using the model described above, an ionosphere with the electron density and T_e/T_i profiles displayed in Figure 14 was considered. In addition, the magnetic field was computed using the International Geomagnetic Reference Field (IGRF) model (e.g., Olsen et al., 2000). Finally, the simulations were performed for a 50 MHz radar at the location of the Jicamarca ISR in Peru and antenna polarizations $\hat{x}$ and $\hat{y}$ were taken to point in SE and NE directions as at Jicamarca.

Let us first analyze magnetoionic propagation effects on the simulated radar voltages, disregarding scattering effects. For this purpose, polarization coefficients Γ_1 and Γ_2 are displayed in Figure 15 as functions of distance and altitude from the radar (in the plots, the positive horizontal axis is directed north). Note that, at low altitudes, where there is

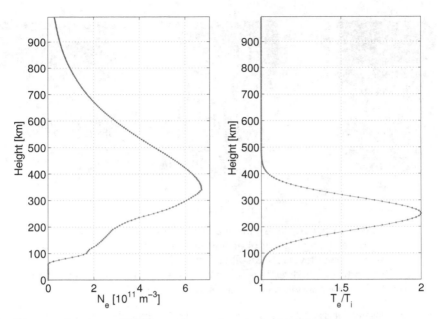

Fig. 14. Electron density and T_e/T_i profiles as functions of height.

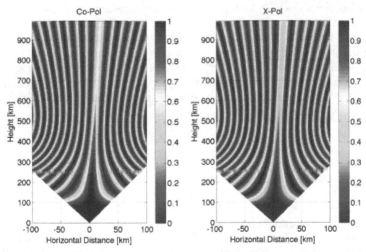

Fig. 15. Polarization coefficients for the mean square voltages detected by a pair of orthogonal linearly polarized antennas placed at Jicamarca. The antennas have very narrow beams and scan the ionosphere from north to south probing different magnetic aspect angle directions. Note that, for most pointing directions, the polarization of the detected fields rotates (Faraday rotation effect), except in the direction where the beam is pointed perpendicular to **B**, in which case, the type of polarization changes from linear to circular (Cotton-Mouton effect).

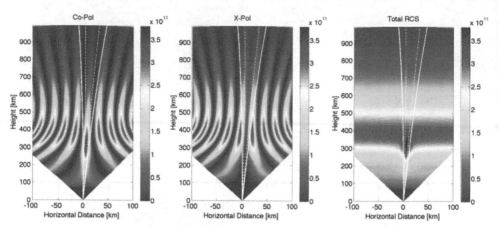

Fig. 16. Co-polarized (left panel), cross-polarized (middle panel), and total (right panel) backscattered power detected by a pair of orthogonal linearly polarized antennas (see caption of Figure 15). Power levels are displayed in units of electron density. In each plot, the dashed white lines indicate the directions half a degree away from perpendicular to **B**, while the continuous lines correspond to the directions one degree off.

no ionosphere, signal returns will be detected only by the co-polarized antenna (i.e., by the same antenna used on transmission). However, as the signal propagates farther through the ionosphere, magnetoionic effects start taking place. We can appreciate that, for most of the propagation directions, the polarization vector of the detected field rotates such that signal from one polarization goes to the other as the radar range increases (Faraday rotation effect). Note, however, that there is a direction in which the wave polarization does not rotate much. In this direction, the antenna beams are pointed perpendicular to the Earth's magnetic field, and it can be observed that the polarization of the detected fields varies progressively from linear to circular as a function of height (Cotton–Mouton effect). Finally, note that at higher altitudes, where the ionosphere vanishes, no more magnetoionic effects take place, and the polarization of the detected signal approaches a final state.

Next, scattering and propagation effects are considered in the simulation of the backscattered power collected by the pair of orthogonal antennas described above. The incoherent scatter volumetric RCS formulated in the previous section is used in the calculations. In Figure 16, the simulated co-polarized (left panel) and cross-polarized (middle panel) power data are displayed as functions of distance and altitude from the radar. In addition, the right panel depicts the total power detected by both antennas. Note that power levels are displayed as volumetric radar cross sections divided by $4\pi r_e^2$ (i.e., power levels are in units of electron density). In each plot, the dashed white lines indicate the directions half a degree away from perpendicular to **B**, while the continuous lines correspond to the directions one degree off.

In the plots, we can observe that there is negligible backscattered power at low altitudes. At higher altitudes between approximately 200 and 700 km (where polarization effects are significant), co- and cross-polarized power maps exhibit features that are similar to the ones observed in Figure 15. Note, however, that there is an enhancement of the detected power in the direction where the antenna beams are pointed perpendicular to **B**; this can be observed more clearly in the plot of the total power (right panel of Figure 16). This feature is characteristic of the incoherent scatter process for probing directions perpendicular to $\mathbf{B}$ and for heights where electron temperature exceeds the ion temperature (i.e., $T_e > T_i$) as described before. At even higher altitudes, scattered signals become weaker and weaker as the ionospheric electron density vanishes.

To model incoherent scatter radar measurements using the propagation model presented in this section, an extra level of complexity has to be considered because, within the range of aspect angles illuminated by the antenna beams, propagation and scattering effects vary quite rapidly. For this reason, the measured backscattered radar signals need to be carefully modeled taking into account the shapes of the antenna beams. A model for the beam-weighted incoherent scatter spectrum that considers magnetoionic propagation and collisional effects is formulated next.

9.3 Soft-target radar equation and magnetoionic propagation

In this section, the soft-target radar equation is reformulated using the wave propagation model described above. Consider a radar system composed of a set of antenna arrays (located in the same area) with matched filter receivers connected to the antennas used in reception. The mean square voltage at the output of the i-th receiver can be expressed as

$$\langle |v_i(t)|^2 \rangle = E_t K_i \int dr \, d\Omega \, \frac{d\omega}{2\pi} \, \frac{|\mathcal{T}(\hat{\mathbf{r}})|^2 \, |\mathcal{R}_i(\hat{\mathbf{r}})|^2}{k^2} \, \Gamma_i(\mathbf{r}) \, \frac{|\chi(t - \frac{2r}{c}, \omega)|^2}{4\pi r^2} \, \sigma_v(\mathbf{k}, \omega), \qquad (78)$$

where t is the radar delay, E_t is the total energy of the transmitted radar pulse, and K_i is the i-th calibration constant (a proportionality factor that accounts for the gains and losses along the i-th signal path). Integrals are taken over range r, solid angle Ω, and Doppler frequency $\omega/2\pi$. In addition, $\mathbf{k} = -2k_0\hat{\mathbf{r}}$ denotes the relevant Bragg vector for a radar with a carrier wavenumber k_0 and associated wavelength λ. Above, $\mathcal{T}(\hat{\mathbf{r}})$ and $\mathcal{R}_i(\hat{\mathbf{r}})$ are the antenna factors of the arrays used in transmission and reception. Note that $|\mathcal{T}(\hat{\mathbf{r}})|^2$ and $|\mathcal{R}_i(\hat{\mathbf{r}})|^2$ are antenna gain patterns and the product $|\mathcal{T}(\hat{\mathbf{r}})|^2 \, |\mathcal{R}_i(\hat{\mathbf{r}})|^2$ is the corresponding two-way radiation pattern. The polarization coefficient $\Gamma_i(\mathbf{r})$ is defined as

$$\Gamma_i(\mathbf{r}) = \left| \hat{\mathbf{p}}_i^\mathsf{T} \, \bar{\bar{\Pi}}(\mathbf{r}) \, \hat{\mathbf{p}}_t \right|^2, \qquad (79)$$

where $\hat{\mathbf{p}}_t$ and $\hat{\mathbf{p}}_i$ are the polarization unit vectors of the transmitting and receiving antennas, and $\bar{\bar{\Pi}}(\mathbf{r})$ is the two-way propagator matrix for the wave field components propagating along $\hat{\mathbf{r}}$ (incident on and backscattered from the range r). Note that $\hat{\mathbf{p}}_t$ and $\hat{\mathbf{p}}_i$ are normal to $\hat{\mathbf{r}}$ because propagating fields are represented as TEM waves. In addition, $\chi(t, \omega)$ is the radar ambiguity function and $\sigma_v(\mathbf{k}, \omega)$ is the volumetric RCS spectrum, functions that have been defined before. Similarly, the cross-correlation of the voltages at the outputs of the i-th and j-th receivers can be expressed as

$$\langle v_i(t) v_j^*(t) \rangle = E_t K_{i,j} \int dr \, d\Omega \, \frac{d\omega}{2\pi} \, \frac{|\mathcal{T}(\hat{\mathbf{r}})|^2 \, \mathcal{R}_i(\hat{\mathbf{r}}) \, \mathcal{R}_j^*(\hat{\mathbf{r}})}{k^2} \, \Gamma_{i,j}(\mathbf{r}) \, \frac{|\chi(t - \frac{2r}{c}, \omega)|^2}{4\pi r^2} \, \sigma_v(\mathbf{k}, \omega), \quad (80)$$

where $K_{i,j}$ is a cross-calibration constant (dependent on gains and losses along the i-th and j-th signal paths), and $\Gamma_{i,j}(\mathbf{r})$ is a cross-polarization coefficient defined as

$$\Gamma_{i,j}(\mathbf{r}) = \left(\hat{\mathbf{p}}_i^\mathsf{T} \, \bar{\bar{\Pi}}(\mathbf{r}) \, \hat{\mathbf{p}}_t \right) \left(\hat{\mathbf{p}}_j^\mathsf{T} \, \bar{\bar{\Pi}}(\mathbf{r}) \, \hat{\mathbf{p}}_t \right)^*. \qquad (81)$$

Note that dispersion of the pulse shape due to wave propagation effects has been neglected in our model.

Denoting by $S_i(\omega)$ the self-spectrum of the signal at the output of the i-th receiver and applying Parseval's theorem, we have that

$$\langle |v_i(t)|^2 \rangle = \int \frac{d\omega}{2\pi} S_i(\omega). \tag{82}$$

Likewise, the cross-spectrum $S_{i,j}(\omega)$ and the cross-correlation of the signals at the outputs of the i-th and j-th receivers are related by

$$\langle v_i(t)v_j^*(t) \rangle = \int \frac{d\omega}{2\pi} S_{i,j}(\omega). \tag{83}$$

Assuming that the ambiguity function is almost flat within the bandwidth of the RCS spectrum $\sigma_v(\mathbf{k}, \omega)$ (which is a valid approximation in the case of short-pulse radar applications), we can use equations (78) and (80) to obtain the following beam-weighted spectrum and cross-spectrum models:

$$S_i(\omega) = E_t K_i \int dr \frac{|\chi(t - \frac{2r}{c})|^2}{4\pi r^2} \int d\Omega \frac{|\mathcal{T}(\hat{\mathbf{r}})|^2 |\mathcal{R}_i(\hat{\mathbf{r}})|^2}{k^2} \Gamma_i(\mathbf{r}) \sigma_v(\mathbf{k}, \omega) \tag{84}$$

and

$$S_{i,j}(\omega) = E_t K_{i,j} \int dr \frac{|\chi(t - \frac{2r}{c})|^2}{4\pi r^2} \int d\Omega \frac{|\mathcal{T}(\hat{\mathbf{r}})|^2 \mathcal{R}_i(\hat{\mathbf{r}}) \mathcal{R}_j^*(\hat{\mathbf{r}})}{k^2} \Gamma_{i,j}(\mathbf{r}) \sigma_v(\mathbf{k}, \omega), \tag{85}$$

where

$$\chi(t) = \frac{1}{T} f^*(-t) * f(t) \tag{86}$$

is the normalized auto-correlation of the pulse waveform $f(t)$. In the radar equations (84) and (85), the polarization coefficients $\Gamma_i(\mathbf{r})$ and $\Gamma_{i,j}(\mathbf{r})$ effectively modify the radiation patterns; thus, the spectrum shapes are dependent not only on the scattering process but also on the modes of propagation. This dependence further complicates the spectrum analysis of radar data and the inversion of physical parameters.

10. Summary

In this chapter we have described the operation of ionospheric incoherent scatter radars (ISR) and the signal spectrum models underlying the operation of such radars. ISR's are the premier remote sensing instruments used to study the ionosphere and Earth's upper atmosphere. First generation operational ISR's were built in the early 1960's — e.g., Jicamarca in Peru and Arecibo in Puerto Rico — and ISR's continue to play a crucial role in our studies of Earth's near space environment. These instruments are primarily used to monitor the electron densities and drifts, as well as temperatures and chemical composition of ionospheric plasmas. The latest generation of ISR's include the AMISR — advanced modular ISR — series which are planned to be deployed around the globe and then re-located depending on emerging science needs. With increasing ISR units around the globe, there will be a larger demand on radar engineers and technicians familiar with ISR modes and the underlying scattering theory. For that reason, in our presentation in this chapter, as well as in our recent papers (Kudeki & Milla, 2011; Milla & Kudeki, 2011), we have taken an "engineering approach" to describe the theory of the incoherent scatter spectrum. Complementary physics based descriptions of the same processes can be found in many of the original ISR papers included in references.

11. Acknowledgements

This chapter is based upon work supported by the National Science Foundation under Grant No. 0215246 and 1027161.

12. References

Bowles, K. L. (1958). Observation of vertical-incidence scatter from the ionosphere at 41 mc/sec, *Physical Review Letters* 1(12): 454–455.

Budden, K. G. (1961). *Radio Waves in the Ionosphere*, Cambridge University Press, Cambridge, United Kingdom.

Callen, H. B. & Welton, T. A. (1951). Irreversibility and generalized noise, *Physical Review* 83(1): 34–40.

Callen, J. D. (2006). *Fundamentals of Plasma Physics*, Chapter 2 – Coulomb Collisions. URL: *http://homepages.cae.wisc.edu/ callen/book.html*

Chandrasekhar, S. (1942). *Principles of Stellar Dynamics*, University of Chicago Press, Chicago.

Chandrasekhar, S. (1943). Stochastic problems in physics and astronomy, *Reviews of Modern Physics* 15(1): 1–89.

Chau, J. L. & Kudeki, E. (2006). First E- and D-region incoherent scatter spectra observed over Jicamarca, *Annales Geophysicae* 24(5): 1295–1303.

Dougherty, J. P. (1964). Model Fokker-Planck equation for a plasma an its solution, *The Physics of Fluids* 7(11): 1788–1799.

Dougherty, J. P. & Farley, D. T. (1963). A theory of incoherent scattering of radio waves by a plasma 3. Scattering in a partly ionized gas, *Journal of Geophysical Research* 68: 5473–5486.

Farley, D. T. (1964). The effect of Coulomb collisions on incoherent scattering of radio waves by a plasma, *Journal of Geophysical Research* 69(1): 197–200.

Farley, D. T. (1966). A theory of incoherent scattering of radio waves by a plasma 4. The effect of unequal ion and electron temperatures, *Journal of Geophysical Research* 71(17): 4091–4098.

Gillespie, D. T. (1996a). The mathematics of Brownian motion and Johnson noise, *American Journal of Physics* 64(3): 225–240.

Gillespie, D. T. (1996b). The multivariate Langevin and Fokker–Planck equations, *American Journal of Physics* 64(10): 1246–1257.

Gordon, W. E. (1958). Incoherent scattering of radio waves by free electrons with applications to space exploration by radar, *Proceedings of the IRE* 46(11): 1824–1829.

Hagfors, T. & Brockelman, R. A. (1971). A theory of collision dominated electron density fluctuations in a plasma with applications to incoherent scattering, *The Physics of Fluids* 14(6): 1143–1151.

Holod, I., Zagorodny, A. & Weiland, J. (2005). Anisotropic diffusion across an external magnetic field and large-scale fluctuations in magnetized plasmas, *Physical Review E* 71(4): 1–11.

Kudeki, E., Bhattacharyya, S. & Woodman, R. F. (1999). A new approach in incoherent scatter F region $E \times B$ drift measurements at Jicamarca, *Journal of Geophysical Research* 104(A12): 28145–28162.

Kudeki, E. & Milla, M. A. (2011). Incoherent scatter spectral theories–Part I: A general framework and results for small magnetic aspect angles, *IEEE Transactions on Geoscience and Remote Sensing* 49(1): 315–328.

Mathews, J. D. (1984). The incoherent scatter radar as a tool for studying the ionospheric D-region, *Journal of Atmospheric and Terrestrial Physics* 46(32): 975–986.

Milla, M. A. & Kudeki, E. (2006). *F*-region electron density and T_e/T_i measurements using incoherent scatter power data collected at ALTAIR, *Annales Geophysicae* 24(5): 1333–1342.

Milla, M. A. & Kudeki, E. (2009). Particle dynamics description of "BGK collisions" as a Poisson process, *Journal of Geophysical Research* 114(7): 1–4.

Milla, M. A. & Kudeki, E. (2011). Incoherent scatter spectral theories–Part II: Modeling the spectrum for modes propagating perpendicular to **B**, *IEEE Transactions on Geoscience and Remote Sensing* 49(1): 329–345.

Olsen, N., Sabaka, T. J. & Tøffner-Clausen, L. (2000). Determination of the IGRF 2000 model, *Earth, Planets and Space* 52(12): 1175–1182.

Rosenbluth, M. N., MacDonald, W. M. & Judd, D. L. (1957). Fokker-Planck equation for an inverse-square force, *Physical Review* 107(1): 1–6.

Sulzer, M. P. & González, S. A. (1999). The effect of electron Coulomb collisions on the incoherent scatter spectrum in the *F* region at Jicamarca, *Journal of Geophysical Research* 104(A10): 22535–22551.

Uhlenbeck, G. E. & Ornstein, L. S. (1930). On the theory of the Brownian motion, *Physical Review* 36(5): 823–841.

Woodman, R. F. (1967). *Incoherent scattering of electromagnetic waves by a plasma*, PhD thesis, Harvard University, Cambridge, Massachusetts.

Yeh, K. C. & Liu, C. H. (1972). *Theory of Ionospheric Waves*, Vol. 17 of *International Geophysics*, Elsevier.

9

Aperture Synthesis Radar Imaging for Upper Atmospheric Research

D. L. Hysell and J. L. Chau

Earth and Atmospheric Sciences, Cornell University, Ithaca, New York
Jicamarca Radio Observatory, Lima
U.S.A., Peru

1. Introduction

Radars used for upper-atmospheric applications can be engineered to measure the Doppler spectra of their targets adequately for most intents and purposes, the spectral resolution being limited only by the observing time and the constraints of stationarity. Likewise, they can measure the range to their targets adequately for most intents and purposes, range resolution being limited by system bandwidth, the power budget, and the constraints of stationarity. Problems arise for "overspread" targets, where range and frequency aliasing cannot simultaneously be avoided using pulse-to-pulse methodologies, and more complicated pulse-to-lag or aperiodic pulsing methods are required (see for example (Farley, 1972; Huuskonen et al., 1996; Lehtinen, 1986; Sulzer, 1986; Uppala, 1993)). Important examples of this situation include incoherent scatter experiments (Farley, 1969), observations of meteor head echoes (Chau & Woodman, 2004), and observations of plasma density irregularities present in certain rapid flows, as are found in the equatorial ionosphere during so-called "equatorial spread F" (Woodman, 2009; Woodman & La Hoz, 1976).

Where capabilities are most limited is in bearing determination and the associated problems of imaging in the directions transverse to the radar beam. Electronic beam steering using phased-array radars offers a means of radar imaging (e.g. Semeter et al. (2009)), but the number of pointing positions that can be used is limited by the incoherent integration time required for each position. If the power budget permits, transmission can be done using a broad beam, and beam forming can be done "after the fact", such that all pointing positions are examined simultaneously (e.g. Kudeki & Woodman (1990)). Even so, the angular resolution will be limited by the size of the antenna array unless the diffraction limit is removed through numerical deconvolution. The half-power beamwidth of large-aperture radars used for upper atmospheric research is usually of the order of one degree. At ionospheric altitudes, this translates to a transverse resolution of a few to a few tens of kilometers, which may be larger than the scales at which primary plasma waves are excited. The resolution of medium-sized and small research radars with their relatively smaller antenna arrays is relatively poorer still. In applications involving coherent scatter from plasma density irregularities, targets of interest may exhibit backscatter intensities spanning 30 dB or more of dynamic range. For such targets, the 3 dB beamwidth of the antenna is essentially irrelevant, and even targets in the sidelobes of the antenna radiation pattern can contribute to the power assigned to a given pointing direction. This poses challenges for observing plasma irregularities with important scale sizes of a kilometer or less, which is often the case in ionospheric research.

Aperture synthesis radar imaging utilizes spaced-antenna data to construct true images of the scatterers versus bearing. Some approaches are adaptive, and some achieve "super resolution" by incorporating the effects of diffraction in the analysis. All information about range and Doppler shift can be retained, meaning that the images can be four dimensional, not counting the time axis. As the techniques can synthesize large apertures from small, sparsely-distributed sensors, they may be especially beneficial for small and medium-sized radars, although some of the benefits can only be realized when high signal-to-noise ratios are available.

In this paper, we review the formulation of the radar imaging problem, which is based on concepts and language derived from radar interferometry and radio astronomy. As radar imaging belongs to the class of problems known as inverse problems, some of the ideas from that domain are also reviewed. The factors that govern the resolution achievable in practice will be described, and optimal strategies for sensor placement will be discussed. Error analysis in radar imaging is treated, and some extensions to the basic imaging procedure are outlined. Finally, examples of radar imaging implementations are drawn from upper atmospheric and ionospheric applications. Application in the lower atmosphere exist as well but will not be covered here (Chau & Woodman, 2001; Hassenpflug et al., 2008; Palmer et al., 1998).

2. Imaging problem

The imaging problem has been formulated by (Thompson, 1986), and we follow his treatment below. We consider the far-field problem only and regard the backscatter in a given range gate as a random process constituted by plane waves with sources that are statistically uncorrelated and distributed in space. Imaging data have the form of complex interferometric cross spectra obtained from spaced antenna pairs separated by a vector distance $\mathbf{d}$. Such "visibility" measurements $V(k\mathbf{d}, f_D)$ are related to the "brightness" distribution $B(\hat{\sigma}, f_D)$, the scattered power density as a function of bearing and Doppler frequency, by

$$V(k\mathbf{d}, f_D) = \int A_N(\hat{\sigma})B(\hat{\sigma}, f_D)e^{jk\mathbf{d}\cdot\hat{\sigma}}d\Omega \tag{1}$$

where k is the wavenumber, f_D is the Doppler frequency, and $\hat{\sigma}$ is a unit vector in the direction of the bearing of interest and where the integration is over all solid angles in the upper half space. As different Doppler spectral components of the data are treated independently, we omit f_D in the formalism that follows.

In (1), A_N is the normalized two-way antenna effective area. In radar imaging, the antennas used for reception are typically much smaller than the antennas used for transmission, and A_N is consequently dominated by the characteristics of the transmitting antenna array. Together, the product $A_N B$ is the effective brightness distribution, B_{eff}, which represents the angular distribution of the received signals. It is this quantity that are interested in recovering from the data, the antenna radiation patterns being known. The radiation pattern need only be treated explicitly when heterogeneous receiving antennas are used (see below).

Equation (1) resembles a Fourier transform between baseline and bearing space. In Cartesian coordinates, (1) is

$$V(kd_x, kd_y, kd_z)$$

$$= \int \frac{B_{\text{eff}}(\eta, \xi)}{\sqrt{1 - \eta^2 - \xi^2}} e^{jk\left(d_x\eta + d_y\xi + d_z\sqrt{1-\eta^2-\xi^2}\right)} d\eta d\xi \tag{2}$$

where η and ξ are the direction cosines of $\hat{\sigma}$ with respect to the x and y coordinates, which can be oriented arbitrarily. If the field of view of the sky being considered is sufficiently restricted, the radical in the denominator of the integrand can be regarded as a constant. Then, if the spaced receivers are coplanar so that all d_z can be made to be zero, or if the brightess is finite only where η and ψ are small, (2) becomes a two-dimensional Fourier transform. This condition is required for "fringe stopping," the practice of calibrating the complex gains of the sensor channels so as to remove the fringes in the visibility spectrum of a point calibration target (see phase calibration section below). If B_{eff} is limited by the finite width of the radar radiation pattern, the visibilities can be completely represented by a discrete set of periodic visibility measurements.

3. Inverse methods

Equation (2) shows that the radar imaging problem is actually the linear problem of inverting a Fourier transform. Since visibility data are generally acquired sparsely and incompletely, inversion of (2) by means of a discrete Fourier transform algorithm is generally impossible and would be undesirable in any case unless that algorithm were followed by another one to deconvolve the radar radiation pattern from the resulting image. A number of somewhat ad-hoc approaches to Fourier analysis of sparsely sampled data have emerged, including the Lomb periodogram (Lomb, 1976) and CLEAN (Högbom, 1974), which amount to linear least-squares fitting of representer functions to the data. Performance of such algorithms is uneven and suffers from pathologies inherent in inverse problems that have not been accounted for. The pathologies arise in part from the fact that the measured visibilities contain noise that is not incorporated in (2), which is therefore only an approximation.

As (2) is a linear transformation, the discrete visibility data could be mapped to a discretized version of the brightness distribution through a matrix (G) whose properties would describe the characteristics of the inverse solution. For example, if the column space of G is incomplete, there may be no model brightness that can reproduce the measured visibilities, the problem would be over-determined, and no solution would exist. If the row space of G is incomplete, then it would be possible to add features to the brightness without altering the predicted visibilities, the problem would be under-determined, and solutions to the inverse problem would not be unique.

More generally, both the column and row spaces of G may be incomplete, meaning that the problem is simultaneously over- and under-determined (mixed determined or rank deficient) and that any number of candidate brightness distributions might give acceptably close approximations to the desired solution in terms of a chi-squared prediction error metric. If G is also poorly conditioned, then those candidates could vary widely, and the inversion would be unstable.

Inverse methods are required for mixed-determined, poorly-conditioned problems (see e.g. Aster et al. (2005); Menke (1984); Tarantola (1987) for a review). The strategy generally amounts to reducing the candidate space of model solutions by imposing a priori information. The information may involve expectations about the variance of the model solution (its "roughness" or regularity) or something more specific, such as the range of admissible numerical values it can assume. A priori information may be incorporated implicitly through the inclusion of damping terms in the inversion algorithm (e.g. damped least squares, Tikhonov regularization, etc.) or explicitly using a Bayesian formalism. Other desirable properties, including model-data consistency, model resolution, and data resolution, can also be optimized.

A common approach to radar imaging is based on the linear constrained minimum variance (LCMV) (or sometimes minimum variance distortionless response (MVDR)) principle and was introduced by Capon (1969). Consider the column vector x with n entries corresponding to complex voltage samples from n sensors, each at a coordinate r_n measured from some reference point. Suppose the objective is to discriminate echoes arriving along a wavevector k from other echoes, noise, interference, etc., by forming an appropriated weighted sum of the voltage samples, $y = w^t x$, prior to detection. If the weights are all unity, the signals detected by the sensors from a point source designated by k would be proportional to the column vector with elements $e^{ik \cdot r_1}, e^{ik \cdot r_2}, \cdots, e^{ik \cdot r_n}$. After incoherent integration, the output of the detector with arbitrary weights will be

$$\langle |y^2| \rangle = w^t \langle xx^t \rangle w = w^t R w \tag{3}$$

where w is the column vector composed of the weights, R is the signal covariance matrix, constructed from the measured visibilities, and t denotes the complex conjugate transpose.

Capon's LCMV strategy is to optimize the weights by minimizing the output of the detector while maintaining unity gain in the direction of the point source, viz.

$$w = \arg\min_{w,\gamma} : w^t R w + \gamma(w^t e - 1) \tag{4}$$

where γ is a Lagrange multiplier. The unity-gain constraint is imposed to prevent the trivial solution. The output of the optimal detector using the weights thus found for a given bearing can readily be shown to be $\langle |y|^2 \rangle = (e^t R^{-1} e)^{-1}$. Imaging then is performed by computing the optimized detector output for all possible bearings. The algorithm is essentially a linear beam former, where nulls are adaptively aligned with sources that are not aligned with the bearing of interest.

Capon's LCMV method is simple to implement and execute computationally. While there is no provision for error handling in the algorithm posed above, the remedy is to precompute the visibility error covariance matrix (see below) and then transform (3) through similarity transformation into a space where that matrix is the identity. The method is equally well suited for imaging continua and point targets, making it a superior choice for geophysical remote sensing compared to point-targeting algorithms like MUSIC (Schmidt, 1986) or CLEAN. However, there is no guarantee that the brightness distribution found will be consistent with the visibility data within the tolerance of the specified error bounds. The a priori information contained within the method is moreover far from explicit, making it hard to assess its validity.

Any number of alternative imaging methods exist that can minimize or constrain the model prediction error while managing issues arising from the mixed determined or ill conditioned nature of the inverse problem. In the next section of the paper, we turn our attention to the

MaxEnt algorithm, which does not suffer from the limitations of Capon's method and which possesses a number of other desirable features as a consequence of the incorporation of rather informative prior information.

4. MaxEnt formulation

The algorithm described below derives from the MaxEnt spectral analysis method, a Bayesian method based on maximizing the Shannon entropy of the spectrum (Shannon & Weaver, 1949). The method should not be confused with the maximum entropy method (MEM or ME) or similar autoregressive models, with which it has only a remote connection (Jaynes, 1982). MaxEnt was originally applied to spaced-receiver imaging by Gull & Daniell (1978) and also by Wernecke & D'Addario (1977) in a more generalized way. Variations on the technique were published later by Wu (1984), Skilling & Bryan (1984), and Cornwell & Evans (1985). MaxEnt is now applied to a wide array of problems, including natural language processing (NLP), quantum physics, and climate science, to name a few.

Expansive rationales for MaxEnt have been given by Ables (1974), Jaynes (1982; 1985), Skilling (1991), and Daniell (1991), among others. MaxEnt is a Bayesian optimization technique that maximizes the MAP (maximum a posteriori) probability of an image given prior probability rooted in Shannon's entropy and constraints related to the model prediction error, error bounds, image support, certain normalization, and other factors. As entropy admits only globally positive brightness distributions, it rejects the vast majority of candidate solutions in favor of a small, allowable solution subspace. The entropy metric favors uniform images in the absence of contrary information but is nevertheless edge preserving. Moreover, the use of entropy for prior probability makes the algorithm minimally dependent on unknown quantities and, in that sense, bias and artifact free. It is a formalization of Occam's razor.

The algorithm described here is based on one developed by Wilczek & Drapatz (1985) (WD85) for radio astronomy. The real valued brightness will be represented by the symbol $f_i = f(\theta_i)$. The visibility data come from normalized cross-correlation estimates

$$V(k\mathbf{d}_j) = \frac{\langle v_1 v_2^* \rangle}{\sqrt{\langle |v_1|^2 \rangle - N_1}\sqrt{\langle |v_2|^2 \rangle - N_2}} \tag{5}$$

where the $v_{1,2}$ represent quadrature voltage samples from a pair of receivers spaced by a distance $\mathbf{d}_j$ and $N_{1,2}$ are the corresponding noise estimates. The angle brackets above are the expectation. We will represent the visibility data by the symbol $g_j = g(k\mathbf{d}_j)$ and assign two real values for each baseline; one each for the real and imaginary part of (5). Given M interferometry baselines with nonzero length, there will be a total of $2M + 1$ distinct visibility data. (The visibility for the zero baseline is identically unity.)

In matrix notation, (2) may be expressed as

$$g^t + e^t = f^t h \tag{6}$$

where g, e, and f are column vectors and t represents the transpose. Here, the elements of the matrix h (h_{ij}) are the real or the imaginary part of the point spread function $\exp(ik\mathbf{d}_j \cdot \hat{\sigma}_i)$, depending on whether g_j is the real or imaginary part of (5), and e_j represents the corresponding random error arising from the finite number of samples used to estimate (5). The elements f_i of vector f represent the brightness distribution evaluated across the defined image space.

MaxEnt explicitly associates the prior probability of a candidate image with the Shannon entropy of the brightness distribution, $S = -\sum_i f_i \ln(f_i/F)$. Here, $F = f^t 1 = g_0 = g(0)$ is the total image brightness ("1" being a column with unity elements). Of all distributions, the uniform one has the highest entropy. In that sense, entropy is a smoothness metric. The entropy of an image is also related to the likelihood of occurrence in a random assembly process. All things being equal, a high-entropy distribution should be favored over a low entropy one. The former represents a broadly accessible class of solutions, while the latter represents an unlikely outcome that should only be considered if the data demand it. Finally, only non-negative brightness distributions are allowed by S. In incorporating it, we reject the vast majority of candidate images in favor of a small subclass of physically obtainable ones.

Neglecting error bounds for the moment, the brightness distribution that maximizes S while being constrained by (6) is the extremum of the functional:

$$E(f(\lambda, L)) = S + (g^t - f^t h)\lambda + L(f^t 1 - F) \tag{7}$$

where the λ is a column vector of Lagrange multipliers introduced to enforce the constraints by the principles of variational mechanics and L is another Lagrange multiplier enforcing the normalization of the brightness. Maximizing (7) with respect to the f_i and to L yields a model for the brightness, parametrized by the λ_j:

$$f_i = F\frac{e^{-[h\lambda]_i}}{Z} \tag{8}$$

$$Z = \sum_i e^{-[h\lambda]_i} \tag{9}$$

where we note how Z plays the role of Gibbs' partition function here.

Statistical errors are accounted for in WD85 by adapting (7) to enforce a constraint on the expectation of χ^2. The constraint is incorporated with the addition of another Lagrange multiplier (Λ). The constraint regarding the normalization of the brightness is enforced by the form of f resulting from (8) and need not be enforced further.

$$\begin{aligned} E(f(e, \lambda, \Lambda)) \\ = S + (g^t + e^t - f^t h)\lambda + \Lambda\left(e^t Ce - \Sigma\right) \\ = (g^t + e^t)\lambda + F \ln Z + \Lambda\left(e^t Ce - \Sigma\right) \end{aligned} \tag{10}$$

where the last step was accomplished by substituting (8) and (9) into S. The Σ term constrains the error norm, calculated in terms of theoretical error covariance matrix C, which we take to be diagonal. Rather than finding the brightness with the smallest model prediction error which also has a high entropy, WD85 finds the brightness which deviates from the data in a prescribed way so as to have the highest possible entropy consistent with experimental uncertainties.

Maximizing (10) with respect to the Lagrange multipliers yields $2M + 1$ algebraic equations:

$$g^t + e^t - f^t h = 0 \tag{11}$$

which merely restates (6). Maximizing with respect to the error terms in e yields equations relating them to the elements of λ:

$$\lambda + 2\Lambda C^{-1} e = 0 \tag{12}$$

(no sum implied). Maximizing with respect to Λ yields one more equation relating that term to the others.

$$4\Sigma\Lambda^2 - \lambda^t C\lambda = 0 \tag{13}$$

The resulting system of $2M + 1$ coupled, nonlinear equations for the Lagrange multipliers can be solved numerically. (The algorithm implemented here uses the hybrid method of Powell (1970).) Finally, (8) yields the desired image. The algorithm is robust and converges in practice when provided with data uncontaminated by interference. An analytic form of the required Jacobian matrix can readily be derived from (11).

4.1 Error analysis

Defining ρ_{12} as the normalized cross-correlation of the signals from receivers 1 and 2, an obvious estimator of ρ_{12} is:

$$\hat{\rho}_{12} = \frac{\frac{1}{m}\sum_{i=1}^{m} v_{1i}v_{2i}^*}{\frac{1}{m}\sum_{i=1}^{m}|v_{1i}|^2 \frac{1}{m}\sum_{i=1}^{m}|v_{2i}|^2} \tag{14}$$

where the numerator and denominator are computed from the same m statistically independent, concurrent samples. The error covariance matrix for interferometric cross-correlation or cross-spectral visibility estimates derived from this estimator was given by Hysell & Chau (2006):

$$\langle e_{r12}e_{r34}\rangle = \Re(\delta^2 + \delta'^2)/2 \tag{15}$$

$$\langle e_{i12}e_{i34}\rangle = \Re(\delta^2 - \delta'^2)/2 \tag{16}$$

$$\langle e_{r34}e_{i12}\rangle = \Im(\delta^2 + \delta'^2)/2 \tag{17}$$

$$\langle e_{r12}e_{i34}\rangle = \Im(\delta'^2 - \delta^2)/2 \tag{18}$$

where e_{r12} stands for the error in the estimate of the real part of the correlation of the signals from spaced receivers 1 and 2, for example, and where the indices may be repeated depending on the interferometry baselines in question. Also,

$$\delta^2 = \frac{1}{m}\left[\rho_{13}\rho_{24}^* - \frac{1}{2}\rho_{34}^*\left(\rho_{13}\rho_{23}^* + \rho_{14}\rho_{24}^*\right)\right. \tag{19}$$

$$-\frac{1}{2}\rho_{12}\left(\rho_{13}\rho_{14}^* + \rho_{23}\rho_{24}^*\right)$$

$$\left.+\frac{1}{4}\rho_{12}\rho_{34}^*\left(|\rho_{13}|^2 + |\rho_{14}|^2 + |\rho_{23}|^2 + |\rho_{24}|^2\right)\right]$$

and

$$\delta'^2 = \frac{1}{m}\left[\rho_{14}\rho_{23}^* - \frac{1}{2}\rho_{34}\left(\rho_{13}\rho_{23}^* + \rho_{14}\rho_{24}^*\right)\right. \tag{20}$$

$$-\frac{1}{2}\rho_{12}\left(\rho_{13}^*\rho_{14} + \rho_{23}^*\rho_{24}\right)$$

$$\left.+\frac{1}{4}\rho_{12}\rho_{34}\left(|\rho_{13}|^2 + |\rho_{14}|^2 + |\rho_{23}|^2 + |\rho_{24}|^2\right)\right],$$

ρ_{12} representing the complex correlation of the signals from spaced receivers 1 and 2, for example. In practice, these terms must be based on experimental estimates. The overall stability of error estimators based on data with statistical errors themselves has not been considered.

4.1.1 Added noise

The formulas above were derived in the absence of system noise but can easily be generalized to include noise. The normalized correlation function error covariances for signals in the presence of noise are still given by (19) and (20), only substituting the factor

$$\rho_{Sii} \rightarrow \frac{S+N}{S} \tag{21}$$

wherever correlation terms with repeated indices appear. Here, S and N refer to the signal and noise power, respectively.

On the whole, this analysis shows that the error covariance matrix is diagonally dominant only in cases where either the signal-to-noise ratio or the coherence is small. These limits are seldom applicable to coherent scatter, however. Even the longest interferometry baseline at Jicamarca, nearly 100 wavelengths long, very often exhibits high coherence, and even small, portable coherent scatter radars typically run in the high SNR limit. Since the error covariance is not diagonally dominant in general, neglecting off-diagonal terms misrepresents statistical confidence and could lead to image distortion.

In practice, it is expedient to diagonalize the error covariance matrix computed using the formulas above and to apply the corresponding similarity transformation to forward problem stated in (6) (Hysell & Chau, 2006). We find that the error variances that result fall into two groups with relatively smaller and larger values, respectively. The former correspond roughly to errors associated with measuring interferometric coherence, and the latter to errors associated with interferometric phase.

4.1.2 Error propagation

Error propagation through MaxEnt can be treated as follows (see for example Hysell (2007); Silver et al. (1990)). Using Bayes' theorem, we can cast the MaxEnt optimization problem posed in (10) as one of maximizing the posterior probability of a model image, m, based on visibility data d, which are related linearly through $d = Gm$, in the form

$$p(m|d) \propto e^{S/\Gamma} e^{-\frac{1}{2}e^t C^{-1} e} \tag{22}$$

$$\equiv e^{-E} \tag{23}$$

where the entropy S is the prior probability and the chi-squared model prediction error is transitional probability. The constant Γ weights the two probabilities and must be adjusted according to some criteria. In the variational approach to the optimization problem outlined above, the Lagrange multiplier Λ plays the role of Γ. That variable is controlled by Σ, and so there is always an adjustable free parameter.

Consider small departures δm about the maximum probability (minimum E) solution. In the neighborhood of a maximum, the gradient of the argument E vanishes, and we can always expand

$$p(m|d) \propto e^{-\frac{1}{2}\delta m^t H \delta m} \tag{24}$$

$$H = \frac{\partial^2 E(m)}{\partial \delta m \partial \delta m}$$

with H the Hessian matrix. Now, (24) has the form of a probability density function (PDF) for normally distributed model errors δm which we maximize through the minimization of E. We can consequently identify the Hessian matrix with the inverse model covariance matrix C_m^{-1}. Taking the necessary derivatives gives the error bounds on the image:

$$C_m^{-1} = G^t C^{-1} G + [\Gamma Im]^{-1} \tag{25}$$

The Γ term in (12) represents the influence of the data on the final MaxEnt model. The greater the influence, the smaller the uncertainties. The other term comes from the entropy prior and ensures that the model variances will be very small where the model values themselves are small. This is obviously significant in view of the importance of suppressing spurious artifacts.

4.2 Extensions

Hysell & Chau (2006) introduced two improvements to WD85 important for upper atmospheric radar research. The first involves incorporating the overall two-way antenna radiation pattern in the imaging analysis. Rather than attempting to remove the two-way pattern from the effective brightness distribution through division, with the attendant conditioning problems, we just acknowledge the influence of the pattern on the effective brightness and modify the entropy metric accordingly to anticipate it. If Shannon's expression favors a uniform brightness distribution, the expression that favors distributions that resemble the beam shape is (Skilling, 1989)

$$S' = -\sum_i f_i \ln(f_i / p_i F)$$

where p_i represents the two-way radiation power pattern. Propagating this expression through the preceding analysis alters only the brightness model:

$$f_i = \Gamma p_i \frac{e^{-[h\lambda]_i}}{Z} \tag{26}$$

$$Z = \sum_i p_i e^{-[h\lambda]_i} \tag{27}$$

The remaining formalism is unchanged. The only restriction is that p_i should be positive. In practice, the effect of the modification is to suppress spurious brightness outside in regions where the radiation pattern is depressed.

The second improvement applies when heterogeneous antennas are used for reception. In that case, the p_i in (26) can be made to match the radiation pattern of the transmitting antenna array, which is common to all the received signals. The radiation patterns of the receiving antennas are then explicitly incorporated into the expressions for the effective brightness, B_{eff}, associated with each baseline. In view of (1) or (6), this is done by modifying the point spread function for the given baseline j such that $h_{ij} \rightarrow h_{ij}\wp_{1i}\wp_{2i}$, where $\wp_{1,2i}$ are the radiation amplitude patterns for the antennas at either end of the baseline. Given the principle of pattern multiplication, characteristics of the radiation pattern common to all the receiving antennas can equally well be incorporated in p_i instead of $\wp_{1,2i}$.

4.3 Super-resolution

That radar imaging resolution does not need to be diffraction limited can be appreciated by considering coherence (normalized visibility) measurements made with a single interferometry baseline in the high signal-to-noise ratio, high coherence limit. As shown by Farley & Hysell (1996), the mean-squared error for the coherence estimate in this limit is

$$\delta^2 = \frac{1}{m}\left[\frac{N}{S} + \frac{1}{2}\epsilon + \mathcal{O}\left(\epsilon^2, \frac{N^2}{S^2}, \epsilon\frac{N}{S}, \cdots\right)\right] \tag{28}$$

where $\epsilon \equiv 1 - |V|^2$, m is the number of statistically independent samples used, and where S and N are the signal and noise powers, respectively. Even given a finite number of samples, the coherence estimate for a highly coherent target can be arbitrarily accurate given a high enough signal-to-noise ratio. This means that the angular width of narrow targets can be measured arbitrarily well, regardless of the baseline spacing, if S/N is sufficiently high. Insofar as imaging, an inverse method that accounts for the effects of diffraction in the forward model (i.e. the point spread function) need not be diffraction limited.

On the basis of information theory pertaining to the rate of information transmission through a noisy channel, Kosarev (1990) investigated the resolution limit for spectral analysis, deriving Shannon's resolution limit:

$$SR = \frac{1}{3}\log_2(1 + S/N) \tag{29}$$

This metric represents the maximum achievable resolution improvement over the diffraction-limited, noise-free case for non-parametric signal processing methods. Kosarev (1990) argued that there is no contradiction between this limit and the Heisenberg uncertainty principle. Kosarev (1990) furthermore performed numerical tests, comparing a spectral recovery algorithm based on maximum likelihood with entropy prior probability. Over the S/N range from 10–50 dB, the algorithm was able to achieve the Shannon limit. At Jicamarca, the longest interferometry baseline is nearly 100 wavelengths long, and the diffraction limited resolution is consequently about 0.5°. In practice, useful resolution at about the 0.1°-level can be obtained with strong backscatter.

4.4 Optimal sensor placement

The placement of sensors (receiving antennas or antenna groups) on the ground is typically constrained by practical consideration. If the sensors are subarrays of a fixed phased array, as in the case of the Jicamarca Radio Observatory in Peru or the MU Radar in Japan, a number of modules set by the number of receivers available will be selected from the total available in such a way as to avoid redundant baselines. To avoid ambiguity, baseline lengths can be selected such that the interferometry sidelobes are not illuminated by the transmitting antenna. Baseline orientations may be selected to accommodate anisotropies in the scatterers. As a rule, uniform sampling of visibility space seems to be conducive to artifact reduction, although there may be good reasons to deviate from it.

Sharif & Kamalabadi (2008) studies the optimal placement of sensors for different remote sensing applications, including aperture synthesis imaging. They considered data inversion through Tikhonov regularization, which is similar to the methodology discussed here except with Gaussian model statistics replacing model entropy as the prior probability. This permitted an entirely linear formulation of the problem. Included in the imaging problem were allowances for constraints on image smoothness and support. Optimization was in terms of minimizing different variations on the model prediction error as well as some detection performance metrics and the informativeness of the sensors with respect to the image (mutual information). Computationally expedient means of performing the different optimizations were also found. Sharif & Kamalabadi (2008) demonstrated that rather different results could be obtained depending on the criteria for which the array was optimized.

There are generally good reasons, however, for confining all of the sensors to a plane if possible. If the antennas are above imperfect earth, then their vertical phase centers may not be well known, but at least they will be identical if the antennas are at the same height and the dielectric constant of the earth is homogeneous. Moreover, with $d_z = 0$ for all interferometry baselines and with the radical in the denominator incorporated into the effective brightness, (2) becomes a two-dimensional Fourier transform. Translations in the brightness consequently map to phase shifts in the visibilities which further map to linear phase progressions of received signals across the aperture plane. This simple relationship can be useful in establishing the absolute phases of signals from different sensors.

4.5 Phase calibration

Sensor phase biases associated with differential cable lengths and other systemic issues must be removed through calibration in order for visibilities to be estimated and inverted. While a number of calibration methods exist, calibration remains one of the most challenging aspects of aperture synthesis imaging in practice (see for example Chau et al. (2008)). The "gold standard" for calibration involves observing point targets with known bearings, i.e. radio stars, and adjusting the complex gains of the receiver channels until the measured visibilities match a model based on the known source locations. However, this is only possible for large-aperture radars with adequate sensitivity, and the infrequency of radio star conjunctions may pose practical problems. Feeding common signals through the entire signal chain is another effective calibration strategy, but this too may be possible only infrequently.

Specular meteor echoes are quasi point-targets that can be observed frequently, even by small radars. While their bearings are not known individually, they can be estimated collectively. If the antenna array lies in a plane, phase biases can be estimated such that the bearings of all meteor echoes are consistent across all interferometry baselines (e.g. Holdsworth et al. (2004)). Corrections to the phase bias estimated can then be made such that the center of gravity of the echoes is aligned with the effective two-way radar radiation pattern, with allowances made for the expected anticipated altitude distribution of the specular meteor echoes. Other radar targets may be used to fine-tune the calibration. For example, that echoes from plasma irregularities arise from the locus of magnetic perpendicularity affords accurate knowledge of their elevation along a given azimuth.

While not widely used in upper atmospheric research, closure phase measurements offer additional information for phase calibration (Cornwall, 1989; Jennison, 1958). The idea is that the sum of the visibility phases from a triad of sensors, calculated for instance using bispectral (triple-product) analysis, is bias free. If the baseline of the triad are suitably arranged, this can afford some information about sensor phase bias.

A promising class of calibration techniques involves finding the phase biases on the basis of image characteristics through optimization. Uncompensated sensor phase biases tend to degrade images and introduce artifacts that increase the image variance (decrease the smoothness) as well as the overall image entropy. Holding the phase biases of three non-collinear sensors to be arbitrary, the remaining phases could be adjusted to minimize the image entropy, global variance, or some other cost function. Afterward, the phase offsets could be readjusted to "rotate" the artifact-free image into its proper place, taking into account known characteristics of the radar and the target. This optimization can take place outside of the main imaging computation or possibly within it, adopting some of the principles followed by Sharif & Kamalabadi (2008) for optimizing sensor placement. Their smoothness and support metrics, respectively, could be imposed to accomplish the first and second steps of the aforementioned calibration, respectively, only within a unified imaging framework. Since the brightness/visibility mapping is not linear in the phase biases, the procedure would necessarily be iterative.

5. Examples from the upper atmosphere

Here, we present examples of ionospheric phenomena that have been revealed or clarified using aperture synthesis radar imaging. The examples are taken from observations of the Jicamarca Radio Observatory, a 50-MHz phased array radar operated outside Lima, Peru. Aperture synthesis radar imaging was introduced to upper atmospheric research at Jicamarca in 1991 (Kudeki & Sürücü, 1991), and MaxEnt was applied there first five years later (Hysell, 1996). The number of sensors sampled has grown from four to eight in the intervening years. Twelve-sensor experiments are being planned. The longest interferometry baseline available is nearly 100 wavelengths long in the direction perpendicular to the geomagnetic field. A subset of the main antenna array is used for transmission, and a phase taper is applied to broaden the main beam and reduce the sidelobe level. Images are normally computed over a ≈13°-wide azimuth sector. In practice, only the central part of the sector contains echoes and need be plotted.

At Jicamarca, imaging has mainly been applied to coherent scatter from field-aligned plasma density irregularities. Different varieties of irregularities occupy altitudes between about 95–2500 km at the geomagnetic equator and can be detected by the strong, spectrally narrow radar echoes that arise from them. While imaging is generally performed in two dimensions, the echoes arrive from bearings very close to the locus of perpendicularity to the geomagnetic field, and the images in each range gate can consequently be collapsed into a single dimension. Alternatively, the imaging problem can be formulated in one dimension from the start. Two dimensional images in range and azimuth are produced finally. Sequences of sequential images can also be animated. Imaging in three dimensions has been applied in lower atmospheric applications (Palmer et al., 1998). We have plans to apply it to mesospheric echoes as well.

Images are formed for each Doppler bin, and each image pixel or voxel consequently represents a complete Doppler spectrum. Spectral information is conveyed through color according to the example legend shown in Figure 1. Pixel colors represent the first three moments of the spectrum, with the brightness, hue, and saturation specifying the

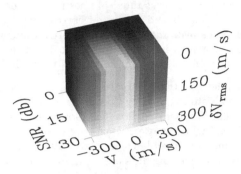

Fig. 1. Scale for interpreting the image pixel coloration. Pixel brightness, hue, and saturation are proportional to the echo signal-to-noise ratio in dB, Doppler shift in m/s, and spectral width in m/s, respectively. Different axes ranges apply to the different images that follow.

signal-to-noise ratio, Doppler shift, and spectral width, respectively. In this example, signal-to-noise ratios between 0–30 dB are represented. The range of Doppler velocities evaluated is controlled by the radar interpulse period and is 300 m/s in this example. By convention, RMS spectral widths between zero and the maximum Doppler shift are portrayed. Incoherent integration times for imaging are typically on the order of a few seconds.

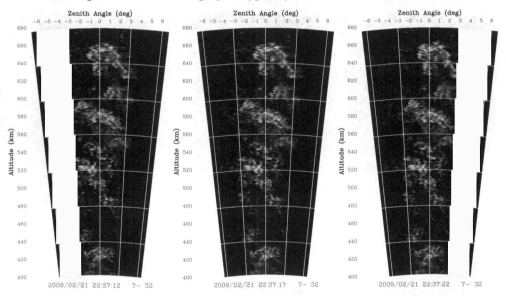

Fig. 2. Coherent scatter from F region plasma density irregularities associated with equatorial spread F depletions on Feb. 21, 2009. The maximum unaliased Doppler velocity is 300 m/s. Signal-to-noise ratios shown span 7–32 dB.

Figure 2 shows radar images of field-aligned plasma density irregularities associated with equatorial spread F (ESF), a nighttime phenomenon characterized by plasma interchange instability and the rapid ascent of depleted plasma wedges from the F region bottomside into the topside. The depletions appear as tilted plumes in the equatorial plane. Here, five plumes between 480–660 km altitude and separated horizontally by about 40 km subtend the Jicamarca beam. No other instrument can provide two-dimensional imagery of ESF plumes with details in the crucial intermediate-scale regime (kilometers to tens of kilometers). Animated sequences of images provide dynamical information with the same detail.

Conventional range-time-intensity (RTI) representations of coherent scatter from ESF demonstrate tremendous qualitative variability from event to event, whereas different plumes generally appear to be similar in radar imagery. RTI plots have been likened to slit-camera images from a "photo finish," which may produce spurious evidence of horses with three or five legs from time to time, for example (Woodman, 1997). Aperture synthesis imaging reduces instrumental distortion, revealing the salient features of the phenomena under study.

Comparisons with in situ observations of ESF have shown that the bright patches in Figure 2 correspond to localized plasma depletions (Hysell et al., 2009). Moreover, the Doppler shifts of those patches correspond closely to the vertical components of the local $\mathbf{E} \times \mathbf{B}$ drifts. However, since the coherent scatter is spatially intermittent and not homogeneous, the Doppler spectrum representing an entire range gate (without imaging) will not be indicative of the average line-of-sight speed of the plasma in that gate. We know this intuitively; even though the Doppler shifts from active ESF predominantly denote ascent, by mass and by volume, the action of interchange instability is to push ionospheric plasma downward.

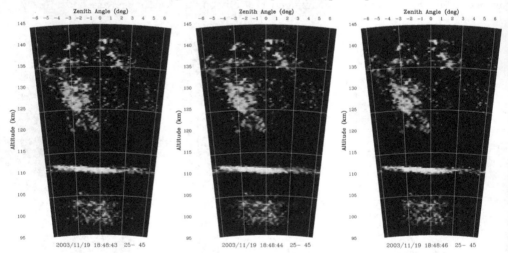

Fig. 3. Coherent scatter from the E and valley regions near twilight on Nov. 19, 2003. The maximum unaliased Doppler velocity is 120 m/s. Signal-to-noise ratios shown span 25–45 dB.

Figure 3 shows coherent scatter imagery from around twilight when strong echoes were observed between about 120–145 km in the equatorial valley region (Chau & Hysell, 2004). The echoes do not appear to be directly connected with the equatorial electrojet, which is also producing irregularities below about 112 km here. The valley echoes are organized into waves with wavelengths of about 10 km which propagate downward and westward. The Doppler shifts are mainly positive (downward) and vary systematically with height. It would have been difficult to distinguish this phenomenon from electrojet-related plasma waves on the basis of RTI information alone.

The cause of the echoes and the underlying source of free energy have not been identified. Recent simulations suggest that the marginal magnetization of the ions at these altitudes is significant and that the irregularities may be due to a class of collisional drift waves. Radar imaging will play a key role in the ongoing investigation of these irregularities.

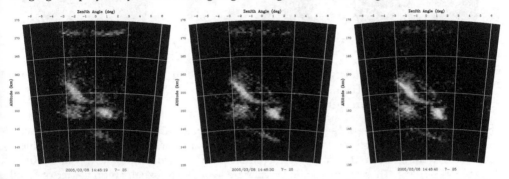

Fig. 4. Radar images of so-called "150-km echoes" observed in the afternoon of Mar. 8, 2005. The maximum unaliased Doppler velocity is 780 m/s. Signal-to-noise ratios shown span 7–25 dB.

Figure 4 presents radar images associated with the so-called "150-km" echoes (Chau, 2004; Kudeki & Fawcett, 1993; Royrvik & Miller, 1981). These daytime echoes are enigmatic, having regular and striking but unexplained patterns in RTI representations. The Doppler shifts of the echoes are known to match the background line-of-sight $\mathbf{E} \times \mathbf{B}$ drift, implying that dielectric plasma polarization probably does not play a significant role in irregularity production. The spectra, in fact, conform in many ways to expectation for incoherent scatter, both looking perpendicular to B and obliquely to B, and part of the 150-km echoes constitute an ion-line enhancement Chau et al. (2009). The most obvious source of free energy for the irregularities is photoelectron production, which peaks nearby, but the mechanisms at work have yet to be articulated.

The 150-km echoes are weak compared to echoes from other equatorial plasma density irregularities. The image in Figure 4 reveals that the echoes are not homogeneous or beam filling but are instead spatially (and temporally) intermittent. Over time, the spatial organization of the echoes in the imagery varies abruptly in a way that does not convey the sense of proper motion.

Lastly, Figure 5 shows images of large-scale waves in the daytime equatorial electrojet (Farley, 1985; Farley & Balsley, 1973; Kudeki et al., 1982). Coherent scatter from the electrojet is the strongest radar target in the upper atmosphere at VHF frequencies and is produced by a combination of gradient-drift and Farley-Buneman instability. Here, large-scale gradient drift waves with wavelengths of 1–2 km can be seen propagating westward under the influence of a sheared zonal electron $\mathbf{E} \times \mathbf{B}$ flow associated with a Cowling conductivity. Echoes come from gradient drift wave turbulence and from small-scale, secondary Farley-Buneman waves, with large-telltale Doppler shifts. At night, the flow and the propagation direction reverse, and the wavelength of the dominant large-scale waves increases.

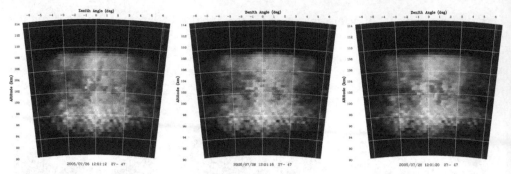

Fig. 5. Coherent scatter from large-scale gradient drift waves in the daytime equatorial electrojet on July 26, 2005. The maximum unaliased Doppler velocity is 600 m/s. Signal-to-noise ratios shown span 27–47 dB.

The dominant wavelength of the waves, their phase speeds, and their dynamical behavior are controlled by a surprisingly complex combination of quasilinear, nonlocal, and nonlinear effects which have been described and simulated by Ronchi et al. (1989) and Hu & Bhattacharjee (1998). Those results could be validated with an unusually high degree of detail using radar imaging experiments (Hysell & Chau, 2002).

6. Summary and future work

Aperture synthesis radar imaging is applied routinely in observations of field-aligned plasma density irregularities at low, middle, and high geomagnetic latitudes, and applications in the lower and middle atmospheres are gradually emerging. Imaging discriminates targets in bearing with resolution limited by the longest interferometry baseline length (rather than by the size of the main antenna) and by the signal-to-noise ratio. It is well suited for heterogeneous sensor arrays with regular or pseudo-random distributions and works alongside other radar modalities like pulse compression and methods for spectral estimation of overspread targets.

We have found the MaxEnt algorithm to be suitable for radar applications in upper atmospheric research. It is an edge-preserving technique, equally applicable to point and continuous targets, and is the embodiment of Occam's razor, suppressing image features for which there is no support in the data. While it was once considered computationally expensive, ongoing improvements in computer performance have made real-time application practical on multi-core systems.

Whereas the example imagery shown above was rendered in two dimensions, imaging is generally performed in three. In the case of scatter from field-aligned plasma density irregularities, information in the direction parallel to the geomegnetic field is limited and is often ignored. However, the magnetic aspect sensitivity of the irregularities, the concentration of the echoes near the locus of perpendicularity, can give insights into the underlying generation mechanism (e.g. Kudeki et al. (1981)). Radar imaging could be used to measure magnetic aspect sensitivity under different conditions in the different regions in the radar field of view. This could be particularly revealing in the auroral electrojet, for example, where different irregularity-producing instabilities may be excited inside, outside, and at the edges of discrete auroral arcs.

An obvious extension of the algorithms described in this paper involves radar imaging of moderately overspread targets, ESF being a good example. Such targets can be investigated using aperiodic or incremental-lag pulses (e.g. Chau et al. (2004); Uppala & Sahr (1994); Virtanen et al. (2009)). This produces temporal lagged-products with nonuniform spacing which can be spectrally analyzed using the same methodologies developed for imaging (Hysell et al., 2008). Moreover, the spatio-temporal lagged products arising from aperiodic aperture synthesis imaging experiments can be analyzed together in one operation, yielding images in range, bearing, and the added dimension of Doppler frequency seamlessly. The total number of distinct lagged products will be given by $N(N-1)/2$, N being the product of the number of sensors and the number of aperiodic pulses considered at a time. Compromises in spectral and angular resolution will be required if the inverse problem is to remain tractable.

Finally, it has been our observation that the backscatter from ionospheric plasma density irregularities in many contexts tends to be "clumpy" rather than diffuse. This information could be exploited in the inversion scheme if the prior probability function were augmented with a component based on an appropriate model Markov chain. Fully exploiting this information would require processing data from multiple ranges and Doppler frequency bins simultaneously, since the spatial organization occurs equally in bearing, range, and Doppler frequency. That the "clumpy" targets often remain organized for long periods of time suggests that this information should be folded in as well.

7. References

Ables, J. G. (1974). Maximum entropy spectral analysis, *Astron. Astrophys. Suppl. Ser.* 15: 383.

Aster, R. C., Borchers, B. & Thurber, C. H. (2005). *Parameter Estimation and Inverse Problems*, Elsevier, New York.

Capon, J. (1969). High-resolution frequency-wavenumber spectrum analysis, *Proc. IEEE* 57: 1408.

Chau, J. L. (2004). Unexpected spectral characteristics of VHF radar signals from 150-km region over Jicamarca, *Geophys. Res. Lett.* 31, L23803: doi:10.1029/2004GL021620.

Chau, J. L. & Hysell, D. L. (2004). High altitude large-scale plasma waves in the equatorial electrojet at twilight, *Ann. Geophys.* 22: 4071.

Chau, J. L., Hysell, D. L., Kuyeng, K. M. & Galindo, F. (2008). Phase calibration approaches for radar interferometry and imaging configurations: Equatorial spread F results, *Ann. Geophys.* 26: 2333–2343.

Chau, J. L., Hysell, D. L., Reyes, P. M. & Milla, M. A. (2004). Improved spectral estimations of equatorial spread F echoes at Jicamarca using aperiodic transmitter coding, *J. Atmos. Sol. Terr. Phys.* 66(1543).

Chau, J. L. & Woodman, R. F. (2001). Three-dimensional coherent radar imaging at Jicamarca: Preliminary results, *J. Atmos. Sol. Terr. Phys.* 63: 253–261.

Chau, J. L. & Woodman, R. F. (2004). Observations of meteor-head echoes using the Jicamarca 50 MHz radar in interferometer mode, *Atmos. Chem. Phys.* 4.1680-7324/acp: 511–521.

Chau, J. L., Woodman, R. F., Milla, M. A. & Kudeki, E. (2009). Naturally enhanced ion-line spectra around the equatorial 150-km region, *Ann. Geophys.* 27: 933–942.

Cornwall, T. J. (1989). The applications of closure phase to astronomical imaging, *Science* 245 (4915): 263–269.

Cornwell, T. J. & Evans, K. F. (1985). A simple maximum entropy deconvolution algorithm, *Astron. Astrophys.* 143: 77.

Daniell, G. J. (1991). Of maps and monkeys, *in* B. Buck & V. A. Macaulay (eds), *Maximum Entropy in Action*, Clarendon, Oxford, chapter 1, pp. 1–18.

Farley, D. T. (1969). Incoherent scatter correlation function measurements, *Radio Sci.* 4: 935–953.

Farley, D. T. (1972). Multiple-pulse incoherent-scatter correlation function measurements, *Radio Sci.* 7: 661.

Farley, D. T. (1985). Theory of equatorial electrojet plasma waves: New developments and current status, *J. Atmos. Terr. Phys.* 47: 729–744.

Farley, D. T. & Balsley, B. B. (1973). Instabilities in the equatorial electrojet, *J. Geophys. Res.* 78: 227.

Farley, D. T. & Hysell, D. L. (1996). Radar measurements of very small aspect angles in the equatorial ionosphere, *J. Geophys. Res.* 101: 5177.

Gull, S. F. & Daniell, G. J. (1978). Image reconstruction from incomplete and noisy data, *Nature* 272: 686.

Hassenpflug, G., Yamamoto, M., Luce, H. & Fukao, S. (2008). Description and demonstration of the new Middle and Upper atmosphere Radar imaging system: 1-D, 2-D, and 3-D imaging of troposphere and stratosphere,, *Radio Sci.* 43, RS2013: doi:10.1029/2006RS003603.

Högbom, J. A. (1974). Aperture synthesis with a non-regular distribution of interferometer baselines, *Astron. Astrophys. Supp.* 15: 417–426.

Holdsworth, D. A., Tsutsumi, M., Reid, I. M., Nakamura, T., T. & Tsuda (2004). Interferometric meteor radar phase calibration, *Radio Sci.* 39, RS5012: doi:10.1029/2003RS003026.

Hu, S. & Bhattacharjee, A. (1998). Two-dimensional simulations of gradient-drift turbulence in the daytime equatorial electrojet, *J. Geophys. Res.* 103: 20,749.

Huuskonen, A., Lehtinen, M. S. & Pirttilä, J. (1996). Fractional lags in alternating codes: Improving incoherent scatter measurements by using lag estimates at noninteger multiples of baud length, *Radio Sci.* 31: 245.

Hysell, D. L. (1996). Radar imaging of equatorial F region irregularities with maximum entropy interferometry, *Radio Sci.* 31: 1567.

Hysell, D. L. (2007). Inverting ionospheric radio occultation measurements using maximum entropy, *Radio Sci.* 42, RS4022: doi:10.1029/2007RS003635.

Hysell, D. L. & Chau, J. L. (2002). Imaging radar observations and nonlocal theory of large-scale waves in the equatorial electrojet, *Ann. Geophys.* 20: 1167.

Hysell, D. L. & Chau, J. L. (2006). Optimal aperture synthesis radar imaging, *Radio Sci.* 41: 10.1029/2005RS003383, RS2003.

Hysell, D. L., Chau, J. L. & Lakshmanan, S. (2008). Improved spectral estimation of equatorial spread F through aperiodic pulsing and Bayesian inversion, *Radio Sci.* 43, RS2010: doi:10.1029/2007RS003790.

Hysell, D. L., Hedden, R. B., Chau, J. L., Galindo, F. R., Roddy, P. A. & Pfaff, R. F. (2009). Comparing F region ionospheric irregularity observations from C/NOFS and Jicamarca, *Geophys. Res. Lett.* 36, L00C01: doi:10.1029/2009GL038983.

Jaynes, E. T. (1982). On the rationale of maximum-entropy methods, *Proc. IEEE* 70: 939.

Jaynes, E. T. (1985). Where do we go from here?, *in* C. R. Smith & W. T. Grandy, Jr. (eds), *Maximum-Entropy and Bayesian Methods in Inverse Problems*, D. Reidel, Norwell, Mass., chapter 2, pp. 21-58.

Jennison, R. C. (1958). A phase sensitive interferometer technique for the measurement of the Fourier transforms of spatial brightness distributions of small angular extent, *Mon. Not. Roy. Astron. Soc.* 118: 276.

Kosarev, E. L. (1990). Shannon's superresolution limit for signal recovery, *Inverse Problems* 6: 55–76.

Kudeki, E., Farley, D. T. & Fejer, B. G. (1982). Long wavelength irregularities in the equatorial electrojet, *Geophys. Res. Lett.* 9: 684.

Kudeki, E. & Fawcett, C. D. (1993). High resolution observations of 150 km echoes at Jicamarca, *Geophys. Res. Lett.* 20: 1987.

Kudeki, E., Fejer, B. G., Farley, D. T. & Ierkic, H. M. (1981). Interferometer studies of equatorial F region irregularities and drifts, *Geophys. Res. Lett.* 8: 377.

Kudeki, E. & Sürücü, F. (1991). Radar interferometric imaging of field-aligned plasma irregularities in the equatorial electrojet, *Geophys. Res. Lett.* 18: 41.

Kudeki, E. & Woodman, R. F. (1990). A post-statistics steering technique for MST radar applications, *Radio Sci.* 25: 591–594.

Lehtinen, M. S. (1986). Statistical theory of incoherent scatter radar measurements, *Technical Report 86/45*, Eur. Incoherent Scatter Sci. Assoc., Kiruna, Sweden.

Lomb, N. R. (1976). Least squares frequency analysis of unevenly sampled data, *Astrophys. Space Sci.* 39: 447.

Menke, W. (1984). *Geophysical Data Analysis: Discrete Inverse Theory*, Academic, New York.

Palmer, R. D., Gopalam, S., Yu, T. Y. & Fukao, S. (1998). Coherent radar imaging using Capon's method, *Radio Sci.* 33: 1585.

Powell, M. J. D. (1970). A hybrid method for nonlinear equations, *in* P. Rabinowitz (ed.), *Numerical Methods for Nonlinear Algebraic Equations*, Gordon and Breach, London, pp. 87–114.

Ronchi, C., Similon, P. L. & Sudan, R. N. (1989). A nonlocal linear theory of the gradient drift instability in the equatorial electrojet, *J. Geophys. Res.* 94: 1317.

Royrvik, O. & Miller, K. L. (1981). Nonthermal scattering of radio waves near 150 km above Jicamarca, Peru, *J. Geophys. Res.* 86: 180.

Schmidt, R. O. (1986). Multiple emitter location and signal parameter estimation, *IEEE Trans. Antennas Propagat.* AP-34: 276–280, March.

Semeter, J., Butler, T., Heinselman, C., Nicolls, M., Kelly, J. & Hampton, D. (2009). Volumetric imaging of the auroral ionosphere: Initial results from PFISR, *J. Atmos. Sol. Terr. Phys.* 71(6-7): 738–743.

Shannon, C. E. & Weaver, W. (1949). *The Mathematical Theory of Communication*, Univ. of Ill. Press, Urbana.

Sharif, B. & Kamalabadi, F. (2008). Optimal sensor array configuration in remote image formation, *IEEE Trans. Image Proc.* 17(2): 155–166.

Silver, R. N., Sivia, D. S. & Gubernatis, J. E. (1990). Maximum-entropy method for analytic continuation of quantum Monte Carlo data, *Phys. Rev.* 41(4): 2380.

Skilling, J. (1989). Classic maximum entropy, *in* J. Skilling (ed.), *Maximum Entropy and Bayesian Methods*, Kluwer Academic Publishers, Dordrecht, pp. 45–52.

Skilling, J. (1991). Fundamentals of MaxEnt in data analysis, *in* B. Buck & V. A. Macaulay (eds), *Maximum Entropy in Action*, Clarendon, Oxford, chapter 2, pp. 19–40.

Skilling, J. & Bryan, R. K. (1984). Maximum entropy image reconstruction: General algorithm, *Mon. Not. R. Astron. Soc.* 211: 111.

Sulzer, M. P. (1986).   A radar technique for high range resolution incoherent scatter autocorrelation function measurements utilizing the full average power of klystron radars, *Radio Sci.* 21: 1033–1040.

Tarantola, A. (1987). *Inverse Theory*, Elsevier, New York.

Thompson, A. R. (1986). *Interferometry and Synthesis in Radio Astronomy*, John Wiley, New York.

Uppala, S. V. (1993). *Aperiodic radar technique for the spectrum estimation of moderately overspread targets*, Master's thesis, University of Washington.

Uppala, S. V. & Sahr, J. D. (1994). Spectrum estimation moderately overspread radar targets using aperiodic transmitter coding, *Radio Sci.* 29: 611.

Virtanen, I. I., Viernen, J. & Lehtinen, M. S. (2009). Phase-coded pulse aperiodic transmitter coding, *Ann. Geophys.* 27: 2799–2811.

Wernecke, S. J. & D'Addario, L. R. (1977). Maximum entropy image reconstruction, *IEEE Trans. Computers* c-26: 351.

Wilczek, R. & Drapatz, S. (1985). A high accuracy algorithm for maximum entropy image restoration in the case of small data sets, *Astron. Astrophys.* 142: 9.

Woodman, R. F. (1997). Coherent radar imaging: Signal processing and statistical properties, *Radio Sci.* 32: 2373.

Woodman, R. F. (2009). Spread F- An old equatorial aeronomy problem finally resolved?, *Ann. Geophys.* 27: 1915–1934.

Woodman, R. F. & La Hoz, C. (1976). Radar observations of F region equatorial irregularities, *J. Geophys. Res.* 81: 5447–5466.

Wu, N. (1984). A revised Gull-Daniell algorithm in the maximum entropy method, *Astron. Astrophys.* 139: 555.

Part 4

Other Advanced Doppler Radar Applications

10

Doppler Radar Tracking Using Moments

Mohammad Hossein Gholizadeh and Hamidreza Amindavar
Amirkabir University of Technology, Tehran
Iran

1. Introduction

A Doppler radar is a specialized radar that makes use of the Doppler effect to estimate targets velocity. It does this by beaming a microwave signal towards a desired target and listening for its reflection, then analyzing how the frequency of the returned signal has been altered by the object's motion. This variation gives direct and highly accurate measurements of the radial component of a target's velocity relative to the radar. Doppler radars are used in aviation, sounding satellites, meteorology, police speed guns, radiology, and bistatic radar (surface to air missile).

Partly because of its common use by television meteorologists in on-air weather reporting, the specific term "Doppler Radar" has erroneously become popularly synonymous with the type of radar used in meteorology.

The Doppler effect is the difference between the observed frequency and the emitted frequency of a wave for an observer moving relative to the source of the waves. It is commonly heard when a vehicle sounding a siren approaches, passes and recedes from an observer. The received frequency is higher (compared to the emitted frequency) during the approach, it is identical at the instant of passing by, and it is lower during the recession. This variation of frequency also depends on the direction the wave source is moving with respect to the observer; it is maximum when the source is moving directly toward or away from the observer and diminishes with increasing angle between the direction of motion and the direction of the waves, until when the source is moving at right angles to the observer, there is no shift. Since with electromagnetic radiation like microwaves frequency is inversely proportional to wavelength, the wavelength of the waves is also affected. Thus, the relative difference in velocity between a source and an observer is what gives rise to the Doppler effect.

Now, suppose that we have received an unknown waveform from the target. This waveform is a result of reflection from a fluctuating target in presence of clutter and noise. The received signal is often modeled as delayed and Doppler-shifted version of the transmitted signal. So not only the Doppler estimation, but the joint estimation of the time delay and Doppler shift provides information about the position and velocity of the target. So we should focus on the joint estimation of both parameters. There are many works for estimating the joint time delay and Doppler shift, with advantages and disadvantages apiece. Among these methods, Wigner Ville (WV) method has proven to be a valuable tool in estimating the time delay and Doppler shift. WV method is a time-frequency processing. It possesses a high resolution in the time-frequency plane and satisfies a large number of desirable theoretical properties [Chassande-Mottin & Pai, 2005]. In fact, these properties are the fundamental motivation for the use of the narrowband(wideband) WV transformation for detecting a deterministic signal with unknown delay-Doppler(-scale) parameters. WV's practical usage is limited by the presence of non-negligible cross-terms, resulting from interactions between signal components. Alternative approaches are proposed for eliminating or at least suppressing the cross-terms [Chassande-Mottin & Pai, 2005; Orr et al., 1992; Tan & Sha'ameri, 2008]. Generally speaking, cross-term suppression may be divided into two categories: signal-independent and signal-dependent paradigm. Coupling the Gabor transformation with the WV distribution is a signal-independent procedure that reveals a cross-term suppression approach through exploitation of partial knowledge about signals to be encountered [Orr et al., 1992]. For signal-dependent method, it is possible to apply an adaptive window over WV distribution where the kernel parameters are determined automatically from the parameters of the input signal. This kernel is capable of suppressing the cross-terms and maintain accurate time-frequency resolution [Tan & Sha'ameri, 2008]. Besides the WV method, there are other time-frequency techniques such as wavelet transform. Wavelet approach combines the noise filtering and scaling together, yielding a reduction in complexity [Niu et al., 1999]. There is also another procedure using the fractional lower order ambiguity function (FLOAF) for joint time delay and Doppler estimation [Ma & Nikias, 1996]. Now another view is presented. It is assumed that the transmitted signal follows an N-mode Gaussian mixture model (GMM). GMM can be used for different transmitted signals. Especially, it presents an accurate modeling for actual signals transmitted in the sonar and radar systems [Bilik et al., 2006]. The received signal is affected by the noise, time delay and Doppler where the conglomerate effects on the signal cause peculiar changes on the moments of received signal. Using moments is a powerful procedure which is used for different applications, specially in parameter estimation. Some people use the moment method to estimate the parameters of a Gaussian mixture in an environment without noise [Fukunaga et al., 1983]. Some apply the method for better parameter estimation in a faded signal transmitted through a communication channel which is suffered from multipath. The method can be implemented using a non-linear least-squares algorithm to represent a parameterized fading model for the instantaneous received path power which accounts for both wide-sense stationary shadowing and small-scale fading [Bouchereau & Brady, 2008]. The most prominent and novel models for the envelope of a faded signal are Rician and Nakagami. There are estimators for the Nakagami-m parameter based on real sample moments. The estimators present an asymptotic expansion which provides a generalized closed-form expression for the Nakagami-m parameter without the need for coefficient optimization for different ratios of real moments [Gaeddert & Annamalai, 2005]. There are also approaches that show the K-factor in Rician model is an exact function of moments estimated from time-series data [Greenstein et al., 1999].

In this chapter, we analyze the effect of noise, time delay, and Doppler on the moments of received signal and exploit them for estimating the position and velocity of the target. We note that in the new method, the noise power is assumed unknown which is estimated along with the time delay and Doppler shift. The new approach exhibits accurate results compared to the existing methods even in very low SNR and long tailed noise. Then, the estimated parameters are used for tracking a maneuvering target's position and velocity. There exist other practical methods for tracking targets such as Kalman filtering [Park & Lee, 2001]. However, when the target motion is nonlinear and/or clutter and/or noise are non-Gaussian, this approach fails to be effective. Instead, unscented Kalman filter (UKF) and extended Kalman filter (EKF) come into use [Jian et al., 2007]. However, in long tailed noise, Kalman filtering results are unsatisfactory. To overcome these difficulties, particle filtering (PF) is utilized [Jian et al., 2007].

Although particle filtering performs better than Kalman filtering in noisy environment, but it also diverges in low SNRs and cannot be trustable in this range of SNR. In addition, this method requires much more processing. We note that Kalman filtering, extended Kalman, unscented Kalman and particle filtering are recursive in nature. The new procedure proposed in this chapter is not recursive and can be used in the non-Gaussian, non-stationary noise, and nonlinear target motion. In here, the target tracking is performed based on the estimated time delay and Doppler. Since the accuracy of the time delay and Doppler estimation are high enough even in the severe noise, the results in tracking are acceptable compared to other rival approaches.

In section II the moment concept is reviewed and moment method is described as the base item in our estimations. Section III provides a model for the received signal. This signal has been influenced by unknown noise, delay and Doppler. It is shown in Section IV that it is possible to estimate Doppler by using the moments of the received random signal. The method is also useful for delay estimation. The noise power and its behavior play a prominent role in our work. So some analysis in this field is presented in this section too. After the parameter estimation, section V is devoted to explain about how the tracking a target is done based on the estimated delay and Doppler. And finally, section VI contains results that illustrate the effectiveness of the proposed method.

2. Moment concept

In probability theory, the moment method is a way in which the moments of a discrete sequence are used to determine its distribution.

Suppose that X is a random variable, and $f_X(x)$ is the probability density function (PDF) of this random variable. The moments of the random variable X is calculated from the following equation:

$$m_n = E\left(X^n\right) = \int_{-\infty}^{\infty} x^n f_X(x)\, dx = \int_{-\infty}^{\infty} x^n\, dF_X(x)\, dx, \tag{1}$$

which $F_X(x)$ is the cumulative distribution function (CDF) of the random variable X, and $E(.)$ is the expectation value.

On the other hand, the moment generating function (MGF) of this random variable is calculated as follows:

$$M_X(u) = E\left(e^{uX}\right), \quad u \in C. \tag{2}$$

Note that the equation will be hold if the expectation value exists.

In here, to obtain the moments of a random variable, the relation between the moment and the moment generating function is use instead of using equation (1). This relation can be demonstrated as follows:

$$M_X(u) = E\left(e^{uX}\right) = \int_{-\infty}^{\infty} e^{ux} f_X(x)\, dx =$$

$$\int_{-\infty}^{\infty} (1 + ux + \frac{u^2 x^2}{2!} + \cdots) f_X(x)\, dx = 1 + u m_1 + \frac{u^2 m_2}{2!} + \cdots . \tag{3}$$

This equation is hold when the moments m_n are finite, i.e. $|m_n| < \infty$.

The moment method claims that using the moment of the random variable X, the PDF of X is completely determined. So if we have:

$$\lim_{n \to \infty} E\left(X_n^k\right) = E\left(X^k\right), \qquad \forall k \tag{4}$$

then, the sequence $\{X_n\}$ has the same distribution as the X. we use (4) for parameter estimation, i.e. The left side of the equation is obtained statically, and the right side is calculated analytically. These two sides should be equal.

To begin our discussion, a model should be considered for our signals. Next section is focused on finding the suitable model.

3. Signal model

We consider the baseband representation of the received signal, which can be expressed as the sum of the desired signal component and non-stationary background noise. The signal component is represented by the linear sum of many non-coherent waveforms whose arrivals at the receiver are governed by a Poisson process [Zabin & Wright, 1994]. The receiver includes two sensors to measure the received signal in presence of background noise:

$$y_1(t) = s(t) + \omega_1(t),$$
$$y_2(t) = s(t - \tau)\exp(j2\pi t\varepsilon) + \omega_2(t), \tag{5}$$

where τ and ε denote the time delay and Doppler respectively, and $s(t)$ is the desired received signal modeled at any time instance t to follow a real N-mode Gaussian mixture distribution [Isaksson et al., 2001]:

$$s(t) \sim \sum_{i=1}^{N} p_i \, N\left(\mu_{s_i}, \sigma_{s_i}^2\right). \tag{6}$$

The processes $\omega_1(t)$ and $\omega_2(t)$ are real zero-mean additive white Gaussian noises (AWGN) with powers of $\sigma_{\omega_1}^2$ and $\sigma_{\omega_2}^2$ respectively. These powers are not constant in practice due to nonhomogeneous environment, but are assumed as random variates which are estimated subsequently. The signal and noise are supposed to be uncorrelated, but the noises $\omega_1(t)$ and $\omega_2(t)$ are possibly correlated.

4. Parameter estimation

In this section, for a random variable X, the moment generating function (MGF), $M_x(u)$, and its asymptotic series are used to determine the moments m_{xi}:

$$M_x(u) = E(e^{uX}) = 1 + u m_{x1} + \frac{u^2 m_{x2}}{2!} + \cdots, \qquad u \to 0. \tag{7}$$

4.1 Time delay estimation

The statistical properties of the signal and noise which are represented in (5) are known. Therefore, their MGF is available, by assuming finite moments of signal and noise. Although the signal follows a Gaussian mixture distribution, the conglomerate effect of the time delay

n	Moment	Central moment	Cumulant
0	1	1	—
1	μ	0	μ
2	$\mu^2 + \sigma^2$	σ^2	σ^2
3	$\mu^3 + 3\mu\sigma^2$	0	0
4	$\mu^4 + 6\mu^2\sigma^2 + 3\sigma^4$	$3\sigma^4$	0

Table 1. Normal distribution moments

and Doppler creates a non-stationary signal, as seen in (5). At first, by using both sensors in the receiver, the time delay is predicted, then this estimated delay facilitates determination of the Doppler shift subsequently. The time delay estimation is described here and discussions about the Doppler estimation are provided in the sequel. It is required to consider the MGF of the normal distributed variate as the starting ground for the next steps:

$$M(u) = \exp\left(\mu u + 0.5\sigma^2 u^2\right),\qquad(8)$$

where μ and σ are the mean and variance of normal distribution. The related moments are depicted in table (I). We suppose the received noise-free signal in the second sensor is denoted by:

$$r(t) = s(t - \tau)\exp(j2\pi t\varepsilon).\qquad(9)$$

First, we assume there is no Doppler i.e. $r(t) = s(t - \tau)$, and the noise variances, $\sigma_{w_1}^2$ and $\sigma_{w_2}^2$, are constant. As mentioned above, we utilize the MGF for the estimation purposes. The noise terms in both sensors have normal distributions. Since the noise terms in (5) and signal $s(t)$ are independent, the difference between MGF of two received signals $y_1(t)$ and $y_2(t)$ in (5) is derived from the noise-free terms $s(t)$ and $r(t)$. Since $r(t)$ is the delayed replica of $s(t)$, it includes two blocks. When the second sensor has not sensed the received signal yet, $r(t)$ merely contains the noise $w_2(t)$ and its MGF can be calculated by (8), but, as soon as the transmitted signal arrives at this sensor, $y_2(t)$ shows a similar behavior to $y_1(t)$. This suitable observation could be used for the time delay estimation.

So, MGF of signal detected at the first sensor is considered as a reference for our estimation in the second sensor. Indeed, the moments of $y_1(t)$ are extractable from this known MGF by using (7). These moments are employed as the reference for comparing among results retrieved from the second sensor. In the second sensor, a rectangular running window is implemented on $y_2(t)$ and this window helps to extract different segments of $y_2(t)$ step by step. The window length depends on two parameters. First, it must be long enough to be trustable in calculating the estimated moments, on the other hand, it should not be so long that damages the real-time characteristics of estimator. Anyway, there is a trade-off between these two factors. The window length is considered constant and moves from the beginning of the signal to the end. Besides the length, the overlap between adjacent frames is another item that is determined according to the required accuracy and tolerable complexity in the time delay estimation. At the beginning of signal block, the windowed signal includes only the noise part of $y_2(t)$, because of delay τ, so it exhibits different moments in comparison with $y_1(t)$. While the first point of window reaches the onset of delayed signal $s(t - \tau)$, the estimated moments become similar to the moments of $y_1(t)$. Mean square error (MSE) criterion is applied for observing the measure of this similarity. At first, we observe large MSE values, but, the window progression leads to a decrease in MSE and after the τ seconds delay point, we get a small amount for MSE nearly equal to zero and will remain constant up to the end of observation time.

Now, Doppler is considered and $r(t)$ is obtained from (9). Doppler changes the constant amount of MSE which had happened after τ seconds. It means that after the delay point, Doppler increases MSE gradually, but this phenomenon is not an annoying event in time delay estimation, even it helps to find the time delay, because this increasing in MSE takes place from the delay point, so it causes the delay point to be the point which has minimum value for MSE.

In figure (1), the Doppler effect on the MSE behavior is showed for three different SNRs. Time delay is equal to 300 microseconds. In SNR=+10dB, the result is clear. In two other SNRs, the minimum point is almost matched well with the actual amount of delay, i.e. 300.

We assume the windowed signal in the k-th step of window moving is denoted by y_{2k} and the i-th moment of this windowed signal is presented as $\hat{m}_{y_{2k},i}$. Therefore, the k-th window whose related moments $\hat{m}_{y_{2k},i}$ are the most similar to those of $y_1(t)$, $m_{y_1,i}$, can be estimated by:

$$\hat{k} = \arg\min_{k} \sum_{i=1}^{L} \left| m_{y_1,i} - \hat{m}_{y_{2k},i} \right|^2, \tag{10}$$

where in here, L is considered 4, and it would reveal a desirable result [Fukunaga et al., 1983]. In fact, when $L=4$, we use 4 moments of signal. So we have 4 equations that are applied to determine the unknown parameter. Although there is only one unknown parameter, but the noise signal does not let us find the parameter by only one equation. But the use of four equations is enough. Note that if more accuracy is needed, L can be considered larger. So, the delay point, $\hat{\tau}$, is the first point of $\hat{k}$-th window.

Despite the presence of Doppler, the proposed moment method estimates the time delay precisely. Consequently, this method can consider the time delay and Doppler simultaneously, and thus, is able to estimate the joint time delay and Doppler accurately.

4.2 Doppler estimation

In this section, we can consider the estimated delay $\hat{\tau}$ as the time origin for the received signal in the second sensor:

$$y_2(t + \hat{\tau}) = r(t + \hat{\tau}) + w_2(t + \hat{\tau}), \quad t \geq 0. \tag{11}$$

According to (9) and (11), we have:

$$y_2(t + \hat{\tau}) = s(t) \exp(j2\pi(t + \hat{\tau})\varepsilon) + w_2(t + \hat{\tau}), \quad t \geq 0. \tag{12}$$

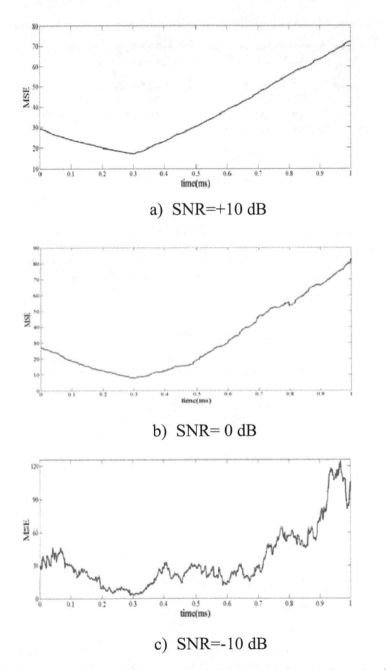

a) SNR=+10 dB

b) SNR= 0 dB

c) SNR=-10 dB

Fig. 1. MSE between the signal $y_1(t)$ moments and the windowed parts of signal $y_2(t)$ moments. a) SNR=+10dB, b) SNR=0dB, c) SNR=-10dB

Doppler and noise effect on the moments of $y_2(t + \hat{t})$ should be noticed. Instead of $y_2(t + \hat{t})$, we work on the real part:

$$y_{2r}(t + \hat{t}) = s(t)\cos(2\pi(t + \hat{t})\varepsilon) + w_2(t + \hat{t}), \quad t \geq 0. \tag{13}$$

$y_{2r}(t + \hat{t})$ includes the noise and signal, and the signal is also affected by Doppler which changes the moments of the signal. Therefore, we prefer to obtain MGF of $y_{2r}(t + \hat{t})$ firstly, then, the moments are obtained from this MGF by (7). The noise-free signal in (13) is independent from the noise $w_2(t + \hat{t})$, so MGF of $y_{2r}(t + \hat{t})$ is:

$$M_{y_{2r}}(u) = M_r(u)M_{w_2}(u), \tag{14}$$

where $M_r(u)$ is MGF of the first term in right side of (13), and:

$$M_{w_2}(u) = \exp\left(0.5\sigma_{w_2}^2 u^2\right). \tag{15}$$

The time varying variance will be comprehensively discussed in the sequel. Here, the problem is to estimate $M_r(u)$. $s(t)$ follows a Gaussian mixture distribution in (6). The presence of the cosine term changes the first term in the right side of (13) to a non-stationary process. Although the cosine term is time variant, fortunately, it is deterministic.

Now, we obtain $M_r(u)$:

$$M_s(u) = \sum_{i=1}^{N} p_i \exp\left(\mu_{s_i} u + 0.5\sigma_{s_i}^2 u^2\right) \quad \Rightarrow$$

$$M_r(u; t) = \sum_{i=1}^{N} p_i \exp\left(\mu_{s_i} u + 0.5\sigma_{s_i}^2 \cos^2(2\pi(t + \hat{t})\varepsilon)u^2\right). \tag{16}$$

Both $M_r(u)$ and $M_{w_2}(u)$ are expressed as the series for $u \to 0$, then by multiplying these two series and ordering their terms, MGF of $y_{2r}(t + \hat{t})$ is asymptotically obtained in the context of (7):

$$
\begin{aligned}
M_{y_{2r}}(u) &= M_r(u)M_{w_2}(u) \\
&= \left(1 + um_{r1} + \frac{u^2 m_{r2}}{2!} + \frac{u^3 m_{r3}}{3!} + \frac{u^4 m_{r4}}{4!} + \cdots\right) \\
&\times \left(1 + um_{w_21} + \frac{u^2 m_{w_22}}{2!} + \frac{u^3 m_{w_23}}{3!} + \frac{u^4 m_{w_24}}{4!} + \cdots\right) \\
&= 1 + u(m_{r1} + m_{w_21}) + \frac{u^2(m_{r2} + m_{w_22} + 2m_{r1}m_{w_21})}{2!} \\
&+ \frac{u^3(m_{r3} + m_{w_23} + 3m_{r1}m_{w_22} + 3m_{r2}m_{w_21})}{3!} \\
&+ \frac{u^4(m_{r4} + m_{w_24} + 6m_{r2}m_{w_22} + 4m_{r1}m_{w_23} + 4m_{r3}m_{w_21})}{4!} + \cdots.
\end{aligned}
\tag{17}
$$

The moments extracted from $M_r(u)$ are shown in Table (II). There exists also another problem. The resulting moments of $y_{2r}(t + \hat{t})$ are time dependent. Since the cosine term is deterministic, the time average of the moments can be substituted instead. Let's define:

$$\zeta_i(\varepsilon) = \frac{1}{T}\int_0^T \cos^i(2\pi(t + \hat{t})\varepsilon)dt, \tag{18}$$

n	Moment
0	1
1	$\sum_{i=1}^{N} p_i \mu_{s_i}$
2	$\sum_{i=1}^{N} p_i(\mu_{s_i}^2 + \sigma_{s_i}^2 \cos^2(2\pi(t+\hat{t})\varepsilon))$
3	$\sum_{i=1}^{N} p_i(\mu_{s_i}^3 + 3\mu_{s_i}\sigma_{s_i}^2 \cos^2(2\pi(t+\hat{t})\varepsilon))$
4	$\sum_{i=1}^{N} p_i(\mu_{s_i}^4 + 6\mu_{s_i}^2\sigma_{s_i}^2 \cos^2(2\pi(t+\hat{t})\varepsilon) + 3\sigma_{s_i}^4 \cos^4(2\pi(t+\hat{t})\varepsilon))$

Table 2. Moments extracted from $M_r(u;t)$

where T is the observation time. Note that for dependency of $\zeta_i(\varepsilon)$ on ε, the moments of $y_{2r}(t+\hat{t})$ are dependent on ε too. Finally, for obtaining the time-independent moments of $y_{2r}(t+\hat{t})$, $m_{y_{2r},i}$, it suffices that all "$\cos^i(2\pi(t+\hat{t})\varepsilon)$" terms in the time-dependent moments to be substituted by $\zeta_i(\varepsilon)$. The final moments are depicted in table (III).

Since now, the moments were obtained analytically, it means we only calculated the right side of equation (4). On the other hand, the moments of the observed signal in the second receiver can be calculated statistically by:

$$\tilde{m}_i = \frac{1}{T} \int_0^T y_{2r}^i(t+\hat{t})dt. \tag{19}$$

Now the left side of the equation (4) is also obtained. Both of these two procedures must yield same results. Thus, ε should be selected in such a way that this equality holds. To do this, MSE criterion is used again:

$$\text{MSE} = \sum_{i=1}^{L} \left| m_{y_{2r},i} - \tilde{m}_i \right|^2. \tag{20}$$

Similar to the previous section, L is considered as 4. So Doppler of the received signal $y_{2r}(t+\hat{t})$ is estimated:

$$\hat{\varepsilon} = \arg\min_{\varepsilon} \sum_{i=1}^{L} \left| m_{y_{2r},i} - \tilde{m}_i \right|^2. \tag{21}$$

4.3 Noise power estimation

The noise power estimation is similar to Doppler estimation. Indeed, these two estimations are done simultaneously. It could be seen that the moments do not merely depend on Doppler.

n	Moment
0	1
1	$\sum_{i=1}^{N} p_i \mu_{s_i}$
2	$\sum_{i=1}^{N} p_i \left(\mu_{s_i}^2 + \sigma_{s_i}^2 \zeta_2(\varepsilon) \right) + \sigma_{\omega}^2$
3	$\sum_{i=1}^{N} p_i \left(\mu_{s_i}^3 + 3\mu_{s_i}\sigma_{s_i}^2 \zeta_2(\varepsilon) \right) + 3\sigma_{\omega}^2 \sum_{i=1}^{N} p_i \mu_{s_i}$
4	$\sum_{i=1}^{N} p_i \left(\mu_{s_i}^4 + 6\mu_{s_i}^2 \sigma_{s_i}^2 \zeta_2(\varepsilon) + 3\sigma_{s_i}^4 \zeta_4(\varepsilon) \right) + 3\sigma_{\omega}^4 + 6\sigma_{\omega}^2 \sum_{i=1}^{N} p_i (\mu_{s_i}^2 + \sigma_{s_i}^2 \zeta_2(\varepsilon))$

Table 3. Final moments extracted from $M_{y_{2r}}(t + \hat{\tau})$

They depend onto the noise power as well. So, in (20), MSE includes two parameters, the noise power and Doppler of the received signal, and should be minimized according to both of them:

$$(\hat{\varepsilon}, \hat{\sigma}_{\omega_2}^2) = \arg\min_{\varepsilon, \sigma_{\omega_2}^2} \sum_{i=1}^{L} \left| m_{y_{2r},i} - \tilde{m}_i \right|^2. \tag{22}$$

Now it is the time to discuss about the variable variance of the noise. This means that in (14) the noise variance is considered unknown. We can estimate the noise variance given N_1 signal-free samples which are at hand occasionally. So, $\sigma_{\omega_2}^2$ becomes a random variate. Since the noise $\omega_2(t + \hat{\tau})$ is assumed Gaussian, the N_1-sample based estimated variance is chi-square distributed with N_1 degrees of freedom:

$$\hat{\sigma}_{\omega_2}^2 = \frac{1}{N_1} \sum_{i=1}^{N_1} \omega_{2_i}^2, \quad \hat{\sigma}_{\omega_2}^2 \sim \chi_{N_1}^2. \tag{23}$$

Hence, the average MGF of the noise over σ^2 is obtained in (14) as:

$$\bar{M}_{\omega_2}(u) = \frac{1}{\sqrt{(1 - \hat{\sigma}_{\omega_2}^2 u^2 / N_1)^{N_1}}}$$

$$= 1 + 0.5\hat{\sigma}_{\omega_2}^2 u^2 + (0.125 + 1/4N_1)\hat{\sigma}_{\omega_2}^4 u^4 + \cdots . \tag{24}$$

In this non-stationary noise scenario due to $\hat{\sigma}_{\omega_2}^2$, the procedure presented for Doppler estimation in the previous part does not change, only MGF and the moments of the normal distribution considered previously for the noise should be substituted by the ones determined in (24).

5. Radar tracking

In the basic section, we said that the proposed parameter estimation can be useful for the tracking of a target. As mentioned, there are various methods for the target tracking which present specific mathematical algorithms. These methods have different performance levels, but most of them are recursive, so that at any time, the data is obtained by using previous data and improving them. Now, some of the most common procedures and their problems are expressed and then, the proposed moment method are described in detail.

5.1 Kalman filter

The Kalman filter is the central algorithm to the majority of all modern radar tracking systems. The role of the filter is to take the current known state (i.e. position, heading, speed and possibly acceleration) of the target and predict the new state of the target at the time of the most recent radar measurement. In making this prediction, it also updates its estimate of its own uncertainty (i.e. errors) in this prediction. It then forms a weighted average of this prediction of state and the latest measurement of state, taking account of the known measurement errors of the radar and its own uncertainty in the target motion models. Finally, it updates its estimate of its uncertainty of the state estimate. A key assumption in the mathematics of the Kalman filter is that measurement equations (i.e. the relationship between the radar measurements and the target state) and the state equations (i.e. the equations for predicting a future state based on the current state) are linear, i.e. can be expressed in the form $y = A.x$ (where A is a constant), rather than $y = f(x)$. The Kalman filter assumes that the measurement errors of the radar, and the errors in its target motion model, and the errors in its state estimate are all zero-mean Gaussian distributed. This means that all of these sources of errors can be represented by a covariance matrix. The mathematics of the Kalman filter is therefore concerned with propagating these covariance matrices and using them to form the weighted sum of prediction and measurement [Ristic et al., 2004].

In situations where the target motion conforms well to the underlying model, there is a tendency of the Kalman filter to become "over confident" of its own predictions and to start to ignore the radar measurements. If the target then manoeuvres, the filter will fail to follow the manoeuvre. It is therefore common practice when implementing the filter to arbitrarily increase the magnitude of the state estimate covariance matrix slightly at each update to prevent this.

5.2 Extended Kalman Filter (EKF)

This method is a class of nonlinear tracking algorithms that provides much better results than the Kalman filter.

Nonlinear tracking algorithms use a nonlinear filter to cope with the following cases:

- The relationship between the radar measurements and the track coordinates is nonlinear.
- The errors are nonlinear.
- The motion model, is non-linear.

In this case, the relationship between the measurements and the state is of the form $h = f(x)$ (where h is the vector of measurements, x is the target state and $f(.)$ is the function relating the two). Similarly, the relationship between the future state and the current state is of the form $x(t + 1) = g(x(t))$ (where $x(t)$ is the state at time t and $g(.)$ is the function that predicts the future state). To handle these non-linearities, the EKF linearizes the two non-linear equations using the first term of the Taylor series and then treats the problem as the standard linear Kalman filter problem. Although conceptually simple, the filter can easily diverge (i.e. gradually perform more and more badly) if the state estimate about which the equations are linearized is poor. The unscented Kalman filter and particle filters are attempts to overcome the problem of linearizing the equations.

5.3 Particle Filtering (PF)

Another example of nonlinear methods is particle filtering. This method makes no assumptions about the distributions of the errors in the filter and neither does it require the equations to be linear. Instead it generates a large number of random potential states ("particles") and then propagates this "cloud of particles" through the equations, resulting in a different distribution of particles at the output. The resulting distribution of particles can then be used to calculate a mean or variance, or whatever other statistical measure is required. The resulting statistics are used to generate the random sample of particles for the next iteration. However, this method also has some problems that restrict the use. This method requires large computational operations and face severe difficulties for real-time applications. On the other hand, this method is also not able to have suitable results in very low SNRs. In these SNRs, PF is not able to bring us to a reasonable particle, and even using Sampling Importance Re-sampling (SIR) method can not lead us to better results [Ristic et al., 2004]. In SIR method, a weighted set of particles is used. These new weighted particles can face and eliminate the noise more powerfully and present better estimation in low SNRs.

5.4 The proposed moment method

In this section, we are going to solve the problems we are faced in PF. This is done based on the time delay and Doppler estimated in the previous section. Three sensors are used. They are located on the vertices of an equilateral triangle. One of the sensors is a transmitter and receiver, the other two sensors only serve as the receiver. The arrangement of the sensors and their positions relative to the target is depicted in figure (2). The target is in the far field of the sensors.

A signal is emitted from the first sensor to the target. When this signal comes into contact with the target, generally speaking, it is scattered in many directions. The signal is thus partly reflected back, hence, all three sensors receive this reflected signal. According to the earlier discussions, the time delay and Doppler of the received signal in each sensor could be estimated.

First, the target position is determined. Suppose the time interval between sending the signal from the transmitter and receiving it in each sensor is shown by T_i for $i = 1, 2, 3$, which i denotes the sensor number. We also use R_i as the distance between the target and the i-th receiver. Since the transmitter is beside the first receiver, we have:

$$R_1 = \frac{1}{2}T_1 \times C_e, \tag{25}$$

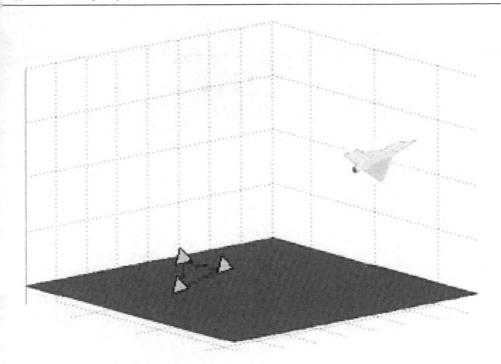

Fig. 2. The arrangement of the three sensors and their positions relative to the target

which C_e is the velocity of the emitted signal that is equal to the light speed. For two other receivers that are not near the transmitter, the distance is calculated as:

$$R_i = (T_i - \frac{T_1}{2}) \times C_e, \quad i = 2, 3. \tag{26}$$

Each sensor provides the locus of the target on a sphere of radius R_i centered at that sensor. As mentioned before, these sensors are located on the vertices of an equilateral triangle.

It can be shown mathematically that the intersection of these three spheres is at two points. To prove this, the equations for the three spheres are considered, and then the intersection of them is obtained. Without losing the generality, we assume that the three points where the sensors are located in, are showed by A, B and C. The points are respectively in $(x_0, 0, 0)$, $(-x_0, 0, 0)$ and $(0, y_0, 0)$ in Cartesian coordinates and are showed in figure (3).

At first, the equations of two spheres with centers A and B and radii R_1 and R_2 are obtained:

$$(x - x_0)^2 + y^2 + z^2 = R_1^2,$$
$$(x + x_0)^2 + y^2 + z^2 = R_2^2. \tag{27}$$

The first equation is subtracted from the second one:

$$2xx_0 - (-2xx_0) = R_2^2 - R_1^2 \quad \Rightarrow$$
$$4xx_0 = R_2^2 - R_1^2 \quad \Rightarrow \quad x = \frac{R_2^2 - R_1^2}{4x_0}. \tag{28}$$

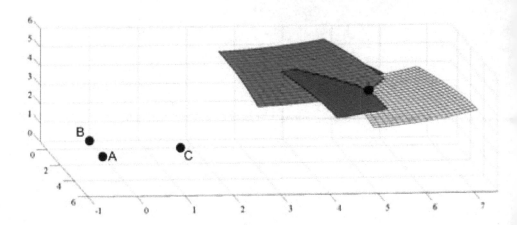

Fig. 3. The position of the three sensors and the intersection of three spheres related to the sensors

Now, the obtained value x is put in the one of the equations (27). We select the first one:

$$\left(\frac{R_2^2 - R_1^2}{4x_0} - x_0 \right)^2 + y^2 + z^2 = R_1^2 \quad \Rightarrow$$

$$y^2 + z^2 = R_1^2 - \left(\frac{R_2^2 - R_1^2}{4x_0} - x_0 \right)^2 . \tag{29}$$

For convenience, the right side of the second equality of (29) is showed by R_{cir}^2. Thus, the intersection of the two spheres is a circle with the following equation:

$$y^2 + z^2 = R_{cir}^2 , \tag{30}$$

Which is located in the plane $x = \frac{R_2^2 - R_1^2}{4x_0}$.

Then the intersection of this circle and the third sphere should be obtained. The third sphere has the center C and radius R_3. So its equation is:

$$x^2 + (y - y_0)^2 + z^2 = R_3^2 . \tag{31}$$

The left side of the equation (31) is extended, and the circle equation is used in it:

$$x^2 + (y - y_0)^2 + z^2 = -2yy_0 + y_0^2 + x^2 + y^2 + z^2$$

$$= -2yy_0 + y_0^2 + \left(\frac{R_2^2 - R_1^2}{4x_0} \right)^2 + R_{cir}^2 \quad \Rightarrow$$

$$y = \frac{y_0^2 + \left(\frac{R_2^2 - R_1^2}{4x_0} \right)^2 + R_{cir}^2}{2y_0} . \tag{32}$$

So, x and y coordinates of the intersection point is:

$$x = \frac{R_2^2 - R_1^2}{4x_0}$$

$$y = \frac{y_0^2 + \left(\frac{R_2^2 - R_1^2}{4x_0}\right)^2 + R_{cir}^2}{2y_0}. \tag{33}$$

Using this two values and the equation (31), the third coordinates is also calculated:

$$z = \pm \sqrt{R_3^2 - \left(\frac{R_2^2 - R_1^2}{4x_0}\right)^2 - \left(\frac{y_0^2 + \left(\frac{R_2^2 - R_1^2}{4x_0}\right)^2 + R_{cir}^2}{2y_0} - y_0\right)^2}. \tag{34}$$

As mentioned, this intersection contains only two points which are located in the two sides of the plane xy and in front of each other. But in reality, only one of these points has a positive height and coincides with the coordinate of a target in sky.

After this proof, we continue our discussion about the tracking. On the one hand, the target position is achievable by using R_is, and on the other hand, the equations (25) and (26) inform about the relation between R_is and T_is. Therefore, the target position can be determined if T_i is known. For calculating this parameter, it should be considered as the signal's time delay to reach to the i-th receiver. Let's assume the first sensor in the section (IV), is the transmitter now, and the second sensor in there is one of the three receivers in here. By using the proposed moment method three times, the time delay can be estimated for all the three receivers. T_i is denoted as the estimated time delay for i-th receiver. Now, all unknowns are obtained, so the position is easily predicted.

Finally, the target velocity should be obtained. The receivers compute three values for Doppler, $\hat{\varepsilon}_i$, by the proposed moment technique. Since the transmitter and the first receiver are at the same sensor, the velocity component along the connecting line between the target and the first sensor is:

$$v_1 = \frac{d}{dt}\|R_1\| = \frac{C}{2f_t}\hat{\varepsilon}_1, \tag{35}$$

where $\|.\|$ represents Euclidean norm, and R_1 is the vector connecting the first sensor to the target. C is the speed of light and f_t is the frequency of the emitted signal. Using v_1, we determine the velocity components along the connecting line between the target and two other sensors (receivers):

$$v_i = \frac{C}{f_t}\hat{\varepsilon}_i - v_1, \quad i = 2,3. \tag{36}$$

In the next section, there are results that compare the different methods available for estimating the time delay and Doppler. There are also some results about tracking a target which has a nonlinear motion. In the parameter estimation results, the proposed moment method is compared with the methods Wigner-Ville (WV), fractional lower order ambiguity function (FLOAF) and wavelet, and in the tracking part, there is a comparison between the proposed method and EKF and PF ones.

6. The results

To prove the procedures were presented in this Chapter, several different tests have been conducted. The results are divided into two categories. At first, the proposed method for estimating the joint time delay and Doppler is examined and compared with other conventional methods. Then, the efficiency of this method in the tracking of the maneuver target is also investigated.

6.1 Parameter estimation results

To estimate the time delay and Doppler parameters, the following assumptions are considered:

- The transmitted desired signal follows a trimodal Gaussian mixture distribution presented in equation (6) with the following mean and standard deviation related to the three modes:
  $\sigma_{s_1} = \sigma_{s_2} = \sigma_{s_3} = 1$,
  $\mu_{s_1} = 2, \mu_{s_2} = 5, \mu_{s_3} = 8$,

  And the probability distribution of the modes is considered as below:
  $p_1 = 0.3, \ p_2 = 0.3, \ p_3 = 0.4$.

- The observation time of the signal is considered 1 millisecond.

- The time delay can be within the observation time of the signal, and in here, it is assumed 300 microseconds.

- Doppler value, $\omega_\varepsilon = 2\pi\varepsilon$, is a number between 0 and 2π that provides a 2π rotation for the frequency shift. Now, Doppler is assumed 0.8π.

The test is done for different SNR values from -10dB to +10dB, and for each SNR, the operation is performed 1000 times. The figure (4) depicts the error existed in the estimation of the time delay for the conventional methods and the proposed moment one. This error is depicted as MSE, calculated from 1000 times of simulation implementation, versus SNR. We have used normalized MSE in our results:

$$MSE\left(\hat{\tau}\right) = E\left[\left(\frac{\hat{\tau} - \tau}{\tau}\right)^2\right], \tag{37}$$

where τ is the actual time delay, and $\hat{\tau}$ is the estimated value of this parameter. The conventional methods are WV [Chassande-Mottin & Pai, 2005], wavelet method [Niu et al., 1999] and FLOAF [Ma & Nikias, 1996].

As shown in Figure (4), all methods are convincing in high SNRs, but in low SNRs, especially negative ones, WV and FLOAF methods are completely unable to estimate the time delay. Wavelet method also has relatively unsuitable results, so that it presents very little reduction in MSE value from SNR=-10dB to SNR=0dB. But the moment method in the both high and low SNRs provides precise answers.

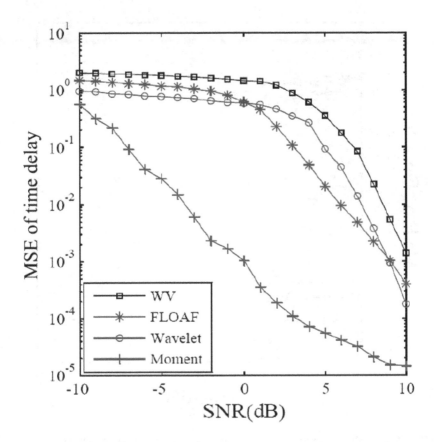

Fig. 4. MSE of estimated time delay in the conventional and proposed methods.

There is a similar observation for Doppler that is showed in figure (5). The error is also as MSE versus SNR. In this figure, the conventional methods are WV [Chassande-Mottin & Pai, 2005] and FLOAF [Ma & Nikias, 1996].

As portrayed in figure (5), WV offers very good results in high SNRs which is expectable. But in the low SNRs, the interaction terms are relatively large and this method fails. So in low SNRs, FLOAF presents more suitable results in comparison with WV. Again in this figure, the power of moment method is absolutely visible.

It is worth mentioning that the obtained results are in an unknown noise power scenario. The moment method also can estimate the noise power. It is important that in addition to parameter estimation, our method can also predict the noise power. This capability helps to recognize the noise environment, and ameliorates noise encountering. To judge the performance of the proposed moment method for estimating the unknown noise power, MSE between the actual and the estimated noise power is portrayed in figure (6). For instance, MSE is 10^{-5} in SNR 8. It means that in this SNR, we have an error between the actual noise power

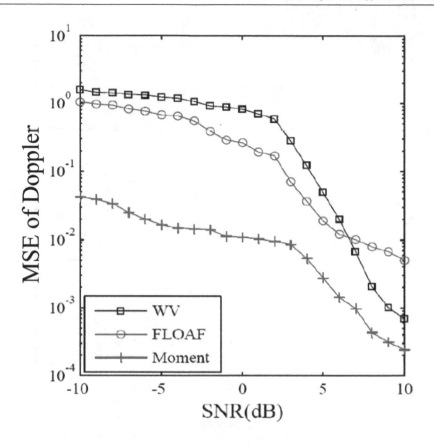

Fig. 5. MSE of estimated Doppler in the conventional and proposed methods.

and the value which our method has estimated for it, and the normalized mean square value of this error is equal to 10^{-5}.

6.2 Radar tracking results

In the following, radar tracking results are presented based on the time delay and Doppler estimations. The original frequency of the signal emitted from the radar, f_t, is considered $10GHz$. A target is at cartesian coordinate $(10000m, 10000m, 10000m)$. It moves with the velocity $v_x = 10m/s, v_y = 10m/s$. In the first 25 sec, $v_z = -10$ m/s and in the following 75 sec, $v_z=+20$ m/s. At first, for SNR=+10dB, test is done for the non-recursive proposed moment method and two recursive conventional methods: EKF [Park & Lee, 2001] and PF [Jian et al., 2007]. The results have been traced for 100 epochs with one second interval and can be seen in figures (7) and (8) as MSE of the estimated position and velocity.

Two points are worth noting in this figures. EKF and PF methods are recursive, so the related curves are decreasing and at first, have not acceptable results. We need some time to have suitable results. In vital application like military, less needed time leads us to a better real-time system and gives the opportunity to react faster. So, a non-recursive method can be valuable.

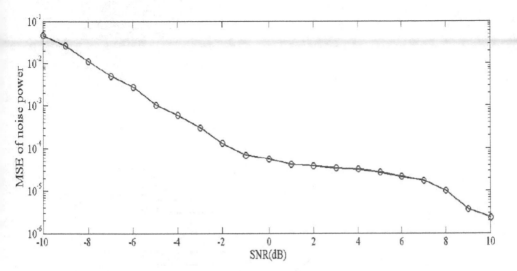

Fig. 6. MSE of estimated noise power in the proposed method

In addition to have a high precision, the moment method is non-recursive, trustable from the beginning, and provides a rapid reaction.

The second point in figures (7) and (8) is the existence of a bulge around the time 25 seconds, where the third component of the speed has changed and made a nonlinear motion. There is no bulge in the curves relating to the moment method, because in this method, the estimation at any time is independent from the other times, so it has no problem in nonlinear motions. In the figures we magnify the results around time 25 seconds and show them in linear scale to depict the bulge obviously. We cannot present all results together in linear scale, because moment results are too small in comparison with EKF and PF results.

To further examine the ability of the proposed method, the test is done at different SNRs. The results of this experiment is showed in figures (9) and (10). In the figures, MSE of the position and velocity estimation is portrayed for our moment method.

In figures (7) and (8), MSE is versus time , and SNR is constant and equal to +10dB. Thus the figures (7) and (8) show the superiority of the proposed method on the two other ones. But in figures (9) and (10), MSE is versus SNR. The power of moment method in the low SNR is quite satisfactory, while the other methods, the EKF and PF, either do not respond or provide answers that are not reliable.

Finally, a necessary point should be noted. We see that our method has much better results in comparison with other ones. The better results are not only because of using moments. Moment method helps us as a tool to encounter the undesired signals logically. In fact, in the first step, we recognize the environment more precisely by a suitable model of noise. Then after the modelling, although the noise is unknown, but the moments of its model are known and used for our estimations. So we can control the noise behaviour. This procedure cannot be found in other methods.

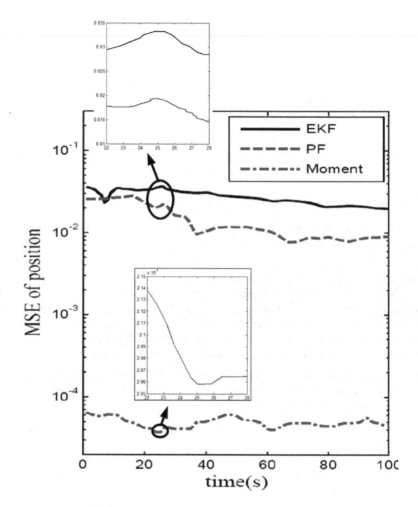

Fig. 7. MSE of estimated position in the conventional and proposed methods for SNR=+10 dB.

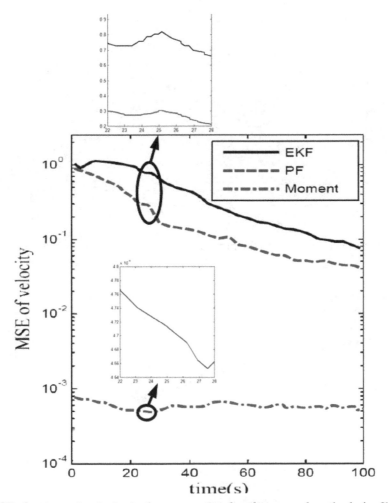

Fig. 8. MSE of estimated velocity in the conventional and proposed methods for SNR=+10 dB.

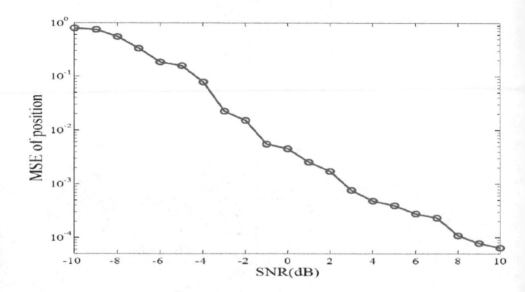

Fig. 9. MSE of estimated position in the proposed method.

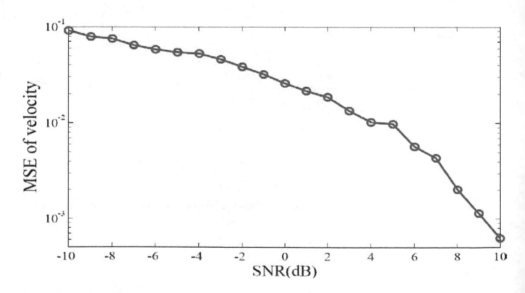

Fig. 10. MSE of estimated velocity in the proposed method.

7. Conclusion

In this chapter, we review different novel methods in joint time delay and Doppler estimation. Each of them has some advantages and disadvantages. The disadvantages are studied and we find a new method which can almost obviate the most of these disadvantages. The new method is based on moment. It exploits the time delay, Doppler, and noise effects exerted onto the moments of the received data. The insight on the moments of the received signal is the criteria for joint estimation of time delay and Doppler. Since the moments of the noise could be obtained, these moments can facilitate separating the main signal from the noise even in a severe noisy environment. So, our estimation in low SNR has suitable results. In addition, we do not encounter with undesirable cross-terms discussed in WV method. After introducing our estimation method, its application in Doppler radar is declared.

The estimated delay and Doppler are used in the target tracking and predicting the position and velocity of the target in a noisy background. So it is applicable in the radar trackers. Test results provide a plausibility of the both estimations and tracking. The estimated position and velocity are completely accurate even in very low SNRs. The tracking can be extended to multiple targets. Based on the features described for mono-target tracking, it is expected to have acceptable results in multiple targets tracking. Multi tracking in low SNRs is one of the most important roles of a Doppler radar which is reachable based on the presented method.

8. References

[1] Bilik, I., Tabrikian, J., Cohen, A. (2006). "GMM-based target classification for ground surveillance Doppler radar," *IEEE Trans. on Aerospace and Electronic Systems*, vol. 42, no. 1, January.

[2] Bouchereau, F., Brady, D. (2008). "Method-of-moments parameter estimation for compound fading processes," *IEEE Trans. Comm.*, vol. 56, no. 2, pp. 166-172.

[3] Chassande-Mottin, E., Pai, A. (2005). "Discrete time and frequency Wigner-Ville distribution: Moyal's formula and aliasing," *IEEE Signal Processing Letters*, vol. 12, no. 7, pp. 508-511, July.

[4] Fukunaga, K., Flick, T. E. (1983). "Estimation of the parameters of a Gaussian mixture using the method of moments," *IEEE Trans. Pattern Analysis and Machine Intelligence*, vol. pami-5, no. 4, pp. 410-416, July.

[5] Gaeddert, J., Annamalai, A. (2005). "Some remarks on Nakagami-m parameter estimation using method of moments," *IEEE Comm. Letters*, vol. 9, no. 4, pp. 313-315.

[6] Greenstein, L. J., Michelson, D. G., Erceg, V. (1999). "Moment-method estimation of the Ricean K-factor," *IEEE Comm. Letters*, vol. 3, no. 6, pp. 175-176.

[7] Isaksson, A. J., Horch, A., Dumont, G. A. (2001). "Event-triggered deadtime estimation from closed-loop data," *In Proc. American Control Conf.*, Arlington, VA, USA, June.

[8] Jian, W., Yonggao, J., Dingzhang, D., Huachun, D., Taifan, Q. (2007). "Particle filter initialization in non-linear non-Gaussian radar target tracking," *Journal of Systems Engineerng and Electronics*, vol. 18, no. 3, pp. 491-496.

[9] Ma, X., Nikias, C. L. (1996). "Joint estimation of time delay and frequency delay in impulsive noise," *IEEE Trans. Signal Processing*, vol. 44, pp. 2669-2687, November.

[10] Niu, X., Ching, P., Chan, Y. (1999). "Wavelet based approach for joint time delay and Doppler stretch measurements," *IEEE Trans. on Aerospace and Electronic Systems*, vol. 35, no. 3, pp. 1111-1119.

[11] Orr, R. S., Morris, J. M., Qian, S. E. (1992). "Use of the Gabor representation for Wigner distribution crossterm suppression," *ICASSP-92*, vol.5, pp. 29-31, March.

[12] Park, S. T., Lee, J. G. (2001). "Improved Kalman filter design for three-dimensional radar tracking," *IEEE Trans. on Aerospace and Electronic Systems*, vol. 37, no. 2, pp. 727-739, April.

[13] Ristic, B., Arulampalam, S., Gordon, N. (2004). "Beyond the Kalman Filter: Particle filters for tracking applications," *Artech House*.

[14] Tan, J. L., Sha'ameri, A. Z. B. (2008). "Adaptive optimal kernel smooth-windowed wigner-ville for digital communication signal," *EURASIP Journal on Advances in Signal Processing*.

[15] Zabin, S. M., Wright, G. A. (1994). "Nonparametric density estimation and detection in impulsive interference channels. I. Estimators," *IEEE Trans. on Communications*, vol. 42, no. 2/3/4, pp. 1684-1697, February/March/April.

11

Volcanological Applications of Doppler Radars: A Review and Examples from a Transportable Pulse Radar in L-Band

Franck Donnadieu
[1]*Clermont Université, Université Blaise Pascal,*
Observatoire de Physique du Globe de Clermont-Ferrand (OPGC),
Laboratoire Magmas et Volcans, Clermont-Ferrand
[2]*CNRS, UMR 6524, LMV, Clermont-Ferrand*
[3]*IRD, R 163, LMV, Clermont-Ferrand*
France

1. Introduction

Many types of radar systems have been applied to the study of a wide range of volcanic features. Fields of application commonly include volcano deformation by interferometric synthetic aperture radar (InSAR), mainly satellite-based (e.g. Froger et al., 2007) but also ground-based like LISA (Casagli et al., 2009), digital elevation model generation using satellite or airborne InSAR measurements, surface products mapping by amplitude images of satellite radars, characterization of unexposed deposits by ground-penetrating radars (Russell & Stasiuk, 1997), monitoring of active lava domes and flows by either ad-hoc ground-based radars (e.g. Malassingne et al., 2001; Macfarlane et al., 2006; Wadge et al., 2005, 2008) or commercial ones (Hort et al., 2006; Vöge & Hort, 2008, 2009; Vöge et al., 2008), and quantitative characterization of explosive activity by means of fixed weather radars (large ash plumes) and transportable radars (Strombolian activity, weak ash plumes). A thorough review of all radar applications in volcanology is beyond the scope of this chapter which, instead, focuses on recent investigations of explosive eruptive regimes enhanced by the developments of dedicated transportable ground-based radars, by the recent advances made in signal interpretation using eruption models, and by the important concerns raised by ash plume hazards.

Below, weather and transportable radar systems used hitherto to monitor and study tephra emissions are first reviewed along with their advantages and limitations. Some differences relevant to the study of radar signals of volcanic origin are also described. Then I present a unique transportable pulse Doppler radar named VOLDORAD, operating in the L-band and dedicated to the study of explosive activity. Because it can be set up close to an eruptive vent and sound at a high rate the interior of even heavily particle-laden plumes within small beam volumes right above the emission source, many features of processes and dynamics of volcanic emissions can be characterized at different time scales from the spatiotemporal analysis of echo signals. Examples are provided of records obtained with VOLDORAD at several volcanoes in different sounding conditions and various types of volcanic activity, from Strombolian lava jets to weak ash plumes. They are meant to illustrate the many capabilities of this type of radar and the interpretation of the variety of Doppler signatures in terms of volcanic processes.

2. Radar monitoring of explosive eruptions

2.1 Ash plume hazards and tephra dispersal forecast

Volcanic ash plumes generate important hazards as widespread ash fallout may cause serious perturbations to surrounding population and infrastructures. In addition, volcanic ash clouds derived from eruptive columns, even of moderate size, can generate direct hazards to aviation, as recently highlighted by the 2010 Eyjafjöll eruption in Iceland. The air traffic was disrupted over Europe for several days, causing a loss of about 1.7 billion dollars to airliners. Although no significant damage to aircraft was reported for this eruption, over 120 aircraft encounters with volcanic ash have nevertheless been documented between 1973 and 2008 (Schneider, 2009). The tracking of large ash clouds has therefore become a main concern in the last decades, as attested by the creation of Volcanic Ash Advisory Centers (VAAC) to provide their expertise to civil aviation in case of significant volcanic eruptions. Volcanic ash transport and dispersion (VATD) models are used to forecast the location and movement of ash clouds over hours to days in order to define hazards to aircraft and to communities downwind. Inputs are eruption source parameters such as plume height, mass eruption rate, duration, and mass fraction of fine ash (Mastin et al., 2009). Values of such parameters are frequently unconstrained in the first minutes or hours after an eruption is detected, and also change during an eruption (e.g. plume height), requiring rapid reevaluation. Dispersion model forecast are routinely validated, verified against all available observations, including field observations, combination of tephra deposits analysis and theoretical models, or in-situ measurements and sampling (aircraft). However remotely-sensed measurements by satellite imagery, ground-based radars and lidars, or better a combination of all, are the most efficient tools for real-time response owing to their continuous data acquisition and potential for automatic processing and rapid parameter quantification (Fig. 1).

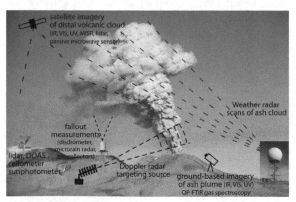

Fig. 1. Synergetic potential of integrated remote-sensing techniques for ash plume monitoring. Photo: Sabancaya volcano, courtesy of J.-L. Le Pennec.

2.2 Ash plume monitoring

Long-range trajectory tracking of ash clouds is achieved primarily by means of satellite imagery. Although Delene et al. (1996) showed the utility of a satellite-based microwave imager passively measuring radiations (19-85 GHz) of millimetric volcanic particles from an ash cloud of Mount Spurr in 1992, satellite visible-infrared radiometric observations from geostationary platforms are usually exploited (e.g., Rose et al., 2000). The evolution of the ash cloud spatial distribution, in particular, can be imaged at intervals of 15-30 min. Important parameters can be further retrieved like the approximate plume height assuming thermal equilibrium with the atmosphere (non unicity of solutions for altitudes above the tropopause), and the concentration and size of distal particles (< 20 microns) transported in the atmosphere, assuming particle sphericity and vertically homogeneous concentration. Using these assumptions, the mass of SO_2 and ash can be integrated on successive images (e.g. Wen & Rose, 1994). Scollo et al. (2010) also showed the potential of Multiangle Imaging SpectroRadiometer (MISR) working in four wavelengths in visible and near-infrared bands, for the 3-D reconstruction of ash plume shape, and for the retrieval of column height, optical depth, type and shape of the finest particles, among the most sensitive inputs for ash dispersal modeling.

Yet, the exploitation of satellite images for monitoring purposes is limited by (1) the presence of clouds at higher levels, (2) an insufficient acquisition rate for event onset detection, (3) a relatively poor spatial resolution, (4) errors of the "split-window" method (brightness temperature difference) when the volcanic plume lies over a very cold surface or when the plume lies above a clear land surface at night where strong surface temperature and moisture inversions exist (Prata et al., 2001). In addition, low ash content and/or small ash plumes might not be clearly observed and near-source emissions are obscured by the emitted tephra. For these reasons, ground-based radar systems represent an optimal complementary solution for real-time monitoring of these phenomena, by providing higher spatial resolution and data acquisition rates, as well as the ability to make observations at night and under any weather conditions. Real-time monitoring of ash plumes is crucial, in particular for the initialization of dispersion models. In this respect, essential input parameters such as plume height, mass flux, and particle concentration can be assessed quantitatively from radar data and directly contribute to improve ash dispersion forecasts.

2.2.1 Radar monitoring of ash plumes

2.2.1.1 Weather radar observations

2.2.1.1.1 Characteristics and advantages

Although ash plume hazards to aviation safety raised concerns early on about the detection capacity of ash clouds by airborne radar (Musolf, 1994; Stone, 1994), most observations of large volcanic ash clouds have been opportunely carried out by fixed meteorological radars of national weather services. Weather radars operate at microwave frequencies from S band (7.5-15 cm wavelength, generally about 10 cm) up to C band (3.75-7 cm wavelength, commonly around 6 cm), X band (2.5-3.75 cm wavelength, commonly around 3 cm) and Ka band (0.75-1.11 cm wavelength, often around 1 cm). With peak powers up to 250 kW or even higher, they have sufficient sensitivity to detect volcanic ash clouds with small particle sizes. Pulsed systems ensure a relatively high spatial range resolution of a few hundreds of meters. Reflector antennas provide half-power beam-widths of a few degrees. These systems usually scan in azimuth and elevation, within a few minutes, allowing the maximum plume height to be tracked through time, along with the spatial variations of its reflectivity. Most have a Doppler capability to measure radial wind velocity, that can be used to infer information on internal velocities of ash clouds and retrieve information on turbulence, which has seldom been used. New generation radars are dual-polarized, which may further help to discriminate ash from hydrometeors.

Powerful weather radars, operating continuously at minute-scale acquisition rate and in all weather, have been used occasionally to track large ash clouds, chiefly since the first radar observations of Hekla eruption in 1970 and Augustine in 1976, because the information held in their records is many fold and potentially very useful for risk mitigation. Scan images of the ash cloud provide reflectivity variations in horizontal and vertical planes. Time evolutions of its height and lateral spreading can then be retrieved, along with its ascent rise rate and lateral transport speed. Mass and volume of radar-detected ash, as well as particle concentrations in the cloud can be estimated provided the grain size distribution can be constrained (from ash fall or other).

2.2.1.1.2 Observations

Harris et al. (1981), Harris & Rose (1983) and Rose & Kostinski (1994) first collected observations of ash plumes from Mt. St. Helens in 1980-1982 using 5 cm and 23 cm radar systems. They tracked the position of the ash cloud of March 19 1982, and estimated its volume (2000 ±500 km^3), the concentration of ash (0.2-0.6 g/m^3), reflectivity factors of 4-5 mm^6/m^3 (6-7 dBZ), and the total mass of ash erupted (3-10×10^8 kg). For the famous paroxysmal eruption of May 18 1980, they obtained a mass of 5×10^{11} kg, an ash volume of 0.2 km^3, and particle concentration of 3-9 g/m^3, for the ash cloud downwind of Mount St. Helens, 1.5-2 h after its eruption (horizontal speed 135 km/h). Reflectivity factors found for these dense (but distal) ash clouds (7-60 mm^6/m^3 or 8-18 dBZ) are several orders of magnitude smaller than those for severe weather considered routinely detectable by airborne weather radar and dangerous for aviation. Eruption-column rise rates and horizontal drift of ash clouds of Mount Pinatubo, Philippines, in 1991, were also tracked using two military C-band weather radars 40 km away (Oswalt et al., 1996). During the second eruption of June 12, 1991, radars indicated an apparent column rise rate in excess of 400 m/s. Radar height measurements were typically 10 to 15 percent lower than ash cloud heights inferred from satellite temperature analyses. Radar observations also suggested that higher eruption columns correlated with greater particle size and density within the column. Using a C-band radar Rose et al. (1995) found that most intense reflections in an ash cloud of Mount Spurr in 1992 came from particles 2 to 20 mm in diameter and with a total particle mass concentration of <0.01 to 1 g/m^3. The radar did not detect distal parts of the ash cloud, which have an atmospheric residence time of longer than 30 minutes, because the larger more reflective ash particles drop out. Maki and Doviak (2001) observed ash plumes of Mount Oyama on Miyake Island, Japan, in 2000, with a 5-cm (C band) radar, and proposed a method to obtain the time-dependent size distribution of ash particles from the time dependence of the reflectivity factor. Lacasse et al. (2004) reported observations of the ash cloud of the Icelandic Hekla volcano in 2000 with a C-band radar at Keflavík international airport. Reflectivity factors in the range 30 to >60 dBZ characterized the eruption column above the vent due to the dominant influence of lapilli and ash (tephra) on the overall reflected signal, whereas values of between 0 and 30 dBz characterized the cloud advected downwind. The plume head had a mean ascent rate of 30 to 50 m/s up to 12 km in altitude (upper limit of the radar). Using the same radar, Marzano et al. (2006a, 2010a) found maximum reflectivities of 34 dBZ for the 2004 ash cloud of Grímsvötn volcano (260 km away), at a height of 6 km (minimum detection altitude). From an inversion technique based on a classification scheme of particles, they estimated ash concentrations of up to 6 g/m³, and ashfall rates of up to 31 kg/h. Likewise, for the 2010 Eyjafjöll eruption, Marzano et al. (2011) determined an ash mass of up to 15×10^8 kg on April 16, and 8×10^7 kg on May 5. Recently, Marzano et al. (2010b) used volume scan data acquired in the S-band by a NEXRAD WSR-88D ground-based weather radar at Augustine vocano in Alaska in 2006 (Wood et al., 2007). From their model-based technique, ash aggregate concentrations of up to 0.2g/m³ were found to correspond to measured reflectivities of up to 55 dBZ at an ash column height of about 4 km. Maki et al. (2001) first reported observations of ash plumes from Mount Oyama in Japan by a 3-cm wavelength polarimetric mobile radar about 40 km away. They discussed the possibility of detecting volcanic ash particles and estimating their size distribution from polarimetric radar parameters such as the differential reflectivity and specific differential phase shift.

2.2.1.1.3 Limitations

As seen previously, ground-based weather radar systems are powerful tools for volcanic ash cloud detection and quantification. Their Doppler capacity has not, so far, been much exploited in the study of ash clouds and could aid understanding of the interplay between their dynamics and their environmental conditions (wind, atmospheric properties such as humidity and temperature profiles, etc). Their main limitations are, in general: (i) their limited sensitivity tending to render invisible to the radar the cloud parts where particle concentration is too low (ultimately all of the ash cloud). This leads to an underestimation of the ash cloud lateral extension, and also of its height because the top of the ash column may be coarse-depleted. Another source of error on column heights, and hence an underestimation of height-derived eruption rates, may come from the incomplete filling of the highest volume scanned by the plume top. The sensitivity of the ground-based radar measurements will decrease as the ash cloud moves farther away. (ii) By using single-polarization weather radar, however, it is fairly difficult to discriminate between ash, hydrometeors, and mixed particles. Ice nucleation and subsequent loss in reflectivity also make ash detection more difficult (Marzano et al., 2006b). These authors suggest that polarimetric radars may improve discrimination of the impact of cloud ice and liquid water on ash aggregates. According to Hannesen and Weipert (2011), however, significant overlap exists between meteorological targets and volcanic ash, so that, even if all polarimetric observables of dual-polarized radars are used, automatic detection might be tricky. With polarimetric data, however, the retrieval of volcanic parameters could be improved by taking into account the mixed particle composition and their shape (Marzano et al., 2012). (iii) Path attenuation effects are not always negligible. According to Marzano and Ferrauto (2003), in the case of hydrometeors, any radar technique above S band should take into account, and possibly remove, path attenuation effects in order to correctly convert measured reflectivity into rain rate. For ash clouds, Marzano et al. (2006b) concluded that C-band may offer some advantages in terms of radar reflectivity response and negligibility of path attenuation. While still tolerable at X-band, the path attenuation cannot be handled at

Ka-band. The advantage of higher frequencies (X-, Ka-band) is the potential diminution of the overall size of the system and a higher sensitivity to fine particles, hence a better detection at low ash concentration. For near-source soundings, path attenuation effects are presumably very important up to X-band and possibly non negligible up to L-band because of the high particle concentrations and sizes (commonly pluri-decimetric), especially in the gas thrust region. Further investigations are needed, even at L-band. (iv) Weather radars cannot track ash clouds over the long-term, due to the low atmospheric residence time of reflective coarse particles. (v) Their maximum detection range is generally within 200-300 kilometers of their fixed location. Portable radar systems overcome the limitation of observing ash clouds from a far distance and always the same volcano. In many respects, the synergetic role of satellite imagery in tracking volcanic ash, particularly after the initial stages of an eruptive event is obvious. (vi) Weather radars are unable to image the lowest few kilometers of the ash column when the volcano is too far away (and the top if above the beam), preventing early detection and retrieval of the near-source ash plume characteristics. To avoid some of these shortcomings, institutes in charge of volcano monitoring have started to integrate nearby dedicated radars into their instrumental networks.

2.2.1.2 Radars dedicated to volcano monitoring

Given the benefits of continuous quantitative retrieval of parameters such as height and mass loading which are crucial to initiate dispersion models, permanent volcano monitoring using weather radars has become more widely used. Ground-based weather radar networks are currently operational at several volcanoes, in Alaska, Iceland, Italy and Guadeloupe. The U.S. Geological Survey first experimented in 1997 with a ground-based Doppler radar at the National Center for the Prevention of Disasters (CENAPRED) in Mexico to track the dispersal of ash plumes of Popocatépetl volcano and at least two eruptions were successfully captured. In addition to the near contiguous network of weather-monitoring Doppler radar NEXRAD operated by the U.S. National Weather Service, the U.S. Geological Survey also deployed a new truck-transportable C-band Doppler radar (MiniMax-250C) during the 2009 eruptions of Redoubt Volcano, Alaska (Hoblitt and Schneider, 2009). Results for 17 ash plumes detected by the radar compared favorably well with those of a nearby WSR-88D NEXRAD operated by the Federal Aviation Administration. The sector-scanning strategy (45°) of the new mobile radar advantageously allowed event onset detection within less than a minute. Heights (9-19 km) and vertical rise rates of the ash columns (25-60 m/s) have been determined. The high radar reflectivity values of the central core of the eruption column (50-60 dBZ) were interpreted as being the result of rapid formation of volcanic ash-ice aggregates (Schneider, 2012).

The X-band is generally preferable providing higher sensitivity with respect to lower frequency bands typically used for weather observations. The Japanese government recently set up an X-band polarimetric radar near Sakurajima volcano, able to monitor its recurrent vulcanian ash plumes (M. Maki, pers. comm.). Since November 2010, the Icelandic Met Office has had on loan from the Italian Civil Protection a mobile X-band dual-polarization radar for volcano monitoring. This radar (75 km from the volcano), along with the fixed weather C-band radar in Keflavík (257 km from the volcano), monitored the ash plumes of Eyjafjoll in 2010 and Grimsvötn in 2011 (Arason et al., 2011, 2012). These authors used in particular the radar time-series of the plume heights to calculate the mean eruptive flow rate. From the polarimetric X-band dataset of this eruption, Hannesen and Weipert (2011) quantified ash concentrations of up to 100 g/m³ and ash fall rates of up to 100 kg/m²/h at a height of 4.5 km from all polarimetric observables. They emphasize, however, the limits of ash quantification, the ambiguity in the separation of precipitation and ash that makes automatic detection still difficult, and the signal weakness from distant ash that prevents radar observations. Vulpiani et al. (2011) explored the benefits of the mobile dual polarization X band radar (DPX 4) operated by the Department of Civil Protection at the airport of Catania Fontanarossa (30 km to the South) to monitor Etna and offer support to the decisions of the authorities that regulate and control air traffic. In an ash plume fed from a lava fountain, maximum reflectivities of 35 dBZ were measured at medium distances of 10-40 km from the volcano. Estimated mass concentrations vary up to a few g/m³, although most are below 1 g/m³. The instrumental monitoring network of Etna operated by the Istituto Nazionale di Geofisica I Vulcanologia (INGV) also comprises, since 2009, and this is unique, a permanent ground-based L-band Doppler radar of the Observatoire de Physique du Globe de Clermont-Ferrand (OPGC, France) targeting the summit craters (Donnadieu et al., 2009a, 2012). Named VOLDORAD 2B, this radar is similar to the transportable volcano Doppler radar (VOLDORAD) successfully applied in several volcanic contexts (Dubosclard et al., 1999, 2004; Donnadieu et al., 2003, 2005), as illustrated later in this chapter (cf. section 7.1, fig. 16). The permanent radar at Etna should complement observations from the INGV monitoring network to constrain the inputs of the tephra dispersal models run automatically to perform tephra dispersal forecast (Scollo et al., 2009).

2.2.1.3 Fallout measurements

A compact X-band continuous wave, low power (10 mW) Doppler Radar (PLUDIX, 9.5 GHz frequency of operation), originally designed as a rain gauge disdrometer, was utilized to measure the terminal settling velocities and infer sizes of plume fallout at Mount Etna in 2002 (Scollo et al., 2005) and Eyjafjallajökull in 2010 (Bonadonna et al., 2011). PLUDIX-derived particle size distributions agree reasonably well with sieve-derived grain size distributions, but only for diameter range above 500 microns, and so should be used within a few kilometers from the source. Such measurements, along with deposit sampling and other methods shown in figure 1, can usefully complement other radar observations of the ash plume/cloud (Fig. 1) by providing the particle size distribution necessary to accurately retrieve the loading parameters (total mass, mass concentrations, mass flux of tephra).

2.2.1.4 Compact portable Doppler radars for near-source measurements

The growing need to get insight into the dynamics of explosive eruptions and to measure eruptive parameters at the source has led to the development of several active remote sensing compact instruments in the last decade or so. The first attempt to bring transportable sounders close to volcanic craters to measure the near-source dynamics was achieved by Weill et al. (1992) who successfully determined vertical velocities in the range 20-80 m/s for over 100 mild Strombolian explosions at Stromboli using a Doppler sodar. This cumbersome acoustic sounder could operate only at a few hundred meters from the vent and, hence, was not well suited to the sounding of larger magnitude, hazardous eruptions. Besides, velocity determinations using sodar require the knowledge of sound velocity at the jet temperature and gas composition, which was not available. Two main types of dedicated portable radars have since been used with the primary goal of studying eruption near-source dynamics through their Doppler capability: commercial micro rain radars, that are continuous-wave frequency-modulated and working at 24 GHz (Seyfried & Hort, 1999; Hort et al., 2003, 2006) and the VOLDORAD system, an L-band pulsed volcano Doppler radar (e.g., Dubosclard et al., 1999, 2004; Donnadieu et al., 2005). Being set up at a chosen location and aiming directly at the emission source (instead of rotation scanning), these compact radar systems can advantageously sound the gas thrust region and provide source eruptive parameters like eruption velocities, but also capture short-lived weak explosive activity, not visible to satellites or weather radars. They have higher temporal (<1 s) and spatial resolutions (tens to hundreds of meters) and higher sensitivity. A comparison of some characteristics of weather radars and transportable volcano Doppler radars is presented in Table 1.

	Location	Max. range	Min. range	Acquisition rate	Volume scanned	Power consumption	Frequency bands
Weather radars	Fixed	100-300 km	Few km	Few min	km^3	100s of kW	S, C, X, Ka
Portable radars	Chosen	10-15 km	10s-100s m	≤ 1 s	10^4-10^8 m^3	Few mW to few 10s of W	L, X, Ka

Table 1. Characteristics of weather and transportable radars for the monitoring of volcanic eruptions. Note in particular the difference in temporal and spatial resolution.

Hort and Seyfried (1998) and Seyfried and Hort (1999) measured mean vertical velocities of about 10 m/s for 12 lava jets during very low activity at Stromboli volcano with a commercial portable FM-CW radar Doppler anemometer 200-300 m away from the eruptive vent. Using the same instrument, Hort et al. (2003) found an increase in eruption duration, much higher velocities and indirect evidence of mean particle size decrease after a rain storm. Gerst et al. (2008) reconstructed the 4D velocity (directivity) of Strombolian eruptions at Erebus and Stromboli from 3 FM-CW radars. FM-CW radars have a narrower field of view (around 1° or so at 3 dB) and can thus target a precise sector of the volcanic emission but, on the other hand, lack the integrated information of longer wavelength pulse radars with a wider beam aperture and deeper range gates. L-band frequency signals are very little attenuated by hydrometeors or volcanic particles and can sound the interior of very dense particle-laden plumes. VOLDORAD also has a higher temporal resolution (<0.1 s).

Donnadieu et al. (2005) showed very detailed time series of power and maximum radial velocities of a Strombolian explosion at Etna and an ash plume at Arenal, acquired at high rate (<0.1 s) with VOLDORAD. Donnadieu et al. (2003, 2005) and Dubosclard et al. (2004) further showed evidence of strong correlation between volcanic tremor and maximum radar velocities for several Strombolian episodes, suggesting the influence of gas bubble dynamics in the conduit on tremor generation at Etna. Using VOLDORAD, Gouhier & Donnadieu (2008) first quantified the mass of tephra of Strombolian explosions at Etna (50-200 tons) from a new power inversion method. From the analysis of the shape of Doppler spectra of 200 Strombolian explosions, Gouhier & Donnadieu (2010) found that 80% of the load is ejected within a 40° dispersion cone and that, for 2/3 of the explosions, ejecta are distributed uniformly within this cone. Using measured maximum radial velocities, at-vent particle and gas velocities can be retrieved, and source gas fluxes estimated when the vent diameter is known (Gouhier & Donnadieu, 2011). Comparing thermal data with records from a FM-CW at Stromboli, Scharff et al. (2008) found a correlation between the radiative energy of Strombolian lava jets and the backscattered energy, suggesting that both methods record the relative variations of mass. They also found pulsations in the power time series of 40% of the eruptions, likely reflecting variations in mass eruption rate and originating in multiple consecutively exploding bubbles.

Scharff et al. (2012) also report the pulsed release (2-5 s) of ash clouds from the dome of Santiaguito with particle radial velocities between 10 and 25 m/s, and preceded by a vertical dome uplift of about 50 cm, as recorded with a FM-CW radar. Using VOLDORAD, Donnadieu et al. (2008) had already reported staccato pressure release in the ash emissions of Arenal volcano, along with a variety of ash plume dynamics from short-lived explosive events with radial velocities of up to 90 m/s, to sustained pulsed ash jetting and to passive dilute ash emissions. Donnadieu et al. (2011) successfully reconstructed the 3D vector of the ash plume transport speed from the echo onsets in contiguous range gates.

3. Specificity of radar signals of volcanic origin

3.1 Examples of meteorological signals

While abundant literature describes the effects of meteorological targets on weather radar signals, few studies characterize volcanic targets from a radar perspective. Not only the dynamics of volcanic eruptions strongly differs from that of common meteorological phenomena but also the target properties. This section points out some differences relevant to the study of radar signals of volcanic origin, for measurements near the emission source and in the distal part of ash clouds.

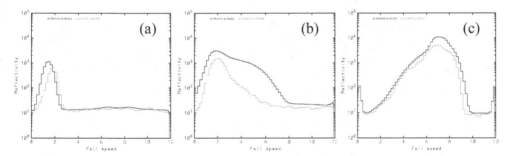

Fig. 2. Examples of meteorological Doppler spectra from a 24 GHz Micro Rain Radar. Reflectivity is shown versus radial velocity (i.e. fall speed with vertical beam) at a 1 hour interval (6:05 U.T. in red and 07:05 U.T. in green) on 21/12/2011: (a) snow crystals at 1430-1730 m a.s.l., (b) mixture of melting snow and water droplets in the radar bright-band (930-1230 m a.s.l.), (c) melt water droplets at 430-730 m a.s.l.. Data of MRR4 at Aulnat Airport (France): courtesy of Yves Pointin (OPGC).

Typical Doppler spectra of meteorological targets showing reflectivity versus fall speeds are presented in figure 2. Because the sounding is vertical, radial velocities (toward the radar) directly indicate fall speeds, unlike in the oblique radar soundings of volcanic emissions. At altitude, low reflectivity snow crystals fall at low speed (Gaussian shape spectrum). At intermediary altitude, a mixture of melting snow and water droplets (radar bright-band) produces more complex spectrum shapes, whereas at lower altitude melt water droplets produce rainfall with high reflectivity (Z>40 dBZ) and higher fall speeds up to 10 m/s. Note the relatively low velocities and spectrum width, as compared with the volcanic emission recorded by VOLDORAD (right panel in figure 3).

3.2 Volcanic features relevant to radar investigations

The contrasted radar signatures from meteorological targets and volcanic emissions are particularly conspicuous in figure 3 showing Doppler spectra recorded by VOLDORAD for rainfall and at the base of a weak ash plume during an explosive event. Whereas the rainfall shows narrow power distribution with a similar weak intensity and a well defined mode at 12 m/s over many range gates, volcanic tephra backscatter much more power spread over a radial velocity range from -30 to +60 m/s in just the two range gates above the eruptive vent.

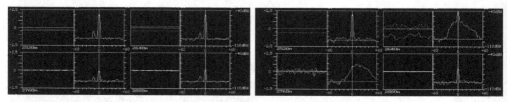

Fig. 3. Doppler spectra recorded by VOLDORAD for rainfall and tephra emission. The snapshots show the signals and power spectral density recorded in 4 range bins (120 m deep) between 2520 and 2880m on 18 February 2004, with a beam elevation angle of 27°. Central ground clutters are not filtered. *Left*: rainfall; *right*: ash plume event.

In addition to the sounding geometry effect, the large contrasts in echo power and velocity distribution (spectrum shape) observed in signals of volcanic origin relative to meteorological ones come at first order from the target sizes, velocity field, trajectories, with huge variations of these in amplitude in time and space, especially near the volcanic source.

The sizes of tephra currently range from decimetric or even metric to micronic particles (6 orders of magnitude) close to the vent while only particles microns to tens of microns in diameter can remain in the atmosphere for days. While Strombolian eruptions consist of the recurrent ejection of small-medium volumes of incandescent lava bombs and lapilli with commonly minor amounts of ash to heights of tens to hundreds of meters, the fragmentation of lava and particle dispersion are much higher in eruptions generating ash plumes. From a radar perspective, the backscattered power is strongly controlled by the larger particles. Consider for instance that an order of magnitude change in particle size leads to an increase by a factor of 10^6 in the reflectivity factor in the Rayleigh domain. However, the relative proportions of each size also counts, so the particle size distribution is of primary importance, albeit challenging to measure accurately over the full size range. Importantly, particles exceeding several centimeters (lapilli and blocks) present in most volcanic tephra emissions, prevent the use of the Rayleigh approximation to quantify near-source products and the Mie formulation must be used at all infra-metric wavelengths.

Because volcanic emissions are made up of a mixture of hot gases, lava and solid fragments of various sizes ejected at high speed (tens to hundreds of m/s) from a vent, their dynamics is complex and rapidly varying in time and space. The moments and shapes of Doppler spectra also fluctuate rapidly, with spatial variations among the range gates related to the dimension of the phenomenon with respect to the sounded volumes. The power spectral density results in particular from the complex combination of the particle velocity field and particle load distribution, i.e. the amount of particles, their velocities and trajectories. Being either short-lived or sustained, volcanic emissions can be roughly viewed as two-phase flows generally oriented vertically upward with a continuum of dynamic behaviors from inertial large blocks mostly following ballistic trajectories soon after their ejection down to the finest low-inertia particles nearly following the gas behavior involving a stronger deceleration soon after their emission. Crosswinds also affect the particles' motion differentially according to their diameter and residual momentum. Gravity further implies that ejecta are propelled upward while others fall out simultaneously and both imprint their signature in near-source radar measurements.

Volcanic emissions are generally highly turbulent close to the source (gas thrust region and convective part) and less turbulent in the distal cloud. The effect of turbulence in the cloud is difficult to assess because the particle behavior is highly size-dependent, from inertial large blocks to gas-entrained fine particles. Although assumed not to be dominant in radar measurements near the source where large particles are present, the turbulence effects affecting overall the small particles should nevertheless tend to increase the spectral width, as observed in radar meteorology.

3.3 Target properties

At second order, the intrinsic properties of the targets, their movements and chemico-physical evolution, also play a role in the measured reflectivity. Because volcanic tephra generally originate from the violent fragmentation of magma by the expanding gas, their shape is also complex and their surface highly irregular at various scales. The effects of shape and roughness of volcanic particles on reflectivity have been little investigated at radar wavelengths. Yet they might be non negligible, at least at short wavelength, as suggested for meteorological targets. In examining the effects of ice crystal shapes on reflectivity at 3 mm wavelength, Okamato (2002) found, for instance, 8 and 5 dB effects of non-sphericity and orientation respectively, for particle sizes approaching the wavelength. In the volcanic case, the analysis is further complicated by in-flight modifications of the ejecta shape and orientation, especially close to the source. Large lava fragments, in particular, deform in-flight due to their plastic nature, as attested by the specific shapes of volcanic bombs (e.g. fusiform), or break up upon impact with other ejecta and because of high strain rates imposed by acceleration, rotation, and drag force. It must be expected that most fragments have a rapidly changing orientation in flight, especially close to the source where turbulence occurs.

Water vapor being the dominant gas species exsolved from magma (commonly >85%), major condensation by the cold atmosphere occurs during eruptions. There is 2.4 factor difference between the dielectric factors of ash (0.39: Adams et al., 1996; Oguchi et al., 2009; Rogers et al., 2011) and liquid water (0.93). According to studies by Oguchi et al. (2009) from 3 to 13 GHz, a water film coating 10-20% of the radius of a sub-millimetric volcanic particle is sufficient to raise the radar cross section to that of a whole liquid water particle (0.93 dielectric factor). Water vapor further promotes the nucleation of ice (0.197 dielectric factor) in ash plumes at high altitude, depending on the vertical atmospheric temperature profile, and favors aggregation of ash particles. Thus, ice formation has a double effect on reflectivity, acting both on the dielectric properties and the size (aggregation). Although the influence of temperature on the dielectric properties of rocks seems rather limited up to 900°K (Campbell & Ulrichs, 1969), possible effects of magmatic temperatures need to be checked. The decrease of rock permittivity with silica content observed by these authors has direct consequences in terms of radar retrievals from eruption products of different composition, from basaltic to dacitic or rhyolitic for instance. As the combined contributions of all these effects might significantly change the reflectivity of ash plumes, they need to be further characterized physically, along with the relevant volcanic particle characteristics, in order to improve the accuracy of volcanic retrievals from radar returns.

4. VOLDORAD, a dedicated volcanological Doppler radar

4.1 Description of the transportable radar

VOLDORAD is a ground-based pulse volcano Doppler radar specifically designed at the Observatoire de Physique du Globe de Clermont-Ferrand (OPGC) for the monitoring of the surface volcanic activity of variable intensity. It can be deployed rapidly near an eruptive vent and target the near-source activity to measure in real-time the eruptive velocities and backscattered power and give information about the amount and rate of tephra emission.

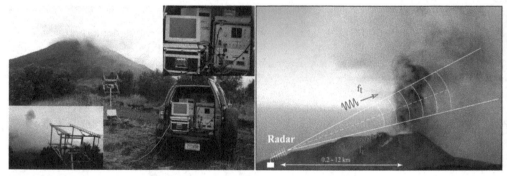

Fig. 4. (*left*): The transportable volcano Doppler radar VOLDORAD 2 deployed at Arenal. The PC and radar in the car trunk are connected to the antenna system via a switch box and fed by a small generator. *Upper inset*: Radar with PC for data storage and real-time monitoring and GPS receiver for time synchronization; *lower inset*: antenna (square array of Yagi) aiming at the summit to sound ash emissions. Photos: courtesy of S. Valade, OPGC (2009).

Fig. 5. (*right*): Principles of near-vent soundings with VOLDORAD.

The signal wavelength (23.5cm) was chosen (i) to sound the interior of dense lava jets and ash-laden plumes, as well as (ii) to avoid attenuation by hydrometeors because cloudy, foggy, rainy, or snowy conditions often occur at volcano summits. It also results from a compromise between transportability (weight, size), variable measurement distances (0.2-12 km) imposed by field conditions, and the HF hardware facilities available at the OPGC.

Acquisition, reception and pre-processing units are mounted on a suspended frame inside a protective metal container (60 cm, 50 kg). A PC controls the radar acquisition, being synchronized to UTC time through a GPS or ethernet connexion, and is used for real-time visualization of Doppler spectra and data storage. The 23 elements' square array antenna is mounted on a tripod adjustable for site and azimuth, and can be easily dismantled for transport. The 3 dB beam width is 9°, equivalent to site and azimuth resolutions of about 160 m at 1 km. The 300 W power consumption is provided through a small electric generator or AC. Owing to its modularity and limited weight (~70 kg), the ensemble is easily transportable, fits in a 4WD vehicle, and can be set up quickly in a volcanic environment. This radar can thus be used for short-term scientific campaigns, as well as over the long term for monitoring purposes.

A number of settings have been designed to be selectable to best adapt to the activity and the sounding conditions. The pulse duration is selectable from 0.4 to 1.5 µs so that the range bin radial resolution can be chosen between 60 and 225 m according to the target dimensions and the type of information searched for. The pulse repetition frequency can be 50, 100 or 200 µs. The non ambiguous maximum range at a 100 microsecond repetition frequency is 12 km. The gain attenuation can be varied by 50 dB through 10 dB steps to best adapt to the eruption intensity. The format of the data stored on the PC hard disk can be chosen in order to adjust the space memory consumption to the duration of the record campaign: either the time series of the raw digitized signal can be recorded, i.e. after coherent integrations in the time-domain, or alternatively only the spectra are saved, i.e. after integrations in the frequency domain.

4.2 Echoing mechanism

A powerful short radio frequency pulse (duration τ) is periodically transmitted into the atmosphere through a switch and a directive antenna which concentrates the energy in a narrow beam. Just after the pulse transmission, the switch connects the antenna to a radio frequency receiver. If targets are located in the antenna beam, part of the pulse energy is backscattered toward the antenna. These radar echoes are fed via a switch to the receiver for amplification and filtering. At the receiver output, the electromagnetic signal is detected and converted into digital data which are then processed and recorded.

Like in the case of atmospheric sounding, two main mechanisms give rise to radar echoes in the case of volcanic targets (Sauvageot, 1992; Doviak and Zrnic', 1993; Dubosclard et al., 1999): (i) Rayleigh ($D < \lambda/4$) or Mie scattering ($D \geq \lambda/4$) from distributed targets, and (ii) Bragg scattering from spatial irregularities of the refractive index induced by turbulent eddies inside the hot jet, and supposedly of secondary importance in the volcanic case because of the large reflectivity of tephra. In addition to the distance of the sounded volumes, the radar reflectivity (η) is deduced from the intensity of the echo signal by using the radar equation:

$$P_r = C_r \frac{\eta}{r^2} \tag{1}$$

where P_r is the echo power measured by the radar receiver and C_r a constant including the radar parameters such as transmitted power, pulse duration, wavelength, antenna characteristics, gain or half power beamwidth. In the simplest case of Rayleigh scattering ($D < \lambda/4$), η is expressed as:

$$\eta = \frac{\pi^5}{\lambda^4}|K|^2 Z \tag{2}$$

where K is the complex dielectric constant of the targets ($|K|^2 = 0.39$ for ash (Adams et al., 1996; Oguchi et al., 2009; Rogers et al., 2011) and Z the radar reflectivity factor. For spherical targets, Z is given by:

$$Z = \int_0^\infty N(D)D^6 \, dD \tag{3}$$

where $N(D)dD$ is the number of targets per unit volume whose diameters are between D and $D+dD$. Z is generally expressed in dBZ units, defined by:

$$Z(dBZ) = 10Log\left[Z(mm^6.m^{-3})\right] \tag{4}$$

For Mie scattering ($D \geq \lambda/4$), one generally uses the so-called equivalent radar reflectivity factor (Doviak and Zrnic', 1993), which is defined as the radar reflectivity factor of a small particle population satisfying the Rayleigh approximation and that would return the same received power. Interestingly, Z characterizes only the target and holds information on the number and size of particles, and thus on the particle concentration. The reflectivity factor obtained from radar measurements is usually calibrated using the dielectric factor of liquid water (0.93) and must be corrected for volcanic ash as the dielectric factor of the latter is lower (0.39):

$$Z_{ash} = \frac{|K|^2_{water}}{|K|^2_{ash}} Z_{water} = 2.38 Z_{water} \tag{5}$$

Finally, the radial velocity (V_r) of the target is calculated from the frequency shift between the transmitted and received signals. The velocity component along the antenna line of sight (toward or away from the radar) causes the returned frequency f_r to be different from the transmitted frequency f_t (Doppler effect), and is proportional to the Doppler shift f_d:

$$f_d = f_r - f_t = -\frac{2V_r}{c}f_t = -\frac{2V_r}{\lambda} \tag{6}$$

Note that a negative Doppler shift ($f_r < f_t$) corresponds to a target with a radial component of motion away from the antenna (positive radial velocity) and *vice versa*. Furthermore, if the velocity vector is normal to the antenna direction, the Doppler shift is zero. In volcanic soundings, the antenna beam can be set either to point upward, e.g. from the volcano slope toward the summit, or downward, for instance aiming toward the eruptive vent from the crater rim. In these cases, contributions of rising and falling particles to the spectra are reversed (Fig. 6).

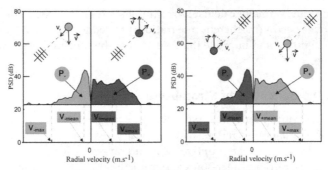

Fig. 6. Volcanic particle contributions to Doppler spectra in different sounding conditions. (a) When the antenna points upward (e.g. at summit craters from the flanks), ascending volcanic particles in red induce echoes with an along-beam velocity component away from the radar (positive radial velocity range, right part of spectrum), whereas falling particles in blue induce radial velocities toward the radar (negative radial velocity range, left part of spectrum). (b) The contributions are reversed when the antenna beam points downward (e.g. down towards a crater from the rim).

For a given range gate and for each component of the complex raw signals, N_c successive digitized samples (coherent integrations) are added together and then averaged in the time domain. This integration process acts as a low-pass filter reducing the high frequency noise and improving the signal-to-noise ratio. In order to avoid aliasing, the value of N_c must be adapted to expected maximum eruption velocities, in such a way that the Nyquist frequency f_N is higher that any Doppler frequency:

$$f_N = \frac{1}{2N_c T_r} \tag{7}$$

where T_r is the pulse repetition period of the radar. From (6) and (7) it results that the maximum radial velocity which can be measured without ambiguity is given by:

$$V_{max} = \frac{\lambda}{4N_c T_r} \tag{8}$$

After N_c pulses have been integrated, the coherent integration stage is repeated until a sequence of 64 integrated complex data is obtained. For each range gate, the 64 coherently integrated complex data are used as a time series input to a FFT (Fast Fourier Transform) algorithm in order to obtain the power spectrum of the radar echo. The frequency resolution (frequency interval between two consecutive spectral lines) is given by:

$$\Delta f = \frac{1}{64 \, N_c \, T_r} \tag{9}$$

Therefore, the corresponding velocity resolution is:

$$\Delta V = \frac{\lambda}{128 \, N_c \, T_r} \tag{10}$$

A mean spectrum can be finally calculated from the averaging in the frequency domain of several consecutive spectra (incoherent integrations) to reduce the noise fluctuations in the resulting Doppler spectra and improve the detection of the spectral line(s) corresponding to the volcanic echoes.

5. Insight into the dynamics of Strombolian activity

In addition to measuring eruptive parameters near the emission source, the main advantage of portable Doppler radars with respect to weather radars resides in the fact that they can be set up close to an eruptive vent and target just one direction at the base of the volcanic flow to retrieve near-source parameters. They are thus able to monitor phenomena of limited spatial dimensions with better spatial resolution and higher acquisition rate. In this respect, they are particularly useful for studying Strombolian activity and small scale ash plumes, generally invisible to remote weather radars and satellites. Although VOLDORAD can also be used to monitor strong ash-laden plumes, given its wavelength, examples of records on such types of activity are provided respectively in this section and the next.

5.1 Information retrieved from the shape of Doppler spectra

5.1.1 Ejecta dispersion

Given the relatively low amount of ash produced during typical Strombolian activity, the echo power is mostly controlled by large blocks that mostly follow ballistic trajectories. An illustration of the strong control of the ejecta's spatial distribution on the shape of Doppler spectra is the discrimination between lava bubble outbursts and lava jets (Fig. 7). When the magma column is high in the conduit, large gas bubbles may deform the lava surface to form lava bubbles several tens of meters in diameter, whose outburst disrupts the surrounding lava film and produces the hemispherical ejection of big lava lumps (decimetric to metric) seen above the crater rim (Fig. 7a). Contrastingly, in the vast majority of events, gas slug explosions occur within the conduit and produce oriented jets of gas and more fragmented lava, generally vertical, that can be captured by the radar beam when (if) pyroclasts go beyond the crater rim (Fig. 7c). Velocity and mass load angular distributions of pyroclasts are, as expected, more uniform for hemispherical lava bubbles that can burst at the free surface and more Gaussian-shaped for lava jets that are subjected to conduit wall friction (Fig. 8). In the case of hemispheric lava bubbles, the uniform mass load and velocity distribution of pyroclasts over the range of ejection angles produce an equal echo power over the range of radial velocities, leading to top-hat shaped spectra (Fig. 7b). In the case of lava jets, the power spectral distribution results from the competing effects of particle velocities and the distribution of ejection angles. As shown in figure 8b, measured maximum radial velocities do not result from particles with the highest velocities in the jet axis, nor from those having the most radial trajectories, but from particles ejected at about 20° to the jet axis (when vertical). Therefore Doppler spectra commonly appear rather triangular when shown in dB (Fig. 7d). Gouhier & Donnadieu (2010) found that, for most Strombolians explosions at Etna, 80% of the tephra mass is ejected within a dispersion cone of about 40°.

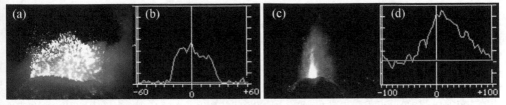

Fig. 7. Shapes of Doppler spectra associated with Strombolian activity. Outburst of a hemispherical lava bubble (a) and associated top-hat Doppler spectrum (b) at Etna's SE Crater (04/07/2001, 21:43'06). (c) Vertical lava jet at Laghetto (29/07/2001, 21:20'56) and the recorded triangular spectrum (d) (power in dBW vs. radial velocity in m/s).

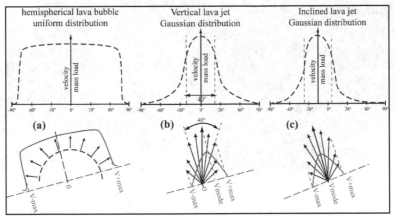

Fig. 8. Ejecta angular distributions (*top*) for lava bubbles (a), vertical jets (b) and inclined jets (c) and associated characteristics of Doppler spectra in red (*bottom*). (a): similar to Fig. 7a & 7b; (b): similar to Fig. 7c & 7d; (c): case of a jet inclined toward the radar.

All intermediary types of pyroclast ejection may exist, including lava jets with low directivity, leading to varied spectrum shapes. The latter can be further complicated by other factors related to the lava state in the conduit and crater, or to gas slug characteristics (overpressure, length) that may enhance fragmentation and produce more fine particles more closely coupled to the gas dynamics.

5.1.2 Inclination of lava jets

Strombolian activity is often imagined as an axi-symmetric dispersion of ejecta around the vertical, but departures are frequently observed from one explosion to another, and ejection in a preferential direction may even persist if conduit conditions are favorable. Relative variations of the ejecta dispersion axis from the vertical, i.e. the inclination of lava jets or lava bubble outburst, can be tracked from the mode of the radial velocities, i.e. the velocity associated with the maximum power in the main range gate (Fig. 8b,c). The velocity mode is shifted toward the maximum negative radial velocities (aiming upward) when the lava jet has an inclination component toward the radar (Fig. 8c) and vice versa. The power distribution in contiguous gates also puts constraints on the ejection geometry and velocities. In figure 9, for instance, ejecta emitted below the 840 m range bin do not reach the 600 m range bin, falling ejecta reaching the 720 m and 960 m bins where the negative parts of the spectra look alike. The absolute inclination angle cannot be retrieved directly firstly because velocities and power distribution vary strongly between explosions and also only the inclination angular component in the beam's vertical plane can be retrieved. Gerst et al. (2008) solved this problem by using simultaneously 3 FM-CW radars to calculate the directivity of Strombolian eruptions and reconstruct time series of the 3-D directivity vector every second. Another solution is to best match the power spectral distribution in the different range gates (e.g., Fig. 9) from ballistic models and Mie scattering theory (Gouhier & Donnadieu, 2010).

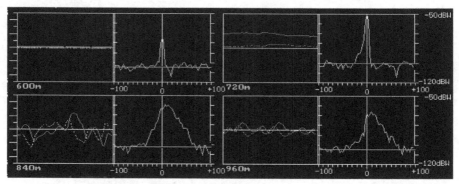

Fig. 9. Doppler spectra recorded in 4 range bins during lava jet activity at Etna. For each gate the raw signals (left) and spectra (right) seen in real-time are shown. The spatial distribution of the power spectral density in the 840 m range gate above the eruptive crater and in contiguous range gates (720 and 960 m) gives information about the dynamics of the lava jets (Laghetto cone, 29/07/2001).

5.2 Information from spectral moments

5.2.1 Mean diameter

Dubosclard et al. (2004) have shown that, in the case of Strombolian activity, the negative power spectral density (when aiming upward) of the range gates above the vent is mostly associated with falling particles. Therefore, the mean negative radial velocity weighted by the power spectral density has been used to retrieve the average particle diameter in the jet, commonly found to be between 1 and a few cm in diameter. Importantly, Gouhier & Donnadieu (2008) pointed out that the radar-equivalent mean diameter retrieved from Doppler spectra differs significantly from the true mode of the particle size distribution which, indeed, corresponds to the most frequent diameter encountered. Thus the radar-equivalent mean diameter cannot be used directly as the true modal diameter. In the gas thrust region of volcanic jets, the power spectrum is very wide, unlike in radar meteorology, and the physical interpretation of the radar-equivalent mean diameter is complex. Because the offset factor depends upon the reflectivity at a given radar wavelength, itself dependent on the number and diameter of particles which vary over a wide range, the conversion requires the Mie scattering formulation.

5.2.2 Eruptive velocities

Initial velocities are of great interest because they control the height reached by the pyroclasts and are related to the gas overpressure. In measuring particle velocities continuously and at high rate, Doppler radars potentially hold information on these parameters and on the detailed kinetics of the jets. Although the variations of measured radial velocities closely reflect the kinetics of the volcanic jet, retrieving absolute initial particle velocities from measured maximum radial velocities is not straightforward, in general, because (i) the latter are associated with oblique trajectories, as illustrated in figure 8b-c, (ii) the distance between the range gate's lower boundary and the emission source must be taken into account; it also controls which particle size induces the measured maximum velocities (Gouhier & Donnadieu, 2011). In order to confidently retrieve initial (at-vent) velocities, eruption models with suitable laws are needed, along with accurate knowledge of the sounding geometry and crater configuration. Velocity calibration can be further supported by an analysis of video-derived velocities. By sounding the volcanic emission near their source, it turns out often that maximum radial velocities, commonly in the range of a few tens to 160 m/s are not very different from eruptive velocities during Strombolian activity at Etna for example. Gouhier & Donnadieu (2011) analyzed 247 Strombolian explosions during the paroxysmal phase of the July 4 2001 eruptive episode of Etna's SE crater and found time-averaged values of 95 ±24 m/s for initial particle velocities, 37.6 ±1.9 m/s for the bulk jet velocity, and 118 ±36 m/s for the initial gas velocity. Note that the initial gas velocity is highly dependent upon the chosen model law, as the gas velocity decrease with height is exponential and more quantitative observations are needed, using complementary techniques like high-rate thermal infrared imagery.

5.2.3 Detailed dynamics at short time scales

Much can be learned from the analysis of time series of echo power and velocities on the eruption dynamics at various time scales, ranging from a single explosion through to an entire eruptive episode, to a series of eruptions. Figure 10 shows such time series for a Strombolian explosion at Yasur volcano, Vanuatu, similar to that shown in figure 11.

The maximum along-beam velocities associated with rising ejecta show a very sharp increase up to a peak near 120 m/s within 0.21 s, attesting to initial accelerations of over 560 m/s^2 (i.e. 57 g). Measured velocities then regularly decrease for 6-7 s, then reach a plateau at 20-30 m/s with sparse, short fluctuations up to 50 m/s. Interestingly, during the strong velocity phase, ample fluctuations associated with short power increases can be seen pulsating every ~1.5 s, suggesting rapid variations in the discharge rate. This could be caused either by the successive explosions of trains of closely packed gas bubbles rather than a single large slug, or by flow oscillations related to the high speed motion of a compressible fluid and conduit irregularities. The sharp onset in echo power in the 344 m range gate for both P- and P+ shows that rising particles with motion components both toward and away from the antenna (pointing downward) contribute to the signal. The former are dominant as the bulk flow is vertically directed. The first break in slope is likely associated with ejecta leaving the gate through its upper boundary. P- then gently increases up to its peak amplitude nearly 3 s later, as more pyroclasts fill up the probed volume. About 6 s after the onset, P+ becomes dominant whereas P- starts to decrease, indicating the contribution from block fallout. Comparing with figure 11, it can be inferred that most of the explosion momentum occurs during the first 6-7 s and corresponds to the main discharge of large ballistic blocks and dense lapilli- and ash-laden plume, with an initial maximum followed by a rapid decrease. The following phase, with lower velocities and echo power, results from the final emptying of the gas slug tail causing the relatively milder release of gas and finely fragmented ash at lower concentrations.

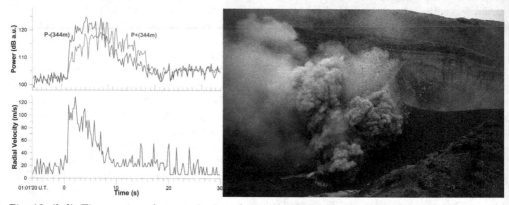

Fig. 10. (*left*): Time series of power (*top*) and maximum radial velocities (*bottom*) of a Strombolian outburst recorded by VOLDORAD aiming downward into Yasur's southern crater on 27/09/2008. Data are smoothed with 3 incoherent integrations (0.21 s).

Fig. 11. (*right*): Strombolian explosion at southern crater of Yasur volcano, Vanuatu. Incandescent pluridecimetric lava blocks resulting from the magma surface disruption by the pressurized gas slug are visible ahead of the ash plume (about a hundred meters high) generated by the magma fragmentation during gas pressure release. Photo courtesy: A. Finizola (2008).

5.2.4 Tephra mass loading

Gouhier & Donnadieu (2008) presented a method to estimate the particle loading parameters (mass, number, volume) of eruptive jets from the inversion of the echo power. The inversion algorithm uses the complete Mie (1908) formulation of electromagnetic scattering by spherical particles to generate synthetic backscattered power values. Assuming a log-normal shape for the particle size distribution, they estimated the total mass of tephra emitted during Strombolian explosions at Etna at around 58 and 206 tons for low and high concentration lava jets respectively. Derived parameters such as mass flux, particle kinetic and thermal energy, and particle concentration can also be estimated. As for particle concentrations, they must be regarded as minima in the case of small scale phenomena (relative to the range gate dimensions) like lava jets, because they are spatially highly heterogenous and might not completely fill the sounded volume. More reliable concentration values could be obtained when several range gates are filled completely, such as one expects from large ash plumes. For instantaneous events like Strombolian explosions, a total mass can be calculated from the echo power maximum amplitude assuming that all particles are present in the beam at the instant of the peak power. The instantaneous mass flux that can be derived in this way differs, however, from the initial mass flux as the power peak does not occur at the signal onset (Fig. 10). The initial mass flux can be estimated from the mass corresponding to the first break in slope of the power curve (cf. 5.2.3.) divided by the time difference between the break in slope and the onset.

The particle size distribution being the main unknown, the tephra mass can be computed as a function of diameter for different reflectivity factors (Z). As seen from the curve shape in figure 12a, the uncertainty on mass is much less for lapilli and blocks than for ash at VOLDORAD's wavelength. This is also demonstrated in figure 12b: for a given tephra volume, Z is much lower for small particles and increases linearly with diameter (in log-log plot), up to particle diameters approaching a quarter of the wavelength (5.9 cm, end of the Rayleigh domain), and beyond this value fluctuates in a lower range. The curve in figure 12a corresponds to the value of Z calculated from the power peak amplitude of figure 10. It can be used to infer a minimum tephra mass of several tons for the high momentum first phase, reasonably assuming an average diameter for ballistic blocks of more than 4 cm. This leads to an initial mass flux of ≥10 tons/s for a period of low activity at Yasur, consistent with values of 26-74 tons/s found by Gouhier & Donnadieu (2008) for larger Strombolian outbursts at Etna.

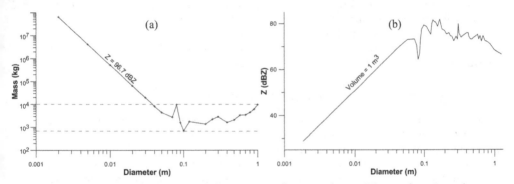

Fig. 12. (a) Tephra mass associated with a given reflectivity factor (Z), as a function of diameter. The curve shown corresponds to the value of Z calculated for the eruption of Yasur volcano in Fig.10. (b) Radar reflectivity factor associated with a tephra volume of 1 m³, as function of diameter, calculated with sounding conditions at Arenal volcano.

For sustained tephra emissions, like ash plumes, the mass flux evolution could be tracked through an integration of the mass of rising particles over time. This is not straightforward, however, because (i) falling particles should be removed, (ii) the particle size distribution must be assumed to be constant with time, and (iii) the integration timestep is not the data acquisition rate that would otherwise cause redundance but is instead related to the average transit time of rising particles through the considered range gates, and (iv) all tephra are not ascending at the same velocity because of their wide range of sizes, and therefore they have different transit times, (v) the particle transit time will vary with time according to velocity variations (e.g. short-lived events). One method to retrieve the time variations of mass flux could be, at first order, to use the P+/P- ratio to find the proportion of rising particles and integrate the power-derived mass curve using as a timestep the range gate dimension along the flow direction divided by the bulk (power-weighted) flow velocity.

5.2.5 Gas flux estimates

Although maximum velocities measured by the radar result from particles, initial gas velocities can be estimated at first order using the corrections seen in section 5.2.2. and adequate model laws for the coupling of gas and particle velocities and gas velocity decrease with height. The volume and mass gas fluxes Q^g can also be estimated from the initial gas velocity (V_0^g) and vent section, the radius (r) of which can be observed in the field or inferred from modeling of acoustic signals:

$$Q_{vol}^g = \pi r^2 \times V_0^g \tag{11}$$

$$Q_{mass}^g = \rho^g \times Q_{vol}^g \tag{12}$$

where ρ^g is the gas density at the considered atmospheric pressure (elevation) and temperature; water vapor being the ultra-dominant species, its density can be used for ρ^g. Note that gas fluxes evolve similarly to the maximum radial velocities and therefore have a large peak at the explosion onset followed by a rapid decrease in a matter of seconds. Averaging over the duration of the emission for a large number of Strombolian explosions, Gouhier & Donnadieu (2011) found volume and mass gas fluxes of $3\text{-}11 \times 10^3$ m³/s and 0.5-2 ton/s during the paroxysmal stage of a Strombolian eruptive episode of the SE crater at Etna. Radar-derived gas flux estimates at the source can then be compared with fluxes inferred from other ground-based techniques, like combined OP-FTIR gas spectroscopy and SO_2 flux measurements by DOAS, or ground-based thermal imagery.

6. Investigations of ash plumes with VOLDORAD

There is a continuum in the types of activity between Strombolian explosions seen previously and ash plumes. Ash plumes display a variety of behaviours depending on whether they are short-lived or sustained (steady state or not), whether they are jet plumes or buoyant plumes according to their momentum (mass loading, particle size distribution, fluxes), and also depending on environmental conditions including crosswind, elevation of the emission point, humidity and atmospheric temperature profiles, among the main ones. As an example, figure 13 shows three ash plumes with varying ash concentration and momentum, and also differently affected by the wind advection.

Fig. 13. Examples of various ash plumes at Arenal and Popocatépetl volcanoes. (a): low concentration buoyant plume bent over by wind advection at Arenal. (b): vertical jet plume, a few hundreds of meters in height, on same day (May 23, 2005). (c): dense ash plume of Popocatépetl buoyantly rising to about 2 km in height and drifting to the North on July 28 2007. Photos courtesy: Hotel Kioro Arenal and CENAPRED.

Ash is the major tephra component and is the main source of the ensuing hazards to humans, infrastructures and aviation, as shown by the 2010 eruption of Eyjafjallajökull in Iceland. This is because fine ash can remain in the atmosphere for hours to days, forming an ash cloud in the distal part. Although termed an ash plume, the proximal part does not comprise only ash, especially in the gas thrust and convective regions sounded by VOLDORAD. Below, examples are described which give insight into the ash plume dynamics close to the source.

6.1 Discriminating ballistics and ash

Tephra emissions are commonly explosive, having initial excess momentum compared with purely buoyant plumes. Therefore the explosive emission driven by the expansion of overpressured gas propels ash, lapilli and blocks in the air. Ash-sized particles closely follow the turbulent gas regime whereas inertial blocks mainly follow ballistic trajectories. So both are strongly decoupled, although a continuum of dynamic behaviors occurs in between for intermediary particle sizes. Because the spatiotemporal distribution of their velocity field and mass loading are contrasted, the dynamics of ballistics and ash can be discriminated when radar targeting the gas thrust region of the volcanic jet. Figure 14 illustrates the distinctive Doppler signatures for a jet plume at Arenal volcano similar to that shown in figure 13b and recorded with the beam aiming upward (27°) toward the summit. Although not obvious from the analysis of the time sequence of Doppler spectra the discrimination becomes particularly conspicuous on velocigrams. The velocigrams represent the power spectral density (dB color scale) as a function of radial velocities (y-axis) and time (x-axis) in 5 contiguous 120 m-wide range bins from 2367 to 2847 m. The 2607 m range bin, located above the vent, first records the jet plume onset. A 3-D representation of the velocigram at 2607 m is shown in the inset (cf. also book cover image). The ballistics are characterized by a short-lived signal (10-15 s) rapidly transiting through the gates. Range gates above the vent show positive radial velocities shifting to negative in a matter of seconds, as a result of the progressive bending of the ballistic trajectories through the radar beam.

Contrastingly, blocks only enter range gates located down-beam with negative radial velocities. So the time evolution of the spectral shape of this signal holds information about the ejection geometry (height, angles, orientation) and mass load spatial distribution, in addition to source parameters retrieved in section 5. Single streaks from individual blocks are sometimes visible on the velocigrams, and power-derived sizes are often decimetric. Considering lapilli to block sizes ranging between 0.04-1 m, Valade & Donnadieu (2011) found a mass of ballistics in the range 0.5-7 tons, i.e. a dense rock equivalent volume of 0.2-2.8 m³, for a similar event at Arenal. The second signal characterizes the ash plume, with lower backscattered power (by 10-20 dB), longer duration (>1 mn), slower transit through the gates, and with only negative velocities because the wind pushes the ash toward the radar. Interestingly, these characteristic maximum radial velocities may be used to constrain the effect of the wind and the buoyant ascent velocity. Although clearly smaller than for ballistics, the particle size distribution in the ash plume is poorly constrained, and so is the ash mass. Also, the longer duration and wider spatial coverage of the ash cloud requires spatial and temporal integration to obtain the total mass, which is nevertheless presumably greater than the mass of ballistics.

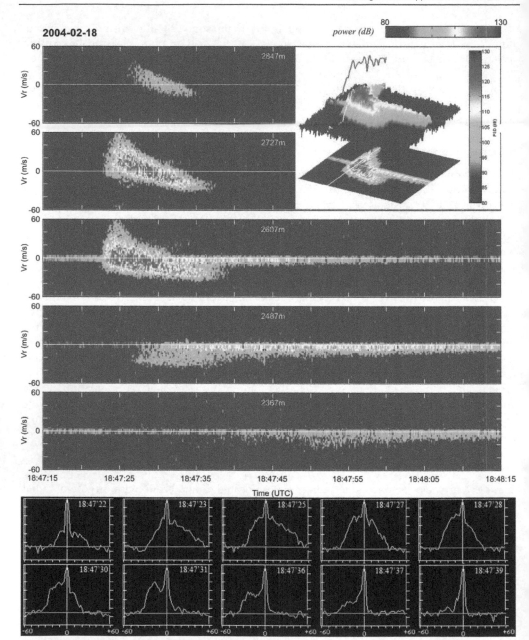

Fig. 14. Distinctive Doppler signatures of ballistics and ash dynamics for a jet plume at Arenal volcano (18/02/2004 at 18:47 U.T.). *Top*: velocigrams of 5 contiguous range bins. Inset: 3-D velocigrams of the 2607 m range bin above the vent. *Bottom*: Doppler spectra at 2607 m, showing power spectral density (dB arbitrary units) versus radial velocity (-60 to + 60 m/s); central peaks with constant power value are non filtered ground echoes.

6.2 Ash plume dynamics

The onset impulsivity of the same jet plume is nicely displayed on the time series of figure 15a, showing a sharp onset in maximum radial velocities, rapidly peaking over 80 m/s and then decreasing exponentially over about 10 seconds. The power maximum amplitude is also at the onset, indicating a mass flux maximum at the beginning, and P+ decreases rapidly after a few seconds and then more gently after about 10 s from the onset. The contribution of falling ballistics keeps P- high for a longer time. The 10 second phase of strong velocities and echo power corresponds to the initial mushroom-like ash plume head heavily charged with large blocks which dominate the signal. After this phase, radial velocities remain steady and low (10-20 m/s) and the power decreases more gently. The ballistics might therefore come from the disruption of the solidified lava plug by the gas overpressure accumulated underneath. This would clear the vent or fracture and open a way to a milder sustained gas release remobilizing variable quantities of ash and possibly fragmenting the lava to form juvenile ash.

Not all tephra emissions are impulsive at Arenal, and there are a wide variety of eruptive behaviors (Mora et al., 2009; Valade et al., 2012). Some emissions comprise mainly ash and are sustained typically for a minute or so (Fig. 15b). In contrast to the impulsive signal of jet plumes (Fig. 15a), the peak in echo power comes about 10 s after the onset. The second striking feature is the large oscillations correlated between P+ and P-, and having a remarkable periodicity of about 3 s. This indicates pulsations in the amount of material emitted, suggesting a staccato pressure release (Donnadieu et al., 2008). This observation supports the clarinet model of Lesage et al. (2006) for the volcanic tremor at Arenal, in which intermittent gas flow through fractures produces repetitive pressure pulses. The repeat period of the pulses is stabilized by a feedback mechanism associated with standing or traveling seismic waves in the magmatic conduit. Moreover, these rhythmic variations might well be a common feature of persistently active volcanoes with intermediate lava composition. In eruptions of Santiaguito volcano, Guatemala, Scharff et al. (2012) also observed multiple explosive degassing pulses occurring at intervals of 3-5 s, with common velocities of 20-25 m/s.

Figure 15c illustrates the signature of a larger ash plume of Popocatépetl, in Mexico, reaching a few kilometers above the volcano. Because its summit culminates at nearly 5450 m a.s.l., even small ash plumes generate hazards to the aviation, to the surrounding infrastructures and airports and the 30 million inhabitants leaving within 100 km of the volcano in important cities like México and Puebla. Its crater is 600 by 800 m wide with a growing lava dome inside. The relatively low velocities (<35 m/s) measured at 5085 m, along with the velocity peak not reached immediately after the onset, suggest low excess momentum and a mainly buoyant uprise, like most ash plumes in 2007 at Popocatépetl. Note that, in the case of ash plumes, radial velocities might reflect plume velocities more closely because convection and turbulence create eddies entraining particles that would tend to generate echoes with radial velocities toward and away from the radar, with comparable amounts of echo power in the absence of wind. From the time lag of P- relative to P+, however, it can be inferred that wind was blowing with some component toward the radar, i.e. to the north. In addition the power backscattered in 3 range gates simultaneously reveals that the horizontal dimension of the plume at the beam level was 2-3 times the radial resolution, i.e. >300 m. The comparable level of echo power at 5085 m and 5235 m, along with the similar evolution of P+ at 5235 m and P- at 5085 m during the first seconds of the main emission indicate that the plume axis was near the boundary between these range gates, i.e. at a slant distance of 5160 m. This plus the fact that the main emission is shortly preceded by a weaker emission at 4935 m suggests that the ash plume originated from the northern part of the dome, closer to the radar.

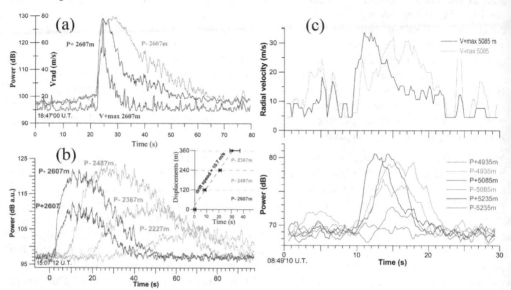

Fig. 15. Time series of echo power and velocities recorded by VOLDORAD (aiming upward) for (a) an impulsive jet plume (18/02/2004, 18:47) and (b) a sustained mild ash emission (11/02/2005, 15:07) at Arenal, and (c) a buoyantly rising ash plume at Popocatépetl (25/01/2007, 08:49).

6.3 Transport speed of ash plumes

The wind has a strong influence on ash plume dynamics, causing it to bend and, importantly for model-derived mass eruption rate estimates, be lower in height. When some wind component exists in the direction of the beam, radar echoes reflect the radial component of the drifting ash plume. As seen in figure 15b for a weak ash plume at Arenal, the near-source displacements of the plume can be tracked through echo onsets induced by ash entering successive probed volumes in the radar beam. When plotting the along-beam displacement versus time, a constant transport velocity is commonly reached within a few seconds of the initial ash emission (10.7 m/s along-beam in this case), as wind advection and buoyancy take over momentum. The departure of the first data point, i.e. the emission onset, from the general trend gives a relative indication of the slant distance of the vent to the radar. As shown by Donnadieu et al. (2011), the plume azimuth and uprise angles can further be constrained by comparing the amplitude decrease of the radar echoes as a function of distance from the source with results from a simple geometric plume model. This allows the three dimensional vector of the ash cloud transport speed to be reconstructed with an accuracy of a few percent. This method may have applications for determining pyroclasts fluxes, for volcano monitoring, for the modeling of tephra dispersal, and for remote measurements of volcanic gas fluxes for which the plume transport speed is needed.

7. Other applications of transportable radars

Beyond their main use to measure near-source eruptive parameters, compact Doppler radars can be utilized for a number of other applications in volcano monitoring. The identification of erupting vents using range gating and the tracking of rockfalls are illustrated in this section; possible investigations on fallout are discussed in the concluding section.

7.1 Discrimination of active vents

The summit areas of active volcanoes have complex and evolving morphologies, often comprising multiple craters, themselves possibly nesting several vents, all potentially active simultaneously with various dynamics. The relatively good spatial resolution (tens of meters) of dedicated ground-based radars often allows the spatial discrimination of the surface activity and, in particular, the identification of the eruptive vents. This information is obviously very useful in volcano monitoring to locate the activity in real-time. Note that large wavelength signals, such as that used by VOLDORAD (L band) can penetrate through dense ash-laden plumes or lava fountains and give information on possible activity occurring in craters behind.

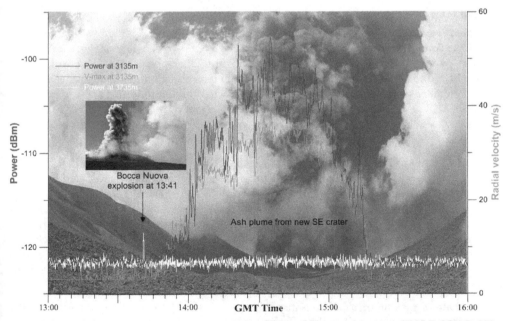

Fig. 16. Spatial discrimination of active craters of Etna using range gating (VOLDORAD 2B). Simultaneous tephra emissions of Bocca Nuova and new SE Crater at Etna on July 9 2011 are discriminated using echoes in range gates at 3735 m and 3135 m respectively (Radar data: OPGC-INGV; Photo courtesy: Tom Pfeiffer, www.volcanodiscovery.com).

At Etna, a ground-based Doppler radar operated jointly by the INGV in Catania and the OPGC in Clermont-Ferrand has been continuously monitoring the tephra emissions of the summit craters since 2009 (Donnadieu et al., 2009a, 2012) from a shelter at La Montagnola, about 3 km to the south. Echoes are recorded in 11 range bins, 150 m deep and 9° wide in azimuth and elevation, defining a 1650 m-long truncated conical volume covering the summit craters. These were very active in 2010-2012, showing different eruptive styles, including short-lived ash plumes (e.g. new SE Crater April 8 2010; Bocca Nuova, August 25 2010) and large ash columns several kilometers high sustained for several hours from lava fountains, such as the eruptive episodes of the new SE Crater in 2011 (Donnadieu, 2012; http://wwwobs.univ-bpclermont.fr/SO/televolc/voldorad/).

Figure 16 presents records from VOLDORAD 2B of simultaneous activity of Bocca Nuova and new SE Crater during cloudy weather. Radar monitoring is not hampered by clouds that sometimes make visual observations impossible. As shown by the echo power curves from the range gates at 3135 m and 3735 m, the short-lived explosion forming a weak plume at Bocca Nuova can be discriminated from the strong and dense column fed from the lava fountain originating in the new SE Crater and sustained for about 1.5 h. While the former cause a power increase of only a few dB, the latter result in much more powerful echoes (> 20 dB), with a progressive onset and more abrupt waning phase. Radial velocities in the convective ash and lapilli plume above the lava fountain commonly reach 30-40 m/s.

7.2 Tracking of rock falls

Not only the explosive activity can be monitored using transportable radars, but also lava flow or dome instabilities (Wadge et al., 2005; Hort el al., 2006). Viscous basaltic andesite lava flows continuously outpour from the summit of Arenal volcano and slowly flow on top of loose pyroclastic material down the steep and unstable upper slopes. Due to the joint actions of cooling and pushing by new lava, instabilities occur and generate repeated rock falls, sometimes evolving into small pyroclastic flows. While monitoring the ash emissions with VOLDORAD 2 from the west between January 26 and March 4 2009, signals from very frequent rock falls could be recorded in several range gates because their lowest part hit the volcano's upper slopes, where destabilizations occurred toward the SW.

The radar signature of rock falls is characterized by echoes with only radial velocity components toward the radar in contiguous bins at slant distances consistent with the location of the volcano's upper slopes (4013 and 3878 m on Fig.17). Radial velocities are typically low (<20 m/s). The amplitude of the backscattered power is less in closer bins, as expected if not all the destabilized material goes all the way down the slope. As seen from the power curves, signal onsets are delayed from the most remote range gates to the closest, as the destabilized material tumbles down. It is interesting that many small rock fall events detected by the radar are not always well recorded on seismograms, and thus both techniques appear complementary as noticed by Vöge et al. (2008).

An estimate of maximum block velocities during rock falls (V_{rf}) can be retrieved from the radial velocities (V_r) measured by the radar on the upper slopes when the geometry is known:

$$V_{rf} = V_r \times \frac{1}{\cos(\alpha)\cos(\theta_{rf} - \theta_{ant})} \tag{13}$$

where α is the azimuth difference between the rock fall direction and the beam axis, θ_{rf} the average angle of rock fall relative to the horizontal and θ_{ant} the elevation angle of the beam.

Approximate angle values determined from field observations ($\alpha\approx45°$, $\theta_{rf}\approx35°$, and $\theta_{ant}\approx22°$ for the lowest part of the beam) lead to rock falls velocities equal to 1.45 times the measured radial velocities. With maximum radial velocities commonly in the range 7-14 m/s, the upper range of rock fall velocities is 10 to 20 m/s. Likewise, average rock fall speeds can be estimated from the time delays of power onsets in successive range gates and the gate radial resolution.

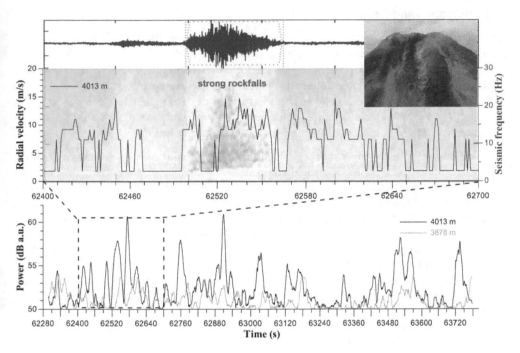

Fig. 17. Radar and seismic signals associated with rock falls at Arenal (02/02/2009). *Bottom*: 25 minute record of power from two range gates hitting the volcano upper slopes, where rock falls are detected by VOLDORAD; *middle*: maximum radial velocity (component towards radar) superimposed on a color spectrogram showing the seismic frequency content associated with rock fall events; *top*: seismic trace of a large rock fall and view of Arenal's summit from north showing lava flows and rock falls producing dust (seismic data courtesy of M. Mora; photo © F. Donnadieu, 14/05/2006).

In theory, the volume of rock falls could be determined from the power measured by the radar. In the case of Arenal, however, the size distribution of the fragments tumbling down the flanks is unknown and challenging to measure. Also, rock falls occurring in corridors can be masked by topographic obstacles or take place in other directions not visible to the radar so that it is likely that not all the falling material is detected and some events can even remain undetected by the radar. Nevertheless, as most rock falls are generated in preferential directions during long periods of activity, transportable Doppler radars can be used to track this type of activity in order to assess the state of the activity, the stability of the lava flow or dome, and better understand the destabilization processes (Hort et al., 2006). They could be further used to correlate the rock fall activity, linked to the lava effusion rate, to the emissions of tephra. As both types of event could be recorded by the radar, such a comparative quantified analysis over a representative time sequence spanning several weeks could provide information on the eruptive behavior of the volcano and the dynamics of its upper plumbing system.

8. Conclusion and future prospects

Illustrations provided in this chapter show the many capabilities of Doppler radars to investigate and monitor volcanic phenomena in real-time and in all weather conditions. The L-band portable volcano Doppler radar of the OPGC is particularly useful for monitoring explosive activities of variable intensity from a chosen location. Owing to its 23.5 cm wavelength, VOLDORAD is able to sound the interior of dense particle-laden volcanic jets. By directly probing the jets near the vent, quantified eruption source parameters can be retrieved in different volumes with a high spatiotemporal resolution.

A major challenge in the mitigation of ash plume-related risks is to determine the relative mass flux of volcanic material propelled into the atmosphere, in particular the mass transport rate in the cloud at the neutral buoyancy level. The mass transport rate represents only a fraction of the total flux of magma erupted at the vent (magma mass eruption rate), which also includes the effusive activity (lava flows, lava dome) as well as all the lava falling back into the crater and its immediate surroundings contributing to the growth of pyroclastic cones (e.g. ballistics). The proportion of material propelled into the atmosphere strongly controls the hazards to humans and infrastructures, the economic costs and environmental consequences, but also represents an essential input to volcanic ash transport and dispersion models. Its estimation in near real-time could allow models to be constantly refined by comparing their predictions with measurables from ground-based and satellite remote sensing methods and ground deposit data. Model inputs from deposit observations (e.g. thickness) to quantify eruption characteristics cannot be done in real-time as this requires the collection of many fallout samples at remote locations. Although models allow the tephra mass flux to be estimated from the ash column height, the latter needs to be accurately measured and defined, in particular with regard to the strong effect of crosswinds. This can be achieved most reliably through a combination of methods, but radars appear particularly relevant in this case. Weather radars can provide the plume height, within the uncertainties discussed in section 2, along with characteristics of the ash cloud including transport speed estimates. Compact Doppler radars sounding the plume base at a high acquisition rate should help discriminate the mass eruption rate from the mass transport rate and link models of plume ascent to models of tephra dispersal by providing crucial source kinetic and mass loading parameters. The radar echo power is also related to the amount of material ejected and can be inverted to retrieve the mass. Among the most stringent assumptions for this is the particle size distribution in the sounded volumes. It can be inferred from direct measurements of fallout (rain radar, disdrometer etc) or deposit sampling, in situ sampling by aircraft, or comparison with data from similar eruptions. We could more ambitiously imagine vertical soundings by transportable Doppler radars from underneath wind-advected ash clouds to provide the vertical distribution of both the particle concentration and fallout velocities, owing to their directive antenna and range gating capability. Tracking the time evolution of these parameters would give access to the internal dynamics of ash clouds and sedimentation processes. Terminal fall velocities could in turn be related to the particle size distribution that could be compared with field deposits and results from tephra dispersal models. Fallout measurements from ground collectors and several continuously operating laser or microwave disdrometers would improve understanding of the relationship between particle size and fall velocity, and the spatial heterogeneities often observed in ground deposits.

Unfortunately, a comprehensive real-time technique that can provide the erupted mass associated with the whole particle-size spectrum does not yet exist. This could only be derived from a combination of complementary techniques. This argues for a synergetic strategy for the assimilation of multiple datasets quantitatively describing the different parts of a plume from the gas-thrust region through the convective region to the buoyant distal cloud. As described in this chapter, the proximal region can be well quantified by radars, and developments are expected with multiple/complementary frequencies and dual polarimetry. The cross-correlation of data from radars and complementary passive remote-sensing methods, particularly ground-based imagery (IR, VIS, UV), is also a potentially powerful tool to retrieve crucial parameters like particle size distribution, gas and tephra mass fluxes. Ground-based radars are not useful for long-term volcanic cloud tracking because the large ash particles, that provide strong radar signals fall out soon after an eruption. C-band radars, for instance, do not detect ash particles with diameter <1-100 μm in drifting volcanic clouds that can persist in the atmosphere for several days or more (Rose et al., 1995). Thus, for the long-range tracking of ash clouds, satellite-based imagers (IR, VIS, UV, microwave) bring an obvious synergetic contribution, along with MISR and lidars, more sensitive to micron- to submicron-sized particles and aerosols, ceilometers, sun photometers, and DOAS mainly for SO_2. An important objective of future works should aim at comparing and calibrating data from different instruments and/or acquired at different wavelengths, always cross-validated with field data (e.g. Bonadonna et al., 2011, 2012; Donnadieu et al., 2009b; Gouhier et al., 2011). Further coupling with other geophysical methods, in particular seismic and acoustic, seems promising to investigate the eruptive behavior of a volcano from down the conduit up through the magma-air interface where explosions occur, up to the surface where the dynamics of the tephra emissions can be recorded. Finally, back to figure 1, the synergetic integration of source-targeting Doppler radars, medium-range weather radars (few tens-hundreds km), and satellite- and ground-based imagery, combined with traditional monitoring networks and field methods, is a promising approach to improve the assessment of ash plume-related hazards, the forecast from tephra dispersal models and the mitigation of associated risks. In order to retrieve accurate eruption parameters, future research should focus on joint measurements of ash plume characteristics at different levels by means of complementary techniques. To this purpose, a good strategy would be to (i) carry out well-targeted multi-method experiments on volcanoes showing either recurrent activity or sudden resumption of activity, and (ii) operate long-term observations at selected laboratory volcanoes having a well-instrumented monitoring network.

9. Acknowledgments

I am much indebted to the OPGC staff for developments of the radar system, in particular G. Dubosclard, R. Cordesses, C. Hervier, J. Fournet-Fayard, P. Fréville, C. Bernard, C. Reymond, and for implementation of databases (S. Rivet, P. Cacault). I am also grateful to C. Hervier, T. H. Druitt, M. Gouhier, S. Valade for assistance in data acquisition in the field, and to E. Bonny, A. Clenet, J. Doloire, M. Gouhier, A. Maillet, L. Perrier, L. Pouchol, N. Rodriguez, S. Saumet, D. Tailpied, P. Tinard and S. Valade for help in data processing of the different campaigns. Y. Pointin kindly provided data on meteorological signals. All these colleagues are warmly acknowledged. External partners facilitated radar soundings: Universidad de Costa Rica (M. Mora, G. Alvarado, L.-F. Brenes, F. Arias and C. Ramirez), DGMWR in Port Vila Vanuatu (E. Garaebiti, C. Douglas), UNAM-CENAPRED in Mexico. Radar measurements from VOLDORAD 2B on Etna were obtained in the frame of a collaboration between OPGC and INGV Catania (M. Coltelli, S. Scollo, M. Prestifilippo). Field campaigns were supported by the French CNRS INSU programs 2003-ACI Risques Naturels, ANR-06-CATT-02 Arc-Vanuatu, 2010-TERMEX-MISTRALS, 2011-2012 CT3, and by the European project VOLUME FP6-018471.

This is Laboratory of Excellence *CLERVOLC* contribution n° 12.

10. References

Adams, R.J., Perger, W.F., Rose, W.I., & Kostinski, A. (1996). Measurements of the complex dielectric constant of volcanic ash from 4 to 19 GHz. *J. Geophys. Res.*, 101, B4, pp. (8175-8185).

Arason, P., Petersen, G. N., Bjornsson, H. (2011). Observations of the altitude of the volcanic plume during the eruption of Eyjafjallajökull, April–May 2010. Earth Syst. Sci. Data, 3, pp. (9–17).

Arason, P., Bjornsson, H., Petersen, G. N., Roberts, M. J., & Collins M. (2012). Resonating eruptive flow rate during the Grímsvötn 2011 volcanic eruption. *30th Nordic geological winter meeting*, 9-12 January 2012, Reykjavík, Iceland.

Bonadonna, C., Folch, A., Loughlin, S., Puempel, H. (2012). Future developments in modelling and monitoring of volcanic ash clouds: outcomes from the first IAVCEI-WMO workshop on Ash Dispersal Forecast and Civil Aviation. *Bull. Volcanol.*, 74, 1, pp. (1-10).

Bonadonna, C., Genco, R., Gouhier, M., Pistolesi, M., Cioni, R., Alfano, F., Hoskuldsson, A., & Ripepe M. (2011). Tephra sedimentation during the 2010 Eyjafjallajökull eruption (Iceland) from deposit, radar, and satellite observations. *J. Geophys. Res.*, 116, B12202.

Casagli, N., Tibaldi, A., Merri, A., Del Ventisette, C., Apuani, T., Guerri, L., Fortuny-Guasch, J., & Tarchi D. (2009). Deformation of Stromboli Volcano (Italy) during the 2007 eruption revealed by radar interferometry, numerical modelling and structural geological field data. *J. Volcanol. Geotherm. Res.*, 182, pp. (182–200).

Campbell, M.J., & Ulrichs, J. (1969). Electrical properties of rocks and their significance of Lunar Radar Observations. *J. Geophys. Res.*, 74, 25, pp. (5867-5881).

Delene, D.J., Rose, W.I., & Grody, N.C. (1996). Remote sensing of volcanic clouds using special sensor microwave imager data. *J. Geophys. Res.*, 101, B5, pp. (11579-11588).

Donnadieu, F., Dubosclard, G., Allard, P., Cordesses, R., Hervier, C., Kornprobst, J., & Lénat, J.-F. (2003). Sondages des jets volcaniques par radar Doppler : applications à l'Etna. *Rapport quadriennal C.N.F.G.G. 1999-2002*, pp. (119-124).

Donnadieu, F., Dubosclard, G., Cordesses, R., Druitt, T.H., Hervier, C., Kornprobst, J., Lénat, J.-F., Allard, P., & Coltelli, M. (2005). Remotely monitoring volcanic activity with ground-based Doppler radar. *E.O.S. Trans.*, 86, 21, pp. (201-204).

Donnadieu F., Gouhier M., Fournet-Fayard J., Hervier C., 2008. Applications of pulsed ground-based Doppler radar to the study and monitoring of volcanoes, Ground-based radar observations for volcanoes workshop, Univ. Reading UK, pp. (6-8).

Donnadieu, F., Hervier, C., Fréville, P., Fournet-Fayard, J., Fournol, J.-F., Menny, P., Reymond, C., & Bernard, C. (2009a). The VOLDORAD 2B radar: Operational handbook, Note OPGC 07.09, Univ. Blaise Pascal Clermont-Ferrand.

Donnadieu F., Roche O., Clarke A., Gurioli L., (2009b). Advances in studies of volcanic plumes and pyroclastic density currents. Report of the IAVCEI Commission on Explosive Volcanism, workshop in Clermont-Ferrand (France), 26-29 October 2009, 18pp.

Donnadieu, F., Valade, S., & Moune, S. (2011). Three dimensional transport speed of wind-drifted ash plumes using ground-based radar. *Geophys. Res. Lett.*, 38, L18310, doi:10.1029/2011GL049001.

Donnadieu, F. (2012). VOLDORAD, In: *Observatoire de Physique du Globe de Cleront-Ferrand web site*, 2012, < http://wwwobs.univ-bpclermont.fr/SO/televolc/voldorad/>.

Donnadieu, F., Coltelli, M., Scollo, S., Fréville, Hervier, C., P., Prestifilippo, M., Rivet, S., Cacault, P., Fournet-Fayard, J., Bernard, C., & Reymond, C. (2012). Doppler radar monitoring of the explosive activity at Etna's summit craters with VOLDORAD 2B: toward an integrated approach. MEMOVOLC meeting, Jan. 17-18, Clermont-Ferrand, France.

Doviak, R. J. and D. S. Zrni´c, 1993: Doppler Radar and Weather Observations. Academic Press. Cambridge University Press, 562 pp.

Dubosclard, G., Cordesses, R., Allard, P., Hervier, C., Coltelli, M., & Kornprobst, J. (1999). First testing of a volcano Doppler radar (Voldorad) at Mount Etna, Italy. *Geophys. Res. Lett.*, 26, pp.(3389-3392).

Dubosclard G., Donnadieu, F., Allard, P., Cordesses, R., Hervier, C., Coltelli, M., Privitera, E., & Kornprobst, J. (2004). Doppler radar sounding of volcanic eruption dynamics at Mount Etna. *Bull. Volcanol.*, 66, 5, pp. (443-456), DOI : 10.1007/s00445-003-0324-8.

Froger, J.-L., Remy, D., Bonvalot, S. and Legrand, D. (2007). Two scales of inflation at Lastarria-Cordon del Azufre volcanic complex, central Andes, revealed from ASAR-ENVISAT interferometric data. *Earth Planet. Sc. Lett.* 255, pp. (148-163).

Gerst, A., Hort, M. Kyle, P. R., & Vöge, M. (2008). 4D velocity of strombolian eruptions and man-made explosions derived from multiple Doppler radar instruments. *J. Volcanol. Geotherm. Res.*, 177 (3), pp. (648–660), doi:10.1016/j.jvolgeores.2008.05.022.

Gouhier, M., Harris, A., Calvari, S., Labazuy, P., Guéhenneux, Y., Donnadieu, F., & Valade, S. (2011). Lava discharge during Etna's January 2011 fire fountain tracked using MSG-SEVIRI. *Bull. Volcanol., in press.*

Gouhier, M., & Donnadieu, F. (2008). Mass estimations of ejecta from Strombolian explosions by inversion of Doppler-radar measurements. *J. Geophys. Res.,* 113, B10202, doi:10.1029/2007JB005383.

Gouhier, M., & Donnadieu, F. (2010). The geometry of Strombolian explosions: insight from Doppler radar measurements. *Geophys. J. Int.,* 183, pp. (1376–1391), doi: 10.1111/j.1365-246X.2010.04829.x

Gouhier, M., & Donnadieu, F. (2011). Systematic retrieval of ejecta velocities and gas fluxes at Etna volcano using L-Band Doppler radar. *Bull. Volcanol.,* 73, pp. (1139–1145).

Hannesen, R., & Weipert, A. (2011). An algorithm to detect and quantify volcanic eruptions using polarimetric X-band radar data. Int. Workshop on X-band Weather Radar, 14-16 Nov. 2011, Delft, Netherlands.

Harris, D. M., Rose, W.I.Jr., Roe, R., & Thompson, M.R. (1981). Radar observations of ash eruptions. In: *The 1980 Eruptions of Mount St. Helens, Washington,* edited by P. W. Lipman and D. R. Mullineaux, U.S. Geol. Surv. Prof. Pap. 1250, pp. (323-333).

Harris, D. M., & Rose, W.I.Jr. (1983). Estimating particle sizes, concentrations, and total mass of ash in volcanic clouds using weather radar. *J. Geophys. Res.,* 88, C15, pp. (10969–10983), doi:10.1029/JC088iC15p10969.

Hoblitt, R. P., & Schneider, D. J. (2009). Radar observations of the 2009 eruption of Redoubt Volcano, Alaska: Initial deployment of a transportable Doppler radar system for volcano-monitoring. American Geophysical Union, Fall Meeting 2009, abstract #V43A-2209.

Hort, M., & Seyfried, R. (1998). Volcanic eruption velocities measured with a micro radar. Geophys. Res. Lett. 25, 1, pp.(113–116).

Hort, M., Seyfried, R., & Vöge M. (2003). Radar Doppler velocimetry of volcanic eruptions: theoretical considerations and quantitative documentation of changes in eruptive behaviour at Stromboli Volcano, Italy. *Geophys. J. Int., 154,* pp. (515–532), doi:10.1046/j.1365-246X.2003.01982.x.

Hort, M., Vöge, M., Seyfried, R., & Ratdomopurbo, A. (2006). In situ observation of dome instabillities at Merapi Volcano, Indonesia: A new tool for hazard mitigation. *J. Volcanol. Geotherm. Res.,* 153, pp. (301–312), doi:10.1016/j.jvolgeores.2005.12.007.

Lacasse, C., Karlsdóttir, S., Larsen, G., Soosalu, H., Rose, W. I., & Ernst, G.G.J. (2004). Weather radar observations of the Hekla 2000 eruption cloud, Iceland. *Bull. Volcanol.,* 66, pp. (457–473), doi:10.1007/s00445-003-0329-3.

Lesage, P., Mora, M.M., Alvarado, G.E., Pacheco, J., & Métaxian J.-P. (2006). Complex behavior and source model of the tremor at Arenal volcano, Costa Rica. *J. Volcanol. Geotherm. Res.,* 157, pp. (49–59).

Macfarlane, D. G., Wadge, G., Robertson, D. A., James, M. R., & Pinkerton, H. (2006). Use of a portable topographic mapping millimeter wave radar at an active lava flow. *Geophys. Res. Lett.,* 33, L03301, doi:10.1029/2005GL025005.

Maki M., & Doviak, R. J. (2001). Volcanic ash size distribution determined by weather radar. *IEEE Int. Geosc. Rem. Sens. Symp., IGARSS '01.*, 2001, vol.4, pp. (1810 - 1811), doi: 10.1109/IGARSS.2001.977079.

Maki M., Iwanami, K., Misumi, R., Doviak, R. J., Wakayama, T., Hata, K., & Watanabe, S. (2001). Observation of volcanic ashes with a 3-cm polarimetric radar. *Proc. 30th Radar Meteorol. Conf.*, 18-24 July, Munich, Germany, P5.13, p. (226-228).

Malassingne, C., Lemaitre, F., Briole, P., & Pascal, O. (2001). Potential of ground based radar for the monitoring of deformation of volcanoes. *Geophys. Res. Lett.* 28, 851–854.

Marzano, F.S., & Ferrauto, G. (2003). Relation between weather radar equation and first-order backscattering theory. *Atmos. Chem. Phys.*, 3, pp. (813–821).

Marzano, F. S., Barbieri, S., Vulpiani G., & Rose W.I. (2006a). Volcanic ash cloud retrieval by ground-based microwave weather radar. *IEEE Trans. Geosc. Remote Sens.*, 44 (11), pp. (3235–3246), doi:10.1109/TGRS.2006.879116.

Marzano, F.S., Vulpiani, G., & Rose W.I. (2006b). Microphysical characterization of microwave radar reflectivity due to volcanic ash clouds. *IEEE Trans. Geosc. Remote Sens.*, 44, pp. (313–327), doi:10.1109/TGRS.2005.861010.

Marzano, F.S., Barbieri, S., Picciotti, E., & Karlsdóttir, S. (2010a). Monitoring sub-glacial volcanic eruption using C band radar imagery. *IEEE Trans. Geosc. Remote Sens.*, 48, 1, pp. (403-414).

Marzano, F. S., Marchiotto, S., Barbieri, S., Textor, C., & Schneider, D. (2010b). Model-based Weather Radar Remote Sensing of Explosive Volcanic Ash Eruption, *IEEE Trans. Geosc. Remote Sens.*, 48, pp. (3591-3607).

Marzano, F. S., Lamantea, M., Montopoli, M., Di Fabio, S., & Picciotti, E. (2011). The Eyjafjöll explosive volcanic eruption from a microwave weather radar perspective. *Atmos. Chem. Phys. Discuss.*, 11, pp. (12367–12409), doi:10.5194/acpd-11-12367-2011.

Marzano, F. S., Picciotti, E., Vulpiani, G., & Montopoli, M. (2012). Synthetic signatures of volcanic ash cloud particles from X-band dual-polarization radar. *IEEE Trans. Geosc. Remote Sens.*, 50, 1, pp. (193-211).

Mastin, L. G., Guffanti, M., Servranckx, R., Webley, P., et al. (2009). A multidisciplinary effort to assign realistic source parameters to models of volcanic ash cloud transport and dispersion during eruptions. *J. Volcanol. Geotherm. Res.*, 186, pp. (10–21).

Mie, G. (1908). Beiträge zur Optik trüber Medien, speziell kolloidaler Metallösungen. *Ann. Phys.*, 330, 3, pp. (377–445).

Mora, M.M., Lesage, P., Donnadieu, F., Valade, S., Schmidt, A., Soto, G., Taylor, W., & Alvarado, G. (2009). Joint Seismic, Acoustic and Doppler Radar observations at Arenal Volcano, Costa Rica: preliminary results. In: *The VOLUME project*, Bean, C. J., Braiden, A. K., Lokmer, I., Martini, F., & O'Brien, G. S. , pp. (330-340), VOLUME Project Consortium, ISBN 978-1-905254-39-2, Dublin.

Musolf,M. (1994). Airborne radar detection of volcanic ash. *In*: First International Symposium on Volcanic Ash and Aviation Safety, Casadevall, T.J. (Ed.), *U.S. Geol. Surv. Bull.*, 2047, pp. (387–390), Seattle, Washington, U.S.A.

Oguchi, T., Udagawa, M., Nanba, N., Maki, M., & Ishimine, Y. (2009). Measurements of dielectric constant of volcanic ash erupted from five volcanoes in Japan. *IEEE Trans. Geosc. Remote Sens.*, 47, 4, pp. (1089 - 1096).

Okamato, H. (2002). Information content of the 95-GHz cloud radar signals: Theoretical assessment of effects of non sphericity and error evaluation of the discrete dipole approximation. *J. Geophys. Res.*, 107, D22, 4628.

Oswalt, J. S., Nichols, W., & O'Hara, J. F. (1996). Meteorological observations of the 1991 Mount Pinatubo eruption, In: *Fire and Mud: Eruptions and Lahars of Mount Pinatubo, Philippines*, Newhall C. G., & Punongbayan R. S., pp. (625–636), Univ. of Wash. Press, Seattle.

Prata, F., Bluth, G., Rose, W.I., Schneider, D., & Tupper, A. (2001). Comments on 'Failures in detecting volcanic ash from a satellite-based technique'. *Remote Sens. Environ.*, 78, 3, pp. (341–346).

Rogers, A.B., Macfarlane, D.G., Robertson, D.A. (2011). Complex permittivity of volcanic rock and ash at millimeter wave frequencies. *IEEE Remote Sens. Lett.*, 8, 2, pp. (298–302).

Rose, W.I., Kostinski, A.B. (1994). Radar remote sensing of volcanic clouds. *In*: First International Symposium on Volcanic Ash and Aviation Safety, Casadevall, T.J. (Ed.), *U.S. Geol. Surv. Bull.*, 2047, pp. (391–395), Seattle, Washington, U.S.A.

Rose, W. I., Kostinski, A. B., & Kelley, L. (1995). Real-time C band radar observations of 1992 eruption clouds from Crater Peak vent, Mount Spurr Volcano, Alaska. *In*: The 1992 eruptions of Crater Peak vent, Mount Spurr Volcano, Alsaka, Keith, T. (Ed.), pp. (19-26), *U.S. Geol. Surv. Bull.*, 2139.

Rose, W. I., Bluth, G. J. S., & Ernst, G. G. J. (2000). Integrating retrievals of volcanic cloud characteristics from satellite remote sensors — A summary. *Phil. Trans. R. Soc. A*, vol. 358, 1770, 1585–1606.

Russell, J. K., & Stasiuk, M. V. (1997). Characterization of volcanic deposits with ground-penetrating radar. *Bull. Volcanol.*, 58, pp. (515–527).

Sauvageot, H. (1992). *Radar meteorology*, Artech House, ISBN 0890063184, Boston.

Scharff, L., M. Hort, A. J. Harris, M. Ripepe, J. Lees, and R. Seyfried (2008), Eruption dynamics of the SW crater of Stromboli volcano, Italy. *J. Volcanol. Geotherm. Res.*, 176, pp. (565–570).

Scharff, L., Ziemen, F., Hort, M., Gerst, A., & Johnson J.B. (2012). A Detailed View Into the Eruption Clouds of Santiaguito Volcano, Guatemala, Using Doppler radar. *J. Geophys. Res.*, in revision.

Schneider, D.J. (2009). Explosive volcanic eruptions: what can radar do for you? 2d Nat. Symp. on Multifunction Phased Array Radar, November 18-20, 2009, Norman, OK, USA.

Schneider, D.J. (2012). The Use of a Dedicated Volcano Monitoring Doppler Weather Radar for Rapid Eruption Detection and Cloud Height Determination. 92nd American Meteorological Society Annual Meeting, Jan. 22-26 2012, New Orleans, LA, USA.

Scollo, S., Coltelli, M., Prodi, F., Folegani, M., & Natali, S. (2005). Terminal settling velocity measurements of volcanic ash during the 2002–2003 Etna eruption by an X-band microwave rain gauge disdrometer. *Geophys. Res. Lett.*, 32, L10302, doi:10.1029/2004GL022100.

Scollo, S., Prestifilippo, M., Spata, G., D'Agostino, M., & Coltelli, M. (2009). *Nat. Hazards Earth Syst. Sci.*, 9, pp. (1573–1585).

Scollo, S., Folch, A., Coltelli, M., & Realmuto, V. J. (2010). Three-dimensional volcanic aerosol dispersal: A comparison between Multiangle Imaging Spectroradiometer (MISR) data and numerical simulations. *J. Geophys. Res.*, 115, D24210.

Seyfried, R., & Hort, M. (1999). Continuous monitoring of volcanic eruption dynamics: A review of various techniques and new results from a frequency-modulated radar Doppler system. *Bull. Volcanol.*, 60, pp. (627–639).

Stone, M. (1994). Application of contemporary ground-based and airborne radar for the observation of volcanic ash. *In*: First International Symposium on Volcanic Ash and Aviation Safety, Casadevall, T.J. (Ed.), U.S. Geol. Surv. Bull., 2047, pp. (391–395), Seattle, Washington, U.S.A.

Valade, S. & Donnadieu, F. (2011). Ballistics and ash plumes discriminated by Doppler radar. *Geophys. Res. Lett.*, 38, L22301, doi:10.1029/2011GL049415.

Valade, S., Donnadieu, F., Lesage, P., Mora, M.M., Harris, A. & Alvarado, G.E. (2012). Explosion mechanisms at Arenal volcano, Costa Rica: an interpretation from integration of seismic and Doppler radar data. *J. Geophys. Res.*, 117, B1, doi:10.1029/2011JB008623.

Vöge, M., & Hort, M. (2008). Automatic classification of dome instabilities based on Doppler radar measurements at Merapi volcano, Indonesia: Part I. *Geophys. J. Int.*, *172*, pp. (1188–1206), doi:10.1111/j.1365-246X.2007.03605.x.

Vöge, M., Hort, M., Seyfried, R., Ratdomopurbo, A. (2008). Automatic classification of dome instabilities based on Doppler radar measurements at Merapi volcano, Indonesia: Part II. *Geophys. J. Int.*, *172*, pp. (1207–1218), doi:10.1111/j.1365-246X.2007.03665.x.

Vöge, M., & Hort, M. (2009). Installation of a Doppler Radar monitoring system at Merapi Volcano, Indonesia. *IEEE Trans. Geosc. Remote Sens.*, 47, 1, pp. (251–271), doi:10.1109/TGRS.2008.2002693.

Vulpiani, G., Montopoli, M., Piccioti, E., & Marzano F.S. (2011). On the use of a polarimetric X-band weather radar for ash clouds monitoring. *Proc. 35th Radar Meteorol. Conf.*, 26-30 September, Pittsburgh, PA, USA, P101.

Wadge, G., Macfarlane, D.G., Robertson, D.A., Hale, A.J., Pinkerton, H., Burrell, R.V., Norton, G. E., & James, M.R. (2005). AVTIS: A novel millimetre-wave ground based instrument for volcano remote sensing. *J. Volcanol. Geotherm. Res.*, 146, 307–318.

Wadge, G., Macfarlane, D.G., Odbert, H.M., James, M.R., Hole, J.K., Ryan, G., Bass, V., De Angelis, S., Pinkerton, H., Robertson, D.A., & Loughlin, S. C. (2008). Lava dome growth and mass wasting measured by a time series of ground-based radar and seismicity observations. *J. Geophys. Res.*, 113, B08210, doi:10.1029/2007JB005466.

Wen, S., & Rose, W.I. (1994). Retrieval of sizes and total masses of particles in volcanic clouds using AVHRR bands 4 and 5. *J. Geophys. Res.*, 99, pp. (5421-5431).

Weill, A., Brandeis, G., & Vergniolle, S. (1992). Acoustic sounder measurements of vertical velocity of volcanic jets at Stromboli volcano. *Geophys. Res. Lett.*, 19, 23, pp. (2357-2360).

Wood, J., Scott, C., & Schneider, D. (2007). WSR-88D radar observations of volcanic ash. WorldMeteorological Organization, *Proc. 4th Int. Workshop Ash*, Mar. 26–30, 2007, Rotorua, New Zealand.

Permissions

The contributors of this book come from diverse backgrounds, making this book a truly international effort. This book will bring forth new frontiers with its revolutionizing research information and detailed analysis of the nascent developments around the world.

We would like to thank Joan Bech and Jorge Chau, for lending their expertise to make the book truly unique. They have played a crucial role in the development of this book. Without their invaluable contribution this book wouldn't have been possible. They have made vital efforts to compile up to date information on the varied aspects of this subject to make this book a valuable addition to the collection of many professionals and students.

This book was conceptualized with the vision of imparting up-to-date information and advanced data in this field. To ensure the same, a matchless editorial board was set up. Every individual on the board went through rigorous rounds of assessment to prove their worth. After which they invested a large part of their time researching and compiling the most relevant data for our readers. Conferences and sessions were held from time to time between the editorial board and the contributing authors to present the data in the most comprehensible form. The editorial team has worked tirelessly to provide valuable and valid information to help people across the globe.

Every chapter published in this book has been scrutinized by our experts. Their significance has been extensively debated. The topics covered herein carry significant findings which will fuel the growth of the discipline. They may even be implemented as practical applications or may be referred to as a beginning point for another development. Chapters in this book were first published by InTech; hereby published with permission under the Creative Commons Attribution License or equivalent.

The editorial board has been involved in producing this book since its inception. They have spent rigorous hours researching and exploring the diverse topics which have resulted in the successful publishing of this book. They have passed on their knowledge of decades through this book. To expedite this challenging task, the publisher supported the team at every step. A small team of assistant editors was also appointed to further simplify the editing procedure and attain best results for the readers.

Our editorial team has been hand-picked from every corner of the world. Their multi-ethnicity adds dynamic inputs to the discussions which result in innovative outcomes. These outcomes are then further discussed with the researchers and contributors who give their valuable feedback and opinion regarding the same. The feedback is then collaborated with the researches and they are edited in a comprehensive manner to aid the understanding of the subject.

Apart from the editorial board, the designing team has also invested a significant amount of their time in understanding the subject and creating the most relevant covers. They scrutinized every image to scout for the most suitable representation of the subject and create an appropriate cover for the book.

The publishing team has been involved in this book since its early stages. They were actively engaged in every process, be it collecting the data, connecting with the contributors or procuring relevant information. The team has been an ardent support to the editorial, designing and production team. Their endless efforts to recruit the best for this project, has resulted in the accomplishment of this book. They are a veteran in the field of academics and their pool of knowledge is as vast as their experience in printing. Their expertise and guidance has proved useful at every step. Their uncompromising quality standards have made this book an exceptional effort. Their encouragement from time to time has been an inspiration for everyone.

The publisher and the editorial board hope that this book will prove to be a valuable piece of knowledge for researchers, students, practitioners and scholars across the globe.

List of Contributors

Masayuki K. Yamamoto
Research Institute for Sustainable Humanosphere (RISH), Kyoto University, Japan

Olivier Bousquet
Météo-France, Centre National de Recherches Météorologiques, France

Shingo Shimizu
National Research Institute for Earth Science and Disaster Prevention/ Storm, Flood, and Land-Slide Research Department, Japan

Chris G. Collier
National Centre for Atmospheric Science, University of Leeds, United Kingdom

Jan Szturc, Katarzyna Ośródka and Anna Jurczyk
Institute of Meteorology and Water Management – National Research Institute, Poland

Lars Norin and Günther Haase
Swedish Meteorological and Hydrological Institute, Sweden

Joan Bech, Bernat Codina and Jeroni Lorente
Dep. Astronomy and Meteorology, University of Barcelona, Spain

Adolfo Magaldi
Institute of Space Sciences, Spanish National Research Council (CSIC), Bellatera, Spain

Erhan Kudeki
University of Illinois at Urbana-Champaign, Peru

Marco Milla
Jicamarca Radio Observatory, Lima, Peru

D. L. Hysell and J. L. Chau
Earth and Atmospheric Sciences, Cornell University, Ithaca, New York, USA
Jicamarca Radio Observatory, Lima, Peru

Mohammad Hossein Gholizadeh and Hamidreza Amindavar
Amirkabir University of Technology, Tehran, Iran

Franck Donnadieu
Clermont Université, Université Blaise Pascal, Observatoire de Physique du Globe de Clermont-Ferrand (OPGC), Laboratoire Magmas et Volcans, Clermont-Ferrand, France
CNRS, UMR 6524, LMV, Clermont-Ferrand, France
IRD, R 163, LMV, Clermont-Ferrand, France

Printed in the USA
CPSIA information can be obtained
at www.ICGtesting.com
JSHW011450221024
72173JS00004B/1013